Advanced Structured Materials

Volume 208

Common engineering materials are reaching their limits in many applications, and new developments are required to meet the increasing demands on engineering materials. The performance of materials can be improved by combining different materials to achieve better properties than with a single constituent, or by shaping the material or constituents into a specific structure. The interaction between material and structure can occur at different length scales, such as the micro, meso, or macro scale, and offers potential applications in very different fields.

This book series addresses the fundamental relationships between materials and their structure on overall properties (e.g., mechanical, thermal, chemical, electrical, or magnetic properties, etc.). Experimental data and procedures are presented, as well as methods for modeling structures and materials using numerical and analytical approaches. In addition, the series shows how these materials engineering and design processes are implemented and how new technologies can be used to optimize materials and processes.

Advanced Structured Materials is indexed in Google Scholar and Scopus.

Azman Ismail · Fatin Nur Zulkipli ·
Mohd Amran Mohd Daril · Andreas Öchsner
Editors

Engineering Frontiers

A Multidisciplinary Odyssey

Springer

Editors
Azman Ismail
Maritime Engineering Technology
and Centre for Women Advancement
and Leadership
Malaysian Institute of Marine Engineering
Technology
Universiti Kuala Lumpur
Lumut, Perak, Malaysia

Mohd Amran Mohd Daril
Malaysian Institute of Industrial
Technology
Universiti Kuala Lumpur
Masai, Johor, Malaysia

Fatin Nur Zulkipli
Information Science Studies
College of Computing,
Informatics and Mathematics
Universiti Teknologi MARA
Machang, Kelantan, Malaysia

Andreas Öchsner
Faculty of Mechanical and Systems
Engineering
Esslingen University of Applied Sciences
Esslingen am Neckar
Baden-Württemberg, Germany

ISSN 1869-8433 ISSN 1869-8441 (electronic)
Advanced Structured Materials
ISBN 978-3-031-56843-5 ISBN 978-3-031-56844-2 (eBook)
https://doi.org/10.1007/978-3-031-56844-2

This Springer imprint is published by the registered company Springer Nature Switzerland AG
The registered company address is: Gewerbestrasse 11, 6330 Cham, Switzerland

Preface

This book describes a diverse collection of engineering research and innovations across 35 chapters. Each chapter unveils a facet of modern engineering excellence. This collection not only presents cutting-edge solutions but also addresses environmental sustainability. Engineering frontiers is an illuminating expedition into the heart of innovation, showcasing the ingenuity and collaborative spirit that define the future of engineering.

Lumut, Malaysia — Azman Ismail
Machang, Malaysia — Fatin Nur Zulkipli
Masai, Malaysia — Mohd Amran Mohd Daril
Esslingen am Neckar, Germany — Andreas Öchsner

Contents

Facile Fabrication of a Low-Cost Electrochemical Temperature Sensor

Shahino Mah Abdullah, Saqib Rafique, Mazwani Abdullah, Mohd Ifwat Mohd Ghazali, Mohd Amran Mohd Daril, Mohamad Ikbar Abdul Wahab, and Khaulah Sulaiman

Abstract The needs of cost-effective electronic devices with easy fabrication procedure have brought the use of organic compound from metal-phthalocyanines (MPc's) derivatives in this work for temperature sensors fabrication. Water soluble MPc's, copper (II) phthalocyanine-tetrasulfonic acid tetrasodium salt (CuTsPc) have been used in the fabrication of electrochemical temperature sensors. The ITO-CuTsPc-ITO electrochemical cells have been prepared and characterized in the temperature range within 25–80 °C. The working principle of this temperature sensor is based on the change in the electrical parameters namely resistance and capacitance of the electrochemical cell due to the temperature variation. Fast response and recovery time were achieved including narrow hysteresis attained from a loop of temperature variation in both increase and decrease directions. The fabricated temperature sensor has shown high consistency and sensitive response towards a specific range

S. M. Abdullah · M. I. M. Ghazali
Faculty of Science and Technology, SMART RG, Universiti Sains Islam Malaysia, Bandar Baru Nilai, 71800 Nilai, Negeri Sembilan, Malaysia
e-mail: shahinomah@gmail.com

M. I. M. Ghazali
e-mail: ifwat.ghazali@usim.edu.my

S. M. Abdullah · S. Rafique · K. Sulaiman
Department of Physics, Faculty of Science, Low Dimensional Research Centre (LDMRC), University of Malaya, 50603 Kuala Lumpur, Malaysia
e-mail: saqibrafique@hotmail.com

K. Sulaiman
e-mail: khaulah@um.edu.my

M. Abdullah · M. A. M. Daril (✉) · M. I. A. Wahab
Quality Engineering Research Cluster, Quality Engineering Section, Universiti Kuala Lumpur Malaysian Institute of Industrial Technology, 81750 Masai, Johor, Malaysia
e-mail: mamran@unikl.edu.my

M. Abdullah
e-mail: mazwani.abdullah@s.unikl.edu.my

M. I. A. Wahab
e-mail: mikbar@unikl.edu.my

A. Ismail et al. (eds.), *Engineering Frontiers*, Advanced Structured Materials 208,
https://doi.org/10.1007/978-3-031-56844-2_1

of temperature values. Further enhancement was made by adding a small amount (1%) of titanium dioxide (TiO_2) nanoparticle (25 nm) into the CuTsPc solution and has increased the sensitivity in capacitance values toward the change in temperature. It has also shown a good linearity and high stability with the change of temperature. This electrolytic nanofluid chemical cell has a great chance to be utilized as capacitive temperature sensor in electronics, energy, automation, IoT, monitoring, and health applications.

Keywords Temperature sensor · Semiconductor · Water soluble · Electrochemical

1 Introduction

A sensor (so-called detector) is a converter that receives any of physical stimulus and turns it into a singal/impulse that can be read in an electrical measurement. It comes from the word 'sense' or 'detect' in which the received physical quantity is able to be collected and converted into an electrical signal. The sensor or detector would give a reading of this electrical signal in several parameters such as current, resistance, capacitance, inductance etc. The performance of the sensor depends on its response and recovery time which is the time it takes to reach 90% of a maximum or mininum reading upon the presence or the absence of respective physical stimulation, respectively.

The raise of new generation technology has driven the use of more efficient operating systems that can save a lot of energy with high speed performance. The sensor could be the best device to be utilized along with the electrical equipement for energy saving and make ease of automated working system. The smaller in size and lighter in weight makes this sensor device easily to be installed and operated with any electrical goods. As such, the temperature sensor could overcome the issue of overheating by detecting the limited temperature for that electical instrument operation, and turning on the cooling system automatically. Other than switching purpose, the sensor could be the tool for giving some specific value of measurement for certain use of applications such as a thermometer.

Temperature is the most important parameter in sensing applications. There are two major types of temperature sensors namely, non-contact and contact. The non-contact type temperature sensors will detect the thermal radiant power of the radiation, while in the case of 'contact' type temperature sensors, it is required both object and the sensor to be in thermal equilibrium until no heat flow occur between them. Both types of sensors need some suppositions in order to determine the influencing factors that control the output since it can make the apparent difference in temperature from the measured temperature. The contact temperature sensors can be further divided into several classes based on their operation mechanisms such as thermal expansion, thermoelectric, resistive and capacitive etc. [1].

The main fundamental operation of many sensors depends on their electrical resistance. The temperature variation is inversely proportional to the change in electrical resistance; however, the variation in electrical resistance with temperature is non-linear in most cases. The degree of linearity, however, does not affect much the sensor sensitivity and accuracy. For instance, the carbon-based resistance–temperature sensors used in cryogenic applications have shown high sensitivity and accuracy like other resistance–temperature based devices used to be.

Recently, the organic electronics technology has emerged to be the new approach of solving several issues by offering a cheaper, easier, and stable production for the fabrication of sensor devices. The use of organic compound, metal-phthalocyanines (MPc's) in this research work is mainly to leverage its versatile tailored properties, especially from its derivatives that fit various fields of applications. The water soluble MPc's derivative was chosen for this study as it is easy to be processed in both solution and thin film-based device fabrications. This material has shown a quick response to most of external physical changes acting on it such as temperature, humidity, light, pressure, electric and magnetic fields etc., thus make it really good in multi-functional sensing applications. The sensors can be prepared in solution form for the fabrication of electrochemical temperature sensor. In this study, the result and discussion will be focusing on the electrochemical temperature sensor with its enhancement into electrolytic nanofluid temperature sensor.

2 Methodology

2.1 Preparation of CuTsPc Solution

Copper (II) phthalocyanine-tetrasulfonic acid tetrasodium salt (CuTsPc) with 988.28 g/mol of molecular weight is a commercially available organic compound obtained from Sigma-Aldrich Malaysia (subsidiary of Merck KGaA). It has a molecular formula of $C_{32}H_{16}CuN_8O_{12}S_4Na_4$ and Fig. 1 (left) shows its molecular structure. The CuTsPc in powder form has been dissolved in de-ionized water to produce a uniform aqueous solution. For composition optimization and selection, it has been prepared in seven different weight percentages (wt%). Firstly, the optimization process was made based on the maximum reach of both resistance and capacitance values with respect to the difference in concentration. Secondly, it was verified by the standard deviation minimum point of both resistance and capacitance of those concentrations. Then, the plots of both concentration and standard deviation were made to correspond with their resistance and capacitance for the selection purpose. The electrochemical cell, or sensor container, was then fabricated by cutting and assembling two pieces silica glass and ITO-coated glass in which they are facing to each other with alternating join in square shape with the ITO inner surfaces separated by a 7 mm gap. The active area of each ITO electrode is 140 mm^2 working when 1 ml CuTsPc aqueous solution was filled inside the cell container. The design of

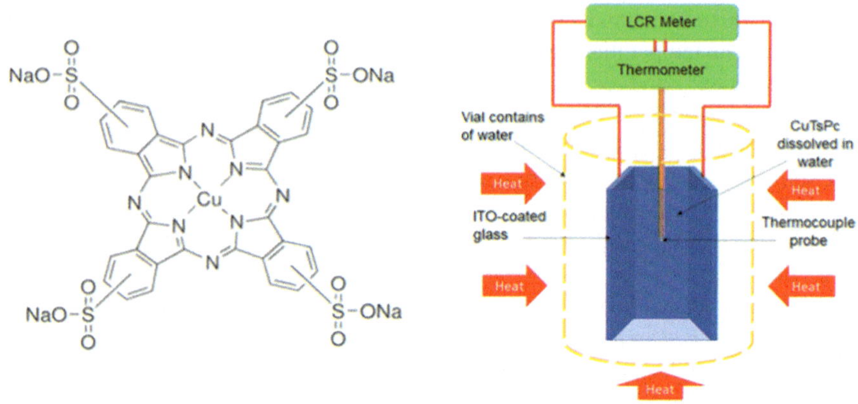

Fig. 1 Molecular structure of the CuTsPc (left), and the temperature sensor (right)

the electrochemical cell container is depicted in a schematic diagram as shown in Fig. 1 (right). The temperature of the electrolytic aqueous solution in this ITO-glass container was measured and recorded using a thermocouple from Fluke 116 HVAC Multimeter. The values of resistance and capacitance of the solution were determined simultaneously using a LCR meter (GW Instek 829). The values of resistance and capacitance were collected every 5 °C in increasing and decreasing directions. In order to observe the change in capacitance and resistance more accurately, the applied temperature was gradually controlled. The experiment has been carried out several times in different ranges of applied temperature to ensure a stable temperature regime while collecting the readings.

2.2 *Preparation of Nanofluids Solution*

The electrolytic nanofluids temperature sensor based on the CuTsPc solution is fabricated as an enhancement of the previous CuTsPc temperature sensor. The additional component of a small amount TiO_2 is used to increase the sensor performance. The selection of the CuTsPc concentration has been made by varying its weight (in mg) and dissolve it into deionized water. The CuTsPc aqueous solution with 0.06 mg/mL concentration has been chosen for the sensor from the previous selection process. The effect of temperature in the range of 25–80 °C has been observed in both resistance and capacitance values. Both values have been recorded in increase and decrease of temperature directions to get a clear picture of its hysteresis. A small amount of TiO_2 of around 5 mg has been added into the CuTsPc aqueous solution to make the electrolytic-nanofluid of TiO_2–CuTsPc mixture. Hot (100 °C) and cold (5 °C) water have been prepared for response/recovery time measurement. The sensor was placed in cold water (as initial state) and immediately placed into hot water for some time. This step has been repeated several times with both resistance and capacitance values were recorded concurrently in obtaining the response/recovery time. The results have been analysed and compared for both sensors without and with the addition of TiO_2 nanoparticles.

3 Result and Discussion

This work has been carried out for the search of potential materials combination for the fabrication of solution-based temperature sensors. It has led to the use of an organic compound copper (II) phthalocyanine-tetrasulfonic acid tetrasodium salt (CuTsPc) that has been dissolved in de-ionized water. Using CuTsPc, an electrochemical cell made of ITO/CuTsPc solution/ITO has been made to characterize its sensing properties in the temperature range of 25–80 °C. The working principle of this temperature sensor is based on the change in the electrical parameters namely resistance and capacitance of the electrochemical cell due to the temperature variation. Fast response and recovery time were achieved including narrow hysteresis attained from a loop of temperature variation in both increase and decrease directions. The fabricated temperature sensor has shown high consistency and sensitive response towards a specific range of temperature values.

The hysteresis curves from the change of resistance and capacitance due to temperature variation are shown in Fig. 2, respectively. The hysteresis value was determined to be less than 3% which is much smaller than that of the previous work that has been achieved [2]. Figure 3 shows the response and recovery time of the CuTsPc temperature sensor. It corresponds to the resistance variation of the temperature sensor when it was measured across increasing temperature from 5 to 100 °C and then it was changed to decrease from 100 to 5 °C. The sensor response and recovery time were determined accordingly to be 20 and 30 s, respectively. It is observed that, the rate of change in resistance value with respect to temperature variation is faster for the response time compared to its recovery time. The presence of oxygen formation was also observed when an applied temperature is increased to nearly reach the water boiling point of 100 °C. It can be one of the factors which later affect the recovery time of the sensor, in which the recovery time will always be longer than the response time. For such case, Lu et al. have further discussed the effect of oxygen to the performance of the sensor in their previous work [3].

In our case, the experiment was done in the temperature range close to the water boiling point and it eventually contributes to the oxygen formation in the form of bubbles on the surfaces of both glass and ITO electrodes. This observation has provided valuable details on the limitation for any solution-based sensor especially using pure water as its medium to have a restricted sensing regime in a temperature range of several degree below 100 °C. This evidence from our previous study when an approximately similar values of response and recovery time were obtained below 95 °C of operating temperature [4, 5]. In the previous work, the hysteresis behaviour of the sensor has shown a stable and consistence value of electrical parameters obtained under the temperature regime below 100 °C.

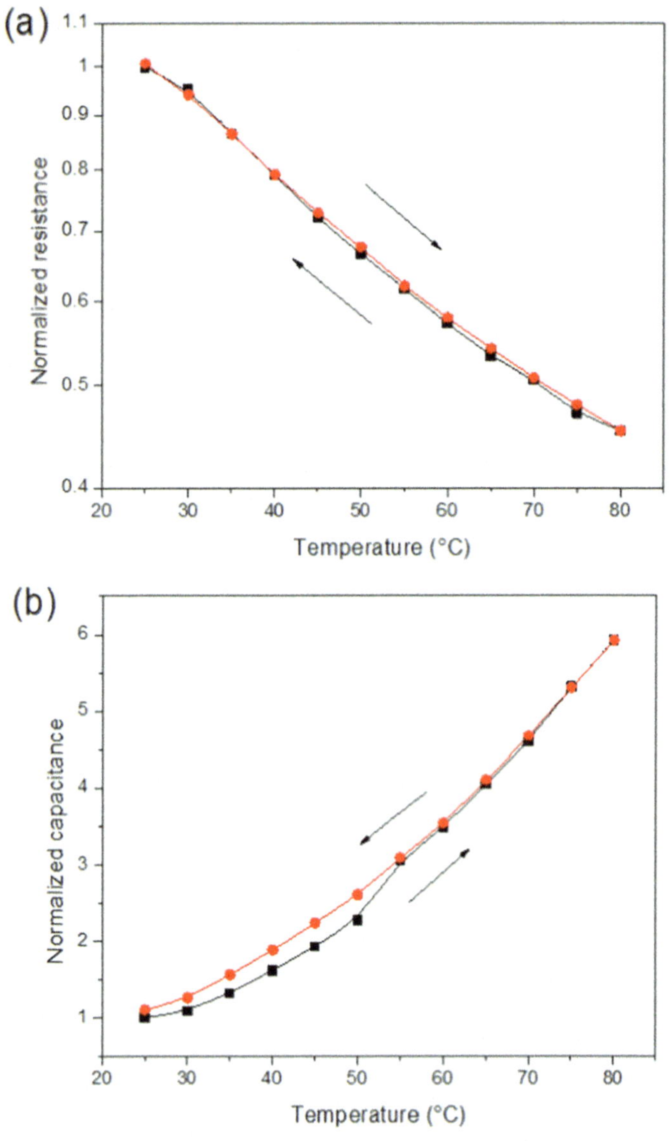

Fig. 2 Hysteresis plot of the normalized resistance (**a**) and capacitance (**b**) towards temperature variation of the cell

In the current work, by using CuTsPc aqueous solution, we have managed to fabricate a highly sensitive and novel electrochemical temperature sensor. A relatively low hysteresis value based on the its gap size that corresponds to the change in resistance has demonstrates a good sensing stability has been achieved and very well-discussed. Based on its hysteresis consistency mainly in normalized resistance

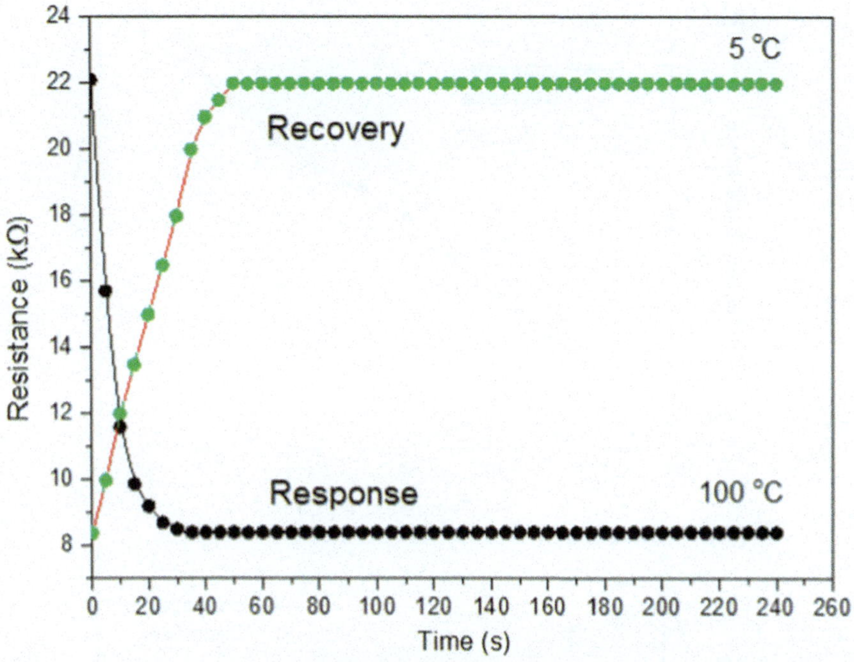

Fig. 3 Response and recovery time of the CuTsPc resistive temperature sensor

curve against the variation in temperature, the sensor is more suitable to be applied as a resistive-type temperature sensor. A good sensitivity behaviour of the sensor was confirmed from its fast response time towards a rapid change of temperature.

Next, this CuTsPc temperature sensor has been further enhanced by using 1% TiO_2 nanoparticles to form an electrolytic nanofluid temperature sensor. All the results are presented below in the similar arrangement like the previous CuTsPc results. Figure 4a, b show the hysteresis for capacitance and resistance, respectively, in both direction of increasing and decreasing temperature. It is observed that, both plots have shown a good hysteresis where the enhancement can be seen from the small deviation between the increase and decrease temperature variation. The linearity of capacitance results is observed to be better than that of the resistance result. At the same time, the range of measured parameters for both capacitance and resistance has changed upon the addition of TiO_2 nanoparticles. Here, the range in capacitance became wider but the range for resistance became smaller. This indicates that the sensitivity of the sensor in capacitance measurement has increased, but the sensitivity in resistance measurement has changed oppositely.

The response and recovery plots for this sensor are shown in Fig. 5a in capacitance and b resistance measurements. The normalized plots were made by C/C_o and R/R_o where C_o and R_o are the initial values of capacitance (137 nF) and resistance (2.7 kΩ), respectively. The response and recovery time in capacitance and resistance are as shown in Table 1. The response and recovery in capacitance are 8 and 15 s

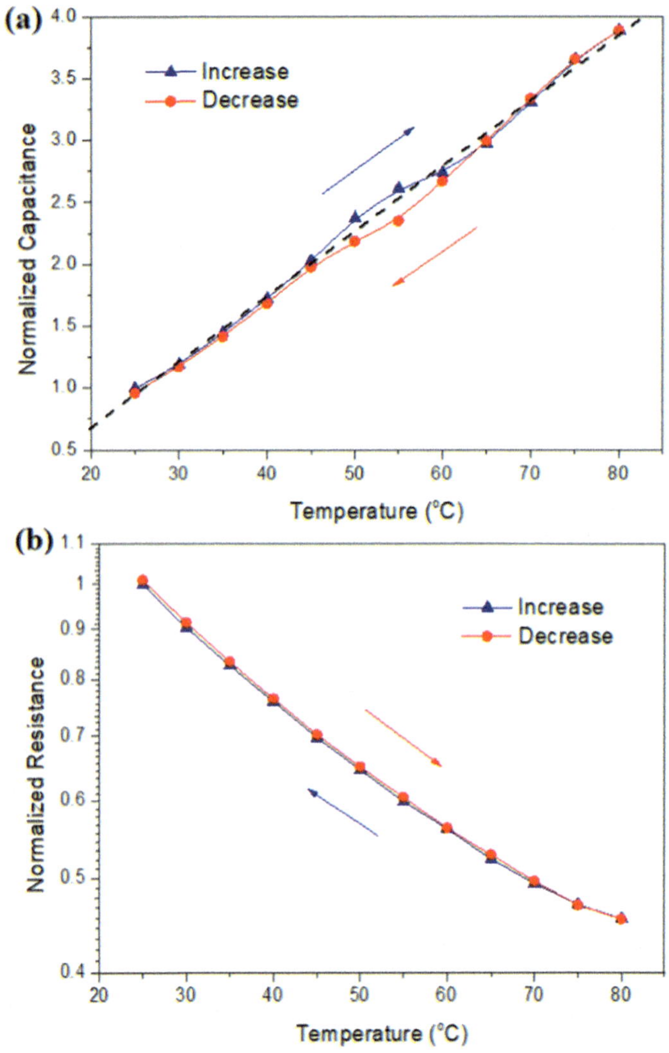

Fig. 4 The hysteresis of the sensor in **a** capacitance and **b** resistance

respectively, while in resistance is > 20 and > 50 s respectively. It can be observed that, the measured capacitance has shown a better and fast response/recovery time compared to the one measured in resistance. Furthermore, the change of capacitance in the CuTsPc temperature sensor with addition of nanoparticles has better performance in linearity, hysteresis, and response/recovery time compared to the CuTsPc temperature sensor without addition of nanoparticles.

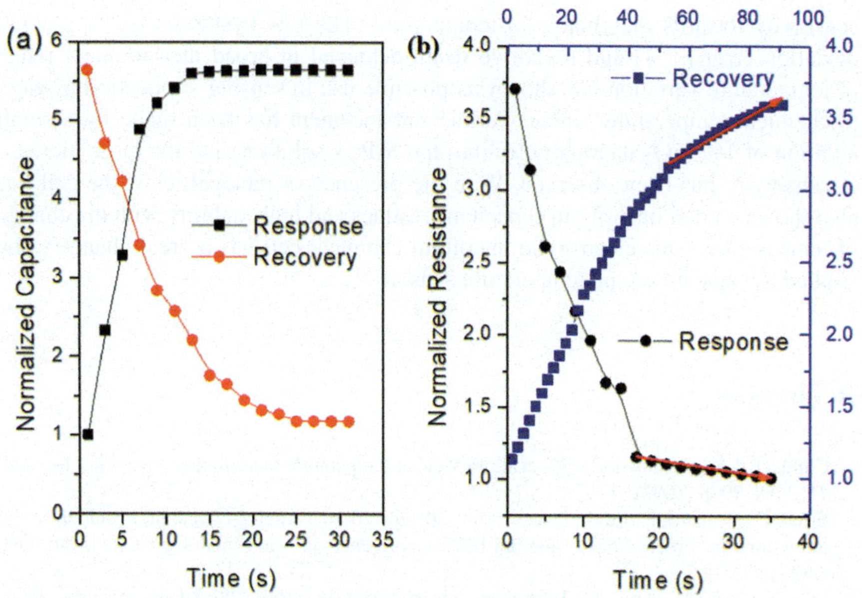

Fig. 5 The response/recovery time in **a** capacitance, and **b** resistance

Table 1 Response/recovery time of TiO_2-CuTsPc electrolytic nanofluid temperature sensor

CuTsPc-TiO_2 temperature sensor	Response (s)	Recovery (s)
Capacitance	8	15
Resistance	> 20	> 50

It can be seen that, from a small addition of TiO_2, large difference has been observed in both sensitivity measured in resistance and capacitance. A lot of enhancements have been obtained from the capacitance value including the sensor sensitivity, response/recovery time, and its linearity. The fabricated TiO_2-CuTsPc temperature sensor has shown a good hysteresis and fast response/recovery time in a measure of capacitance across a certain range of temperature variation [6]. Due to high sensitivity and linearity in capacitance hysteresis, the sensor is suitable to be applied as capacitive-type temperature sensor.

4 Conclusion

A temperature-sensitive chemical cell has been successfully fabricated by using copper (II) phthalocyanine-tetrasulfonic acid tetrasodium salt (CuTsPc), and was properly characterized by means of AC electrochemical measurement using a LCR meter. The sensor cell, was made of ITO-glass as its container, has shown a good sensitivity towards the change in temperature. The low hysteresis (< 2%), small deviation error (± 1) and repetitive result achieved in broad measurement range of temperature variation has shown its possible use in sensing application as electrochemical temperature sensor. Further enhancement has been made by a small addition of 1% TiO_2 nanoparticles into the MPc's solution, and the large increase in sensitivity has been observed. With the presence of nanoparticles, the cell has also shown a good linearity in capacitance values and high stability with the change of temperature. This electrolytic nanofluid chemical cell has a great chance to be applied as capacitive-type temperature sensor.

References

1. Childs, P.R.N., Greenwood, J.R., et al.: Review of temperature measurement. Rev. Sci. Instrum. **71**, 2959–2978 (2002)
2. Erlen, C., Brunetti, F., Lugli, P., et al.: (17–20 June 2006) Trapping effects in organic thin film transistors. In: Proceedings of the 6th IEEE Conference on Nanotechnology, Cincinnati, OH, USA, pp. 82–85
3. Lu, C., Chen, Z., Saito, K.: Hydrogen sensors based on Ni/SiO_2/Si MOS capacitors. Sens. Actuators B Chem. **122**, 556–559 (2007)
4. Ahmad, Z., Abdullah, S.M., Sulaiman, K.: Temperature-sensitive chemical cell based on Nickel (II) phthalocyanine-tetrasulfonic acid tetrasodium salt. Sens. Actuators A Phys. **179**, 146–150 (2012)
5. Abdullah, S.M., Ahmad, Z., Sulaiman, K.: A solution-based temperature sensor using the organic compound CuTsPc. Sensors **14**, 9878–9888 (2014)
6. Abdullah, S.M., Rafique, S., et al.: Highly sensitive capacitive cell based on a novel CuTsPc-TiO_2 nanocomposite electrolytic solution for low temperature sensing applications. Sens. Actuators A Phys. **289**, 94–99 (2019)

Utilizing Perimeter Design in Sandwich Plates for Ship Structures

Rizky Chandra Ariesta, Achmad Zubaydi, Haikal Anjasmara, Roslin Ramli, Mohammad Nurul Misbah, and Wing Hendroprasetyo Akbar Putra

Abstract Sandwich materials are the ship structure design innovations. This material arrangement is based on a core material which is split by two face sheets. Previous research analyzed the application of sandwich material on a tanker's structure. A concern of the selection of a sandwich material is that they have the same strength but reduced weight. Sandwich materials have been widely used in the industrial world, but applications on ships require more detailed calculations. This paper aims to analyze the strength of the sandwich material to be applied to the deck structure. Applying sandwich material on Ro-Ro passenger ship decks is a prospect that can be applied by estimating weight reduction and considering an easy repair process. An analysis is carried out by comparing the existing conditions of the structure and application of the perimeter design benchmarks from rules, also own design. The selected ship is the Roro ship, and this location selection is based on Lloyd's Provisional Rules recommendations. The results of the analyzes were compared with the strength assessment, and the weight loss estimates were presented.

R. C. Ariesta (✉) · A. Zubaydi · H. Anjasmara · M. N. Misbah · W. H. A. Putra
Department of Naval Architecture, Faculty of Marine Technology, Institut Teknologi Sepuluh Nopember, Surabaya, Indonesia
e-mail: chandra@its.ac.id

A. Zubaydi
e-mail: zubaydi@na.its.ac.id

H. Anjasmara
e-mail: haikal.anjasmara98@gmail.com

M. N. Misbah
e-mail: mnmisbah@na.its.ac.id

W. H. A. Putra
e-mail: wing@na.its.ac.id

R. Ramli
Universiti Kuala Lumpur, Malaysian Institute of Marine Engineering Technology, Jalan Pantai Remis, 32200 Lumut, Perak, Malaysia
e-mail: roslin@unikl.edu.my

A. Ismail et al. (eds.), *Engineering Frontiers*, Advanced Structured Materials 208,
https://doi.org/10.1007/978-3-031-56844-2_2

Keywords Sandwich material · Ship structure · Deck · Strength assessment · Weight reduction

1 Introduction

The sandwich material is lightweight, consisting of a core material and two face-plates [1]. The core material has a lower density than the shell material. Production of sandwich material in the structure is made of steel material combined with a perimeter bar at the end of the plate. The core material withstands shear loads, reduces load concentration, and prevents the effects of permanent deformation [2]. One of the significant advantages derived from the utilization of sandwich materials is the reduction in weight, thereby enabling the attainment of higher payloads. Typically, sandwich materials exhibit greater thickness compared to conventional steel materials [3]. This characteristic is advantageous as it provides enhanced structural integrity and load-bearing capacity while still maintaining a lightweight composition.

The use of sandwich materials in car decks has previously been investigated. The use of sandwich material with extruded aluminum profiles has been considered, even though the costs are very high [4]. The investigation results have also been proven admissible based on a strength analysis. Studies on applying sandwich plates to structures on hoistable car decks can reduce the weight by 10%. Weight reduction in sandwich panel systems can generally reduce by 5–15% compared to conventional plates on general cargo barges. In other cases, bulk carriers and SPS container barges are 6–13% and 4–12% [2]. In previous studies, the application of sandwich structures was able to reduce the stress by 14.6% and obtain a weight reduction of 8.87% to 11.6% [3]. Based on these results, it is challenging to obtain a weight reduction of more than 15%; the average result is 5–8%. Moreover, the hybrid sandwich panel was also identified, not to reduce weight but aimed at internal decks and bulkheads because it has a lightweight than a single skin plate structure and can be designed to withstand high pressure [5, 6].

Based on many available sources, analysis of joints in sandwich material structures is rarely used. Fortunately, Lloyd Register issued the rules for applying sandwich plate construction to ship structure [5]. Calculations are carried out for use in new buildings, and the determination of plate thickness can be calculated to obtain equivalent to conventional steel thickness. This study investigated the advantages of sandwich materials using perimeter joints compared to conventional structures. The thickness and perimeter of the sandwich material are designed according to Lloyd's Rules.

2 Methodology

2.1 Description of Ship Structure Model

The initial phase was carried out by selecting the ship model used with an estimated car deck size of 6 m × 1.5 m. The illustration of the joint perimeter model is compared to the two, namely the existing plate and the use of a sandwich panel, shown in Fig. 1.

The car deck structure model being analyzed is in the parallel middle body between the girder frames 36 and 46 with the size of the perimeter model is 6 m in length and 1.5 m in breadth. In this study, the scantling of the car deck consists of profiles measuring 90 mm × 90 mm × 9 mm and a T profile measuring 200 mm × 100 mm × 10 mm thickness of the sandwich plate material designed according to the rules of LR with the equation strength index (R). The previous vehicle's deck plate thickness was 14 mm. In this study, the thickness of the sandwich material used has a faceplate of size 6 mm, and the core is 10 mm. Modifying the model is done by changing the conventional plate system and applying a manufacturing approach by comparing the LR perimeter design (a) and (b) to facilitate the repair process on ships. The configuration of the stiffeners and supports is equal, while the perimeter design varies on the tilt angle of 45° (c) and 63.5° (d) the thickness of 10 mm.

Figure 2 shows a comparison of the perimeter benchmark rules and innovation design. Strength assessment was performed using finite element analysis (FEA) by comparing the values of von Mises stress, normal stress, shear stress, deflection, and the structure's weight. The element used in this structure is of the C3D8R type with six degrees of freedom per node in the core and faceplate material. At the same time, the stiffeners, and girders are modeled with the S4R surface model. Contact between plate surfaces is defined by implementing a node to surface-based tie constraints between deck plates and stiffeners while on plates using surface-to-surface. This

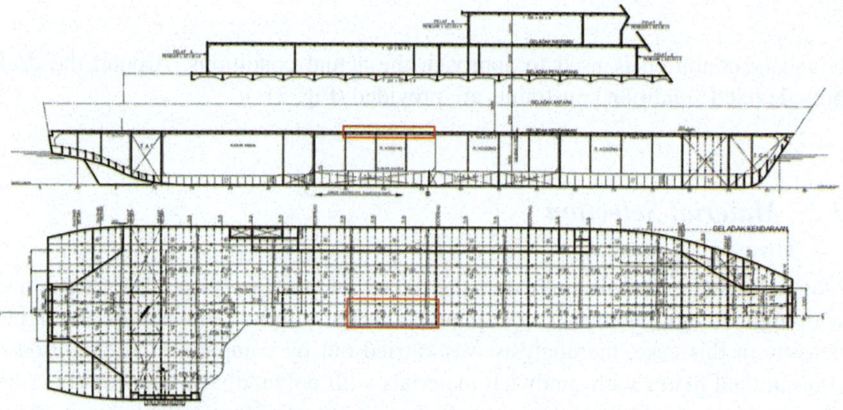

Fig. 1 Details structural arrangement of Roro vessel

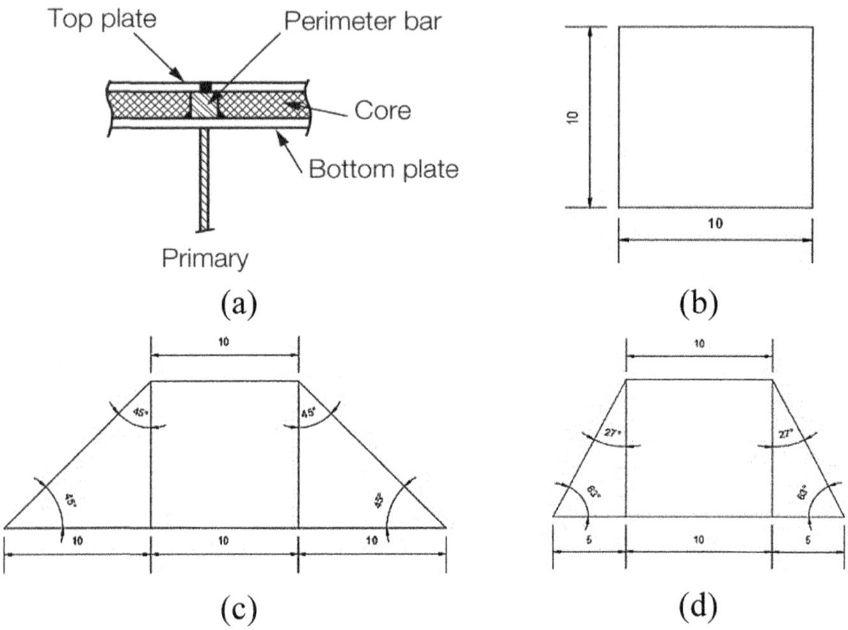

Fig. 2 Perimeter design [7]

Fig. 3 Design of sandwich material [7]

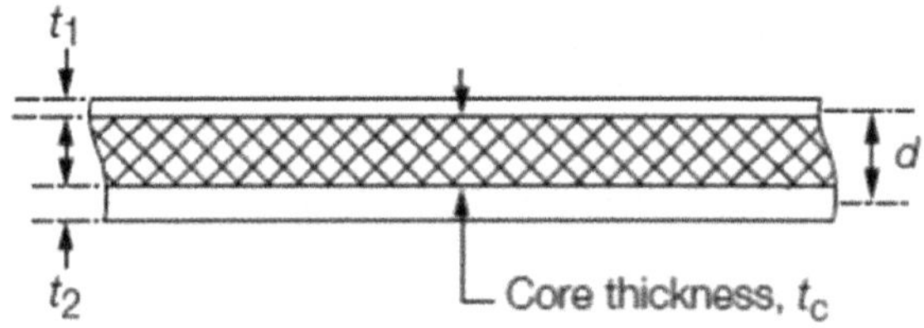

boundary condition is used to approach the actual conditions. Around the deck, provide fixed condition constraints are provided (Fig. 3).

2.2 Material Selection

Material selection in this study uses the theory that the face-plate plates are thinner with high strength, while the core plate is thicker with more flexibility and lighter density. In this case, the analysis was carried out by comparing the condition of conventional plates with sandwich materials with polyurethane fiber matrix cores. The composition of the thickness configuration obtained from calculation is shown in Table 1 and Fig. 1 based on the strength index (R) relation in Eq. (1):

Table 1 Mechanical properties of sandwich material [14]

Properties	Faceplate (Carbon steel)	Core (Polyurethane elastomer matrix fibre)	Unit
Density	7850	1098	kg/m^3
Young's modulus	200e9	901.95e6	MPa
Poisson's Ratio	0.3	0.36	–

$$R = 0.01A_R\left[0.1(b^2/d(t_1 + t_2) + 11.7\left(b.t_c/d^2\right)^{1.3}\right]kP_{eq,R} \tag{1}$$

The mechanical test on the material has met the minimum standard criteria by LR [7]. Where the density value of the material is 1098 kg/m^3. This material has been applied to tanker deck and side hull structures [8] and barge construction in previous studies.

2.3 Design Load

Loading condition scenarios for assessing the deck's strength are carried out based on normal operational loading considering the load on the card deck. The load design conditions are calculated based on the ship's initial structural design rules using the Indonesian Classification Bureau approach to the working car deck load. Pressure load is calculated according to the Biro Klasifikasi Indonesia (BKI) rules. Load conditions are calculated by Eq. (2):

$$P_1 = 9806\ P_w\lambda \tag{2}$$

P_1 is the corrected patch load (N), Pw is the load in tonnes on the tyre print, and λ is the dynamic magnification factor 1.25 in harbour conditions and for seagoing. The magnitude of the load is a truck with size 11.9 m × 2.5 m which produces a wheelprint force in the middle of a span of 88,200 N. This is used as wheel load 1 and 1,76,400 N is used as wheel load 2 acting on the structure. The load print is represented in the red area of the heavy truck shown in Fig. 4.

The present design is purposed for accommodating the loading conditions encountered on the ship's deck during the transportation of the heaviest loads. This assumption is predicated upon the necessity to thoroughly evaluate the structural strength response of the ship deck after the implementation of the novel design.

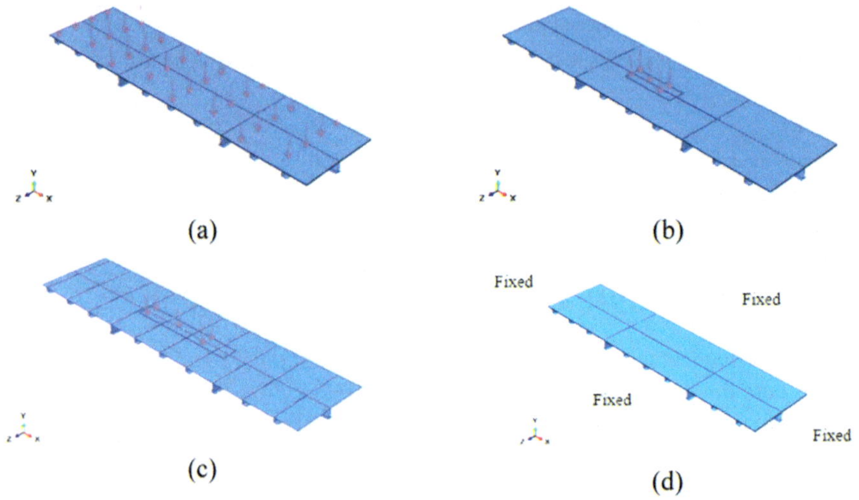

Fig. 4 Reference area model: **a** pressure load, **b** load wheel 1, **c** load wheel 2 and **d** boundary condition

2.4 Convergency Study of Meshing

Convergence study analysis is essential during identification with finite element analysis. During the analysis process, errors can be found; for example, if the mesh needs more details, the results do not make sense. Thus, convergence studies aim to ensure the accuracy of the results and the stability of the running process. The mesh convergence method is carried out by looking at the von Mises stress curve parameter on the *y*-axis and the number of elements on the *x*-axis, which can be seen in Fig. 5. The selected global mesh size is 0.03 m to get the appropriate model verification shown in the convergence mesh area above.

Fig. 5 Convergence study of structure car deck

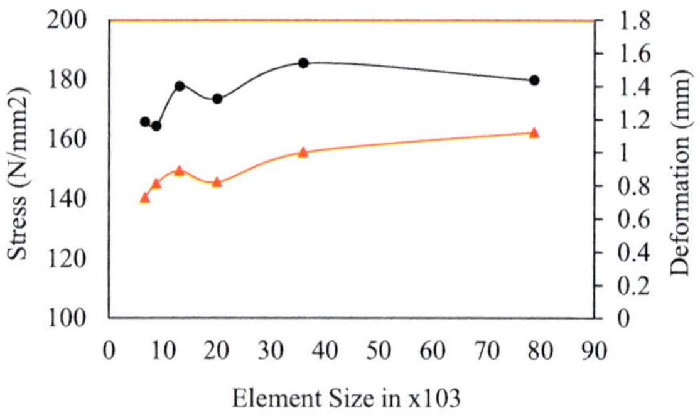

The black and red color lines in the graph in Fig. 5 depict the von Mises stress displacement that manifests in the deck structure. For a single computation, approximately $\pm$ 35,000 elements were employed in this analysis.

2.5 *Acceptance Criteria*

The acceptance criteria of permissible strength is determined by the Biro Klasifikasi Indonesia (BKI) rule where the ship was operating. Maximum allowable stress conditions are 180 MPa for von Mises stress; normal stress is 140 MPa; sheer stress is 90 MPa, and allowable deformation is under 12 mm [9]. This design is an allowable limit when the ship operates, and this aims to ensure the safety of passengers from each deck on the ship.

3 Results and Discussion

The priority for using sandwich materials is in the ship structure by conducting an initial strength analysis and estimating the weight reduction by applying a joint perimeter design according to the LR rules. This investigation ensures that sandwich materials are profitable and suitable for use by the shipyard or the ship owner. This reduction in weight percentage compensates for technical, practical, and financial issues in the design and production of sandwich materials. It can also be used as a reference for ship owners, classification societies and management of shipyards to decide on the use of sandwich materials. Figure 6 compares the von Mises, shear stress and deformation in the model. The model is shown by (a) existing model, (b) parameter LR, (c) parameter 1, (d) parameter 2. Based on the given loading, the results of the finite element analysis results are shown in Figs. 6, 7 and 8.

Using sandwich materials to reduce the structure has main parameters that play an essential role. The weight on the ship is related to the value of the density of the material. Compensating for the density of lighter materials in the core can increase the volume of the ship's cargo. Therefore, this study needs a strength assessment to examine the applied sandwich material's strength directly. This section presents observational data from comparing conventional designs, perimeter based on rules, and design innovations. This investigation was conducted to assess the performance of each variation. In addition, Table 2 presents that stress and deformation are also complied with criteria for acceptance of sandwich materials. One of the challenges of using sandwich materials in structures is that stress concentrations due to loads can cause damage to the core structure [10].

In this case, the weight reduction calculation involves the conversion of the model plate structure in the deck panel. The conventional model plate weighs 989.1 kg, whereas the sandwich model plate has a weight of 946.62 kg, resulting in a reduction in the overall weight of the structure on the plate. In addition to the significant weight

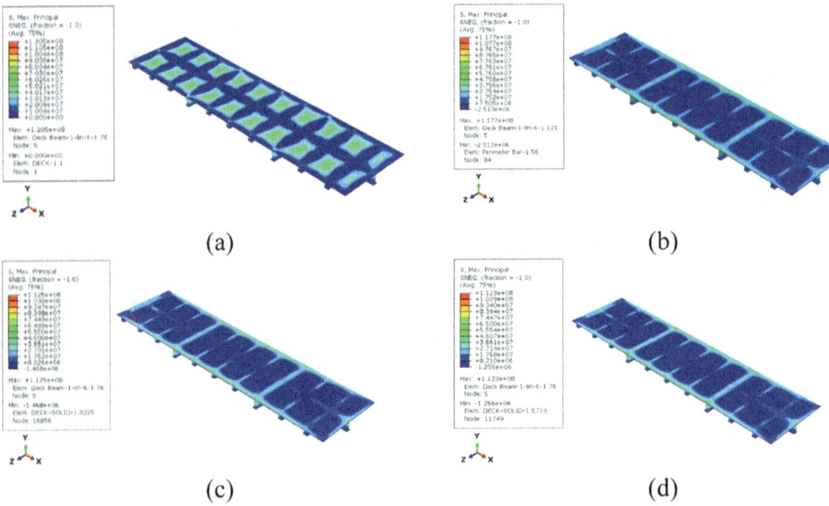

Fig. 6 Comparison of von Mises stress following models

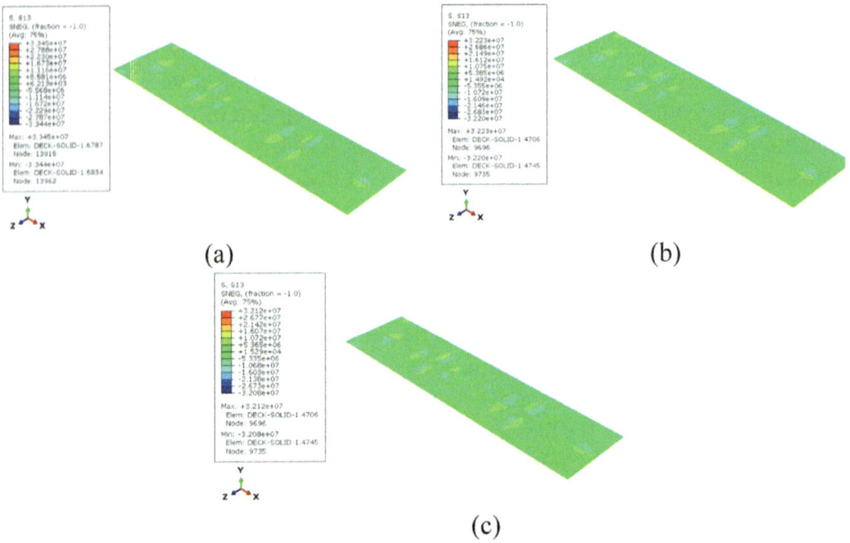

Fig. 7 Comparison of shear stress following models

reduction achieved, the utilization of this structure also brings about noteworthy advantages in terms of repairability and maintenance [11]. The alterations made to the conventional plate design, including the incorporation of a perimeter bar, facilitate easier repairs for the sandwich plate. Notably, this design has been demonstrated to reduce the load on the structure by 4.23% compared to conventional structures.

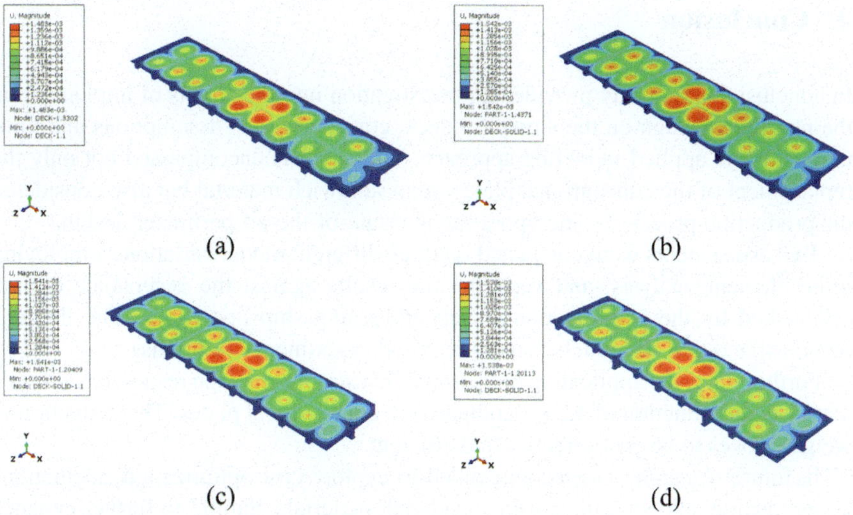

Fig. 8 Comparison of deformation following models

Table 2 Case analysis results

Cases		Von Mises (MPa)	Normal stress (MPa)	Shear stress (MPa)	Deflection (mm)
Existing	Deck load	161.8	120.5	–	1.483
	Wheel load 1	15.8	15.8	–	0.166
	Wheel load 2	44.28	42.39	–	0.464
Perimeter LR	Deck load	166.7	117.7	33.45	1.542
	Wheel load 1	14.26	13.6	3.273	0.200
	Wheel load 2	41.38	37.17	9.716	0.526
Perimeter 1 (45°)	Deck load	155.6	112.5	32.23	1.541
	Wheel load 1	142.4	136	3.258	0.204
	Wheel load 2	41.32	37.18	9.689	0.532
Perimeter 2 (63°)	Deck load	155.1	112.3	32.12	1.538
	Wheel load 1	14.21	13.63	3.252	0.206
	Wheel load 2	41.2	37.21	9.654	0.534

Moreover, when it comes to replacing the plate, the process does not require cutting, but rather allows for the convenient opening of the perimeter panels and the replacement of specific components as needed. This streamlined approach simplifies the inspection of the sub-structure, ensuring efficient maintenance practices [12, 13].

4 Conclusion

In conclusion, this study provided an investigation into the effects of implementing the perimeter design on the structural deck, considering the assumptions and limitations of the applied modeling approach. The analysis encompassed not only the replacement of the conventional plate with a sandwich material but also considered the production process by incorporating the state-of-the-art perimeter design.

Two assessments were conducted on three different model variations, employing finite element analysis, and verifying the results against the acceptance criteria established by the classification society governing ship operations. The findings demonstrated that all models met the required assessment conditions.

Furthermore, the application of the sandwich structure design resulted in a significant weight reduction of 4.23% compared to conventional plates. This reduction in weight proves to be cost-effective and efficient.

In future research, it is recommended to explore cost optimization calculations for producing ship structures using sandwich materials, aiming to further enhance effectiveness and efficiency.

Acknowledgements This research was funded by the Department of Naval Architecture, Faculty of Marine Technology, Institut Teknologi Sepuluh Nopember research scheme.

References

1. Kopp, R., Nutzmann, M., van Santen, J.: Formability of lightweight, vibration damping and medium perfused sandwich sheets. In: Sandwich Structures 7: Advancing with Sandwich Structures and Materials (2005). https://doi.org/10.1007/1-4020-3848-8_73
2. Momčilović, N., Motok, M.: Estimation of ship lightweight reduction by means of application of sandwich plate system. FME Trans. **37**(3), 123–128 (2009)
3. Tuswan, A.Z., Budipriyanto, A., Sujiatanti, S.H.: Comparative study on ferry Ro-Ro's car deck structural strength by means of application of sandwich materials, no. September, pp. 87–96 (2020). https://doi.org/10.5220/0008542800870096.
4. Gunnarson, M.: Beneficial use of aluminium in RoRo equipment. Göteborg (1994)
5. Hakansson, M., Johnson, E., Ringsberg, J.W.: Cost and weight of composite ship structures: a parametric study based on Det Norske Veritas rules. J. Eng. Maritime Environ. 1–20 (2017)
6. Kamaruzaman, A.A.B., Ismail, A., Baharuddin, B.A., Rahman, F.A., Hamid, D.A., Khalid, P.Z.M.: Strength analysis of the hull structure for a submersible drone. In: Advanced Structured Materials, vol. 166 (2022). https://doi.org/10.1007/978-3-030-89992-9_16
7. Lloyds Register: Rules for the application of sandwich panel construction to ship structure. © Lloyd's Register Group Limited 2019 (2019)
8. Ismail, A., Zubaydi, A., Piscesa, B., Tuswan, T., Ariesta, R.C.: Study of sandwich panel application on side hull of crude oil tanker. J. Appl. Eng. Sci. (2021). https://doi.org/10.5937/jaes0-30373
9. BKI: Consolidated edition 2021. In: BKI Rules for Hull Volume II Tahun 2021, vol. II (2021)
10. Ariesta, R.C., Zubaydi, A., Ismail, A.: Identification of damage in a ship hull sandwich plate by natural frequency. no. February, 2021, https://doi.org/10.1088/1757-899X/1034/1/012012

11. Kondratiev, V., Gaidachuk, A.: Weight-based optimization of sandwich shelled composite structures with a honeycomb filler. Eastern-Euro. J. Enterp. Technol. **97**(1), 24–32 (2019). https://doi.org/10.15587/1729-4061.2019.154928
12. SAND: CORe. In: Best Practice Guide for Sandwich Structures in Marine Applications (2013)
13. Zubaydi, A., Budipriyanto, A.: Material Sandwich: Teori, desain, dan aplikasi, 1st edn. Airlangga University Press, Surabaya (2020)
14. Ariesta, R.C., Zubaydi, A., Ismail, A., Tuswan, T.: Damage evaluation of sandwich material on side plate hull using experimental modal analysis. Mater. Today Proc. (2021). https://doi.org/10.1016/j.matpr.2021.04.293

The Water Quality Index: Effect of Antifouling Paint

Norazlina Abdul Nasir, Abdul Rauf Abdullah, Hanisah Johor, Asmalina Mohamed Saat, Fatin Zawani Zainal Azaim, and Muhammad Amir Hussein

Abstract Maintaining or improving water quality is very important to be concerned by both communities and authorities especially on the marine environment which inhabited by the marine life. However, the development of technology keeps increasing, which mostly leads to the pollutions towards the environment including water pollution. One of the pollutants sourcing from the maritime activities is antifouling paint. This may relate to Sungai Manjung's water quality in Perak as there are several maritime activities nearby such as port operations, vessel maintenance and others. The aim of this study was to identify the water quality index and to compare the trend of corrosion defects of aloe vera on capsaicin antifouling additive and standard antifouling paints towards water quality. In this paper, 17 samples of steel plates were submerged at Sungai Manjung for 12 weeks. The initial water parameters, which are pH value, dissolved oxygen (DO), salinity and total dissolved solid (TDS), were being taken. The weight, water parameters, surface condition of samples are recorded. An observation on the conditions of weather and tide level were recorded. A total of 65 data for water parameters collected and analyzed using the procedure of the Canadian Council of Ministers of the Environment (CCME WQI). The weight of the samples was analyzed by using simple regression. The results found that the water quality index for Sungai Manjung is poor after the sample were submerged. On the other hand, the sample steel plate with natural additive gives

N. A. Nasir (✉) · A. R. Abdullah · H. Johor · M. A. Hussein
Maritime Management, Malaysian Institute of Marine Engineering Technology, Universiti Kuala Lumpur, Lumut, Perak, Malaysia
e-mail: norazlinaan@unikl.edu.my

H. Johor
e-mail: hanisah@unikl.edu.my

A. M. Saat
Marine Electrical and Engineering Technology, Malaysian Institute of Marine Engineering Technology, Universiti Kuala Lumpur, Lumut, Perak, Malaysia
e-mail: asmalina@unikl.edu.my

F. Z. Z. Azaim
Student Development, Malaysian Institute of Marine Engineering Technology Lumut, Universiti Kuala Lumpur, Lumut, Perak, Malaysia
e-mail: fatinzawani@unikl.edu.my

A. Ismail et al. (eds.), *Engineering Frontiers*, Advanced Structured Materials 208,
https://doi.org/10.1007/978-3-031-56844-2_3

lower trend of corrosion defects compared to the streel plate with antifouling paint only.

Keywords Water quality · CCME WQI · Trend of corrosion defects · Natural antifouling paint · Sungai Manjung

1 Introduction

About 14,000 people die per day in the world due to the water pollution, which causes diseases and leads to the risk of death [1]. One of the factors that leads to water pollution is the maritime activity. This includes the usage of antifouling paint applied to the surface of vessel or any marine structure to prevent the biofouling process. The antifouling paint affects the water quality by releasing antifouling paint particles (APPs) into the seawater. APPs refer to residues that mostly produced from maintenance, cleaning and repair of vessel activities and also from marinas and fishing harbours [2]. APPs consist of dangerous chemicals and dissolved metals especially zinc and copper, which can affect the water properties negatively thus affecting the marine environment. Using natural additives in antifouling paints, the water pollution rate can be reduced as well as preserving the water quality.

The study on the water quality index at Sungai Manjung found that the river is slightly polluted [3]. Based on the report, the trend of water quality decreased from class ii to class iv. The report by the Department of Environment Malaysia (DOE) 2020 stated that the water quality in Sungai Manjung is still in good condition [4]. However, the station of sample taken was in the upstream of the river. In the estuaries of Sungai Manjung where most of the maritime activities took place, the water quality has not been evaluated. The maritime industries surrounding Sungai Manjung lead to the metal contamination able to change water quality, resulting in changes in the abundance and distribution of sessile and mobile organisms. This will increase metal content in organism tissues [5]. A study in 2019 found that the heavy metals found in Perak River, which are iron (FE), manganese (Mn), lead (Pb), zinc (Zn) and calcium (Ca), are significantly correlated with water properties including temperature, dissolved oxygen, pH and electrical conductivity [6]. This shows that the APPs containing dissolved metals can affect the water quality thus leading to the changes of environmental conditions as well as the habitat of aquatic life. However, there is no such research that has been made at Sungai Manjung which relates to the metal contamination affecting water quality.

Capsaicin is a compound that is responsible for the pungency of capsicums. An element that can be extracted from capsicum is the high-density polyethylene capsaicin (HDPE-capsaicin) powder. High dosage of capsaicin showing encouraging antibacterial property towards gram-negative Escherichia coli and gram-positive marine Bacillus sp. Bacteria. HDPE-capsaicin coatings have been tested on their ability to remove algae adhesion [7]. The antifouling effect and removal rate as high

Table 1 National water quality standards for Malaysia in Class 1

Parameter	Unit	Value
Dissolved oxygen	**mg/l**	**7**
pH	–	6.5–8.5
Salinity	**ppt/%**	**0.5/0.05**
Total dissolved solid	mg/l	500

as 90%. Another natural ingredient used is aloe vera, which is used in the biosynthesis in nanoparticles as a reducing agent [8]. Olive oil is also utilised as natural component in antifouling to improve the coating performance.

This study investigates the water quality index at one station around Sungai Manjung and trend of corrosion defects for different antifouling additive used. The aims is to determine the Sungai Manjung water quality and the trend of corrosion rate of steel plate within different additive antifouling paints. The National Water Quality Standards for Malaysia in Class 1 show in Table 1 was referred to determine the water quality level.

2 Methodology

The samples of A36 steel plate with different mixing ratios of antifouling paint and natural additives were submerge in Sungai Manjung. Table 2 shows the details of steel plate samples. In this study, 17 samples of steel plates with different types of antifouling additives including primer, capsaicin, aloe vera and olive oil were submerged at jetty area in Sungai Manjung as shown in Fig. 1. The base metal is a steel plate that was not layered with any antifouling paint. The bare metal steel plate was required to be polished since there were some black deposits found on its surface. This may affect the corrosion rate of the steel plate surface (Fig. 2).

Steel plate primer is layered with antifouling primer, which is commonly used in the ship hull, coastal structures as well as underwater area in any structures related to marine. Steel plate A was not applied with both standard and natural antifouling paint. Steel plate B and the others were applied with single, two or three types of natural antifouling additive in different ratios according to [9]. Before the samples were immerse, the river water sample was taken. In this paper, 17 samples of steel plates with different types of antifouling additives including primer, capsaicin, aloe vera and olive oil, were submerged at Sungai Manjung.

The frames and pipes were used to support the steel plates underwater. The frames were used to hold the steel plates in the water and prevent them from being swayed and loss into the water. The pipes served as the buoy for the frames and steel plate to stay afloat in the water. This can avoid the steel plates to touch the seabed and stay in the water although the water level is decreasing during the low tide. The frames used were made from chengal wood because of its high durability and good water resistance. Three frames held four steel plates each and frame was holding five

Table 2 List of steel plate with natural additive

Label	Percentage of antifouling additive (%)			
	Antifouling	Capsaicin	Aloe vera	Olive oil
BM	–	–	–	–
Pri	100	0	0	0
A	100	0	0	5
B	100	5	0	0
C	100	5	25	0
D	100	5	5	0
E	100	5	15	0
F	100	5	25	0
G	100	0	5	0
H	100	5	0	5
I	100	5	5	0
J	100	5	5	5
K	100	5	5	15
L	100	10	0	0
M	100	10	5	0
N	100	10	5	5
O	100	10	5	15

Note BM = Bare Metal; Pri = Primer

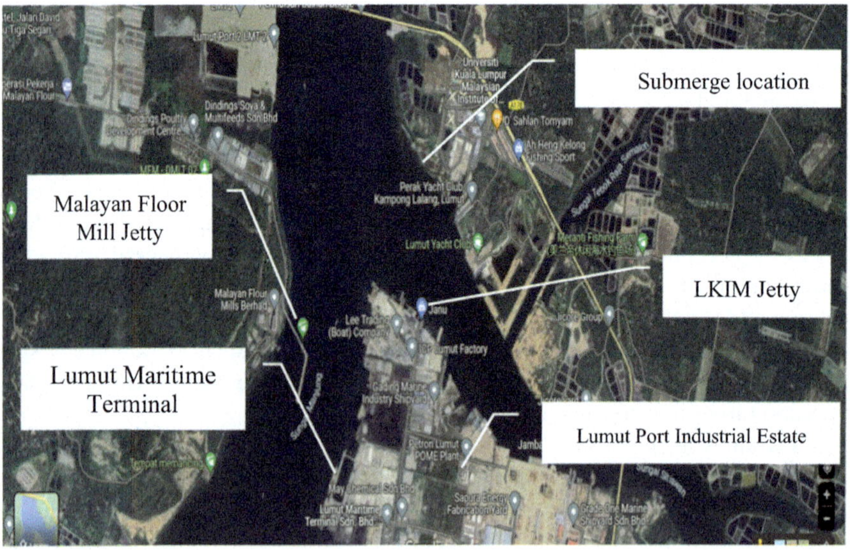

Fig. 1 Study area

Fig. 2 Comparison of surface before immersing and at week 12

steel plates. The pipe used is made of polyvinyl chloride (PVC), which is suitable to become a buoy for the wood frame. The initial data of water sample had been collected in a week before sample were immersed, i.e. observation of the weather condition and tide level. The data of water quality parameters, the weight of steel plates and their surfaces' pictures were taken weekly for 12 weeks. The data collected then were analyzed by using CCME WQI and simple linear regression.

3 Results and Discussion

The results of water quality index are shown in Table 3. The three main values were based on the National Water Quality Standards for Malaysia in Class 1. Class I refers to the standard limit for the conservation of natural environment which is suitable to determine the degree of Sungai Manjung contamination from its original state. The result of CCME WQI score is 19.69. This shows that the water quality around the immersion area is poor. The CCME WQI value obtained is within the scale of 0–44 according to the CCME WQI index categories [10, 11]. Compared to the water before submerging, DO, TDS and salinity increased after 12 weeks. This proves that the river is almost contaminate fully or its state is not within the desirable quality or its natural water quality. Figure 3 shows the surface of steel plate after 12 weeks immersed in the river. The surface condition is mostly full of the barnacles. Under the National Water Quality Standards for Malaysia in Class I, the total dissolved solid (TDS) and the salinity of the river were in critical state as all the parameters' value taken failed or exceeded the limit standards for the conservation of natural environment.

Simple linear regression was used to determine the trend of corrosion defects for the steel plates. The trend of corrosion rate used the steel plates' weight in linear regression is show in Fig. 4. Table 4 shows the weight of each sample in gram. The

Table 3 Results of water parameter

Parameter	pH	TDS (ppm)	DO (mg/L)	Salinity (%)	Temperature (°C)
Initial	7.0	2633	7.00	24	26.1
W1	7.0	2064	7.84	26	27.0
W2	7.0	2143	7.89	24	29.5
W3	7.0	2221	8.30	24	28.1
W4	7.1	1986	8.60	25	28.4
W5	7.1	1907	8.67	28	28.1
W6	7.2	1986	7.00	29	28.4
W7	7.2	1829	7.90	29	26.5
W8	7.1	1671	8.80	26	29.8
W9	6.9	1514	8.53	22	27.9
W10	6.9	1671	8.20	28	27.4
W11	6.9	1750	8.02	27	27.1
W12	6.9	1593	8.00	26	27.6

Fig. 3 Sample K surface in week 0 and week 12

results show that the trend of corrosion defects for natural antifouling paint in sample K is lower compared to the standard antifouling paint, in sample primer steel plate. However, there were some steel plates with natural antifouling paint applied, which were sample plates A, D, F, E, B and J, that showed a higher trend compared to the sample primer. This shows that the ratio or percentage of natural additives is very crucial in affecting the antifouling paint's performance even though the steel plates used the same type of natural additives. This study found that the lowest trend of corrosion rate was sample steel plate K which consist of 5% of capsaicin, 5% of aloe vera and 15% olive oil as its natural additive in the antifouling paint. Based on the trend, sample K shows slower rate in increasing its weight as well as the accumulation of the barnacles on its surface (Fig. 5).

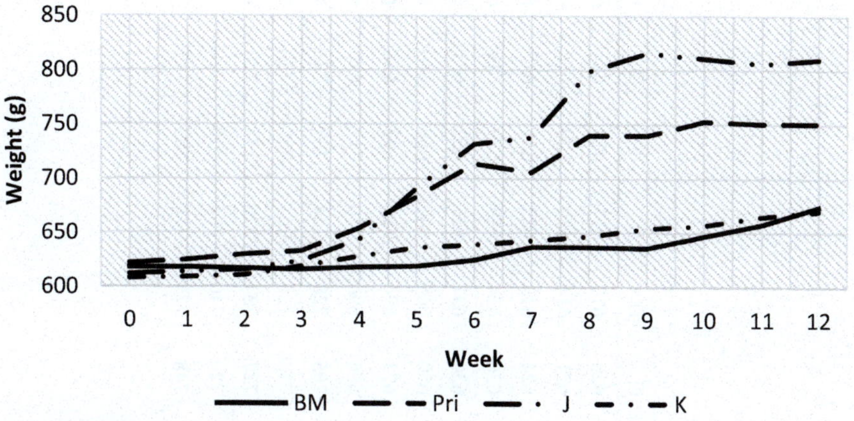

Fig. 4 Trend of corrosion rate in 4 samples

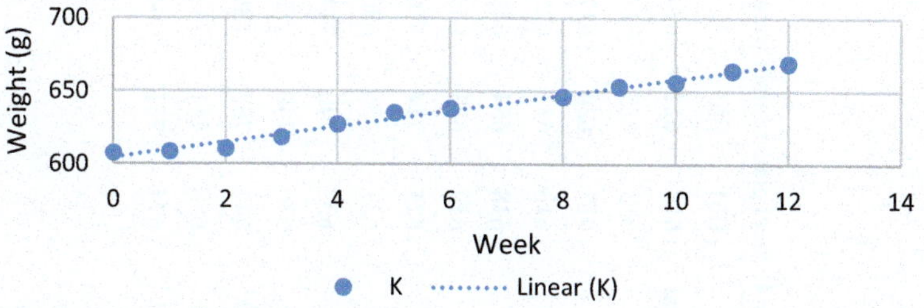

Fig. 5 Trend of corrosion rate sample K

Table 4 Steel plate sample weight

	BM	Pri	A	B	C	D	E	F	G	H	I	J	K	L	M	N	O
Initial	**618**	**622**	**614**	**615**	**620**	**612**	**617**	**628**	**614**	**612**	**619**	**612**	**608**	**621**	**619**	**609**	**617**
W1	618	625	617	617	621	614	619	629	616	613	621	614	609	623	620	610	619
W2	**617**	**630**	**617**	**618**	**624**	**615**	**622**	**631**	**617**	**615**	**624**	**616**	**611**	**624**	**625**	**613**	**622**
W3	616	633	621	622	626	624	629	639	619	617	627	624	619	630	633	625	629
W4	**618**	**654**	**629**	**628**	**634**	**637**	**635**	**652**	**631**	**626**	**641**	**644**	**628**	**639**	**645**	**643**	**644**
W5	619	683	648	653	650	668	640	669	646	639	656	691	636	653	665	672	651
W6	**625**	**714**	**670**	**681**	**672**	**696**	**718**	**698**	**658**	**657**	**666**	**732**	**639**	**672**	**675**	**702**	**667**
W7	637	706	690	704	686	696	710	712	680	665	686	738	643	674	675	678	672
W8	**637**	**740**	**711**	**738**	**705**	**720**	**744**	**737**	**723**	**684**	**699**	**799**	**647**	**685**	**694**	**704**	**692**
W9	636	740	720	752	718	726	763	757	723	691	708	816	654	693	699	705	702
W10	**647**	**753**	**739**	**763**	**727**	**734**	**776**	**758**	**729**	**695**	**712**	**811**	**657**	**701**	**702**	**707**	**704**
W11	658	751	747	768	729	746	764	770	719	699	735	806	665	712	700	702	702
W12	**674**	**751**	**764**	**783**	**748**	**775**	**778**	**783**	**717**	**709**	**763**	**810**	**670**	**715**	**707**	**713**	**702**

4 Conclusion

Increase in environmental impact of antifouling paint pollution urges the need to shift into a more environmentally friendly antifouling solution. The usage of natural additives in antifouling shows better improvement in the reducing ecological impact from conventional antifouling paints. The results only showed the effect in 12 weeks. These results may be due to the industrial activities near the immersion location. Not only the maritime operations, but also related to the presence of the ships near the jetty as there were some maintenances on ship conducted occasionally. Agricultural activities also may contribute to the increasing TDS as excessive fertilizer and pesticides are leeching from the land. This may relate to the highest concentration of nitrate in the water body causing the increases of TDS [12]. Long-term brackish water tidal inundation and enhanced salt sediment may be the main causes of rising water salinity [13].

References

1. Chaudhry, F.N., Malik, M.F.: Factors affecting water pollution: a review. J. Ecosyst. Ecography **7**(1), 225–231 (2017)
2. Torres, F.G., De-la-Torre, G.E.: Environmental pollution with antifouling paint particles: distribution, ecotoxicology, and sustainable alternatives. Mar. Pollut. Bull. **169**, 112529 (2021)
3. Mohd Rohani, N.A.B.M.R.: Study on water quality at Manjung River and its tributaries (2012)
4. Yong, Z.J., Bashir, M.J., Hassan, M.S.: Biogas and biofertilizer production from organic fraction municipal solid waste for sustainable circular economy and environmental protection in Malaysia. Sci. Total. Environ. **776**, 14596 (2021)
5. Kyei, S.K., Darko, G., Akaranta, O.: Chemistry and application of emerging ecofriendly antifouling paints: a review. J. Coat. Technol. Res. **17**, 315–332 (2020)
6. Salam, M.A., Kabir, M.M., Yee, L.F., Khan, M.S.: Water quality assessment of Perak River, Malaysia. Pollution **5**(3), 637–648 (2019)
7. Liu, Y., Shao, X., Huang, J., Li, H.: Flame sprayed environmentally friendly high density polyethylene (HDPE)–capsaicin composite coatings for marine antifouling applications. Mater. Lett. **238**, 46–50 (2019)
8. Rasli, N.I., Basri, H., Harun, Z.: Biosynthesized nanoparticles from aloe vera: a brief review towards membrane technology. Malaysian J. Fundam. Appl. Sci. **15**(6), 895–902 (2019)
9. Saat, A.M.: Performance of Rosin modification antifouling coasted on mild steel surface at various immersion orientation. Int. J. Innov. Technol. Explor. Eng. **8**(2) (2019)
10. Akhtar, N., Ishak, M.I.S., Ahmad, M.I., Umar, K., Md Yusuff, M.S., Anees, M. T., Ali Almanasir, Y.K.: Modification of the water quality index (WQI) process for simple calculation using the multi-criteria decision-making (MCDM) method: a review. Water **13**(7), 905 (2021)
11. Alam, M., Ashfaq, M., Hassan, S., Ali, A.: The analysis of groundwater quality and the impact of remedial measures adopted by the wheat growers: using endogenous switching regression model approach. J. Econ. Impact **3**(3), 144–151 (2021)
12. Sanou, A., Coulibaly, S., Guéi, A.M.L., Baro, M., Tanon, E.F.T., Méité, N., Atsé, B.C.: Assessment of some physico-chemical parameters of the fish farm water in Abengourou, Côte d'Ivoire. Egypt. J. Aquatic Biol. Fish. **26**(5) (2022)
13. Akter, S., Ahmed, K.R., Marandi, A., Schueth, C.: Possible factors for increasing water salinity in an embanked coastal island in the southwest Bengal Delta of Bangladesh. Sci. Total. Environ. **713**, 136668 (2020)

Experimental and Finite Element Analysis of Fibre-Reinforced Polymers Based on Tensile Testing

Roslin Ramli, Mohd Hisbany Mohd Hashim, Anizahyati Alisibramulisi, Suhailah Mohamed Noor, Noorlee Boonadir, Nurain Jainal, and Sjarief Widjaja

Abstract This paper presents the experimental and finite element analysis (FEA) results of a fiber-reinforced polymer (FRP) tensile test. The experimental tensile test is using the universal testing machine "Quasar 100" while FEA is using the ANSYS workbench software. The tensile test has two groups of specimens in which, each has five samples in it. Each group has a different orientation angle of + 45° and -45°. The FRP material which affects the angle orientation is the woven roving mat. In addition to woven roving mat, tissue mats and chopped strand mats are used. The polymer (resin) is used as unsaturated polyester (Norsodyne 3338W), and the catalyst is methyl ethyl ketone peroxide (MEKP). The stacking sequence of the fibres is nine

R. Ramli (✉) · N. Boonadir
Maritime Engineering Technology Section, Universiti Kuala Lumpur Malaysian Institute of Marine Engineering Technology, 32200 Lumut, Perak, Malaysia
e-mail: roslin@unikl.edu.my

N. Boonadir
e-mail: noorlee@unikl.edu.my

M. H. M. Hashim · A. Alisibramulisi
Faculty of Civil Engineering, Universiti Teknologi MARA (UiTM), 40450 Shah Alam, Selangor, Malaysia
e-mail: hisbany@uitm.edu.my

A. Alisibramulisi
e-mail: aniza659@uitm.edu.my

S. M. Noor
Faculty of Civil Engineering, Universiti Teknologi MARA (UiTM), 13500, Permatang Pauh, Pulau Pinang, Malaysia
e-mail: suhailahmn@uitm.edu.my

N. Jainal
Business School, Universiti Kuala Lumpur, 54000 Kuala Lumpur, Malaysia
e-mail: nurain@unikl.edu.my

S. Widjaja
Department of Naval Architecture, Faculty of Marine Technology, Institut Teknologi Sepuluh Nopember, Surabaya, Indonesia
e-mail: sjarief@na.its.ac.id

A. Ismail et al. (eds.), *Engineering Frontiers*, Advanced Structured Materials 208,
https://doi.org/10.1007/978-3-031-56844-2_4

layers of the same size (250 mm × 25 mm × 8 mm) in a flat bar that is balanced and symmetric according to ASTM D3039. The results were presented in the form of experimental and FEA results, showing that [A + 45° and AA + 45°] orientation has a higher tensile strength as a function of angle orientation. The results were significant for future studies and the implementation of the authentic boat.

Keywords Finite element analysis · E-glass · Angle orientation and tensile

1 Introduction

The strength of a structure is important to avoid cracking and damage. In an efficacious composite, there must be adequate bonding between fibre and matrix, and this bonding may be either physical or chemical [1]. The main function of the matrix material is to hold the fibres in the correct position so that, they carry the stress applied to the composite as well as, provide adequate rigidity. At the same time, the matrix protects the fibre from surface damage and from the action of the environment.

Fiber-reinforced polymer (FRP) is another material which could be used to build the hull of boats. FRP can provide a solution in terms of a better strength capacity, anti-corrosion, and possibly less maintenance during service. The FRP is the constituent within a polymeric matrix of a composite which contributes to the high tensile strength, boosting properties in the final part such as strength, rigidity, matrix interaction, conductivity, heat resistance, and resistance against chemical and physical corrosion while minimising weight [2, 3]. On top of preserving a high strength-to-weight ratio as compared to conventional materials, FRP composites can meet diverse design requisites with a relatively low weight. FRPs have high strength-to-weight ratios and can be easily adapted to almost any shape and size of a structure [4].

The mechanical properties of FRP composites are dependent upon the ratio of fibre to matrix material, which is determined by the mechanical properties of the constituent materials, the fibre orientation in the matrix, and ultimately the processing and methods of fabrication [4–6]. In order to achieve multidirectional isotropic behaviours, proper fibre orientation at a different angle is necessary, which can only be done by multiplying laminate preparation [7–9].

Nevertheless, moisture absorption is a crucial factor in polymers and composite materials. This material is especially beneficial for the construction of ship hulls and ensuring stability in marine environments. High water absorption of the material will affect the mechanical properties and stability of the composite [10].

However, in order to determine the mechanical properties of FRP, the researchers use the tensile test. The tensile test is one of the most fundamental and common types of mechanical testing. The process involves the application of force to a material and measuring the response of the specimen to the stress. The experimental approach for the test uses the universal testing machine "Quasar 100," and the FEA approach uses ANSYS Workbench software.

The aim of the present research is to focus on the tensile strength of the different approaches and the angle orientation (+45° and -45°) of the woven roving mats. The parameters of the specimen are used according to ASTM D3039. ASTM D3039 has a constant rectangular cross section which is 25 mm wide and 250 mm long [11]. The tabs are optional and can be bonded to the ends of the specimen to prevent gripping damage. However, a specimen which does not have a tab is similar to the rectangular shape.

2 Methodology

The hull laminate method is used in this research, which is a stacked sequence with a ± 45° woven roving and the layer orientation with different types of fibre mats such as tissue mat and chopped strand mat. Thus, the method of testing which will be used is the tensile test, which is conducted using two distinct approaches: Experiment and finite element analysis (FEA). The methodology has four parts such as angle orientation and stack sequences, specimen preparation, tensile test experiment and finite element analysis.

2.1 Angle Orientation and Stack Sequences

The angle orientation and stack sequence are divided into three, i.e. isotropic, anisotropic, and orthotropic. The isotropic has the same material properties in all directions and normal load. An anisotropic has not the same properties in different directions and the properties vary with rotation at a point. The orthotropic specimen encloses three mutually orthogonal planes of material property symmetry which are almost equivalent to an anisotropic material.

The characteristic of the isotropic material is that it is producing pure normal strain and shear stress produces shear strain.

For the isotropic laminate, the orientation angle and stack sequence is balanced and symmetric. The orientation angle of the woven roving mats is + 45° and -45°. Figure 1 shows the example of the angle orientation.

The stack sequence contains nine layers in which, the materials used are tissue mat (T), chopped strand mat 450 (CSM) and woven roving 600 (WR). The orientation angle and stack sequence involved are shown in Table 1.

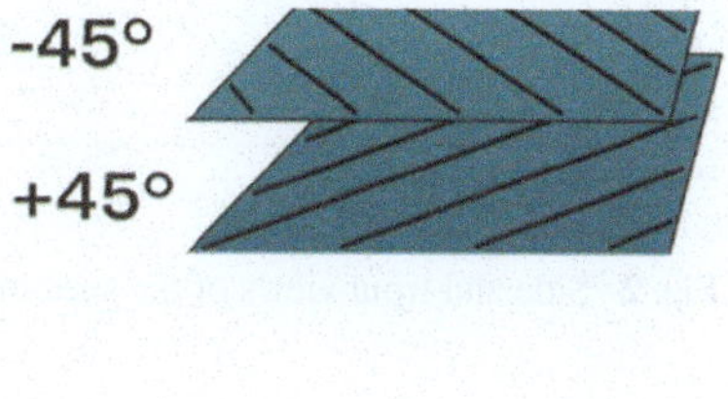

Fig. 1 Example of angle orientation

Table 1 The material of FRP laminate

Material	Code	Types
Tissue mat		E-Glass
Chopped strand mat	CSM 450	E-Glass
Woven Roving Mat	WR 600	E-Glass
Unsaturated polyester resin	Norsodyne 3338W	Polymer
Catalyst (hardener)		Methyl ethyl ketone peroxide

2.2 Specimen Preparation

The specimen has two groups in which, each group has 5 samples. The test specimen requires at least five specimens for each test condition unless a useable result can be gained through the use of less specimens [11]. This test specimen used the FRP material which affects the orientation angle of woven roving mat (+45° and -45°). In addition to the woven roving mat, the tissue mats and chopped strand mat are used. The polymer (resin) used is unsaturated polyester (Norsodyne 3338W) and the catalyst used is the methyl ethyl ketone peroxide (MEKP).

The standard tensile specimens were prepared to have dimensional and tolerance for two different angle orientations of materials. The stack sequence of fibre is nine layers with the same size 250 mm × 25 mm × 8 mm is a flat bar, balanced and symmetric according to ASTM D3039 as shown in Fig. 2.

Glass plates of dimension 304 mm X 304 mm were prepared. The glass of plate was used to obtain a smooth and flat surface. Prior to the process, Mirror Glaze® mold released wax was applied onto the glass to prevent the unwanted bonding between the materials and the glass surfaces. The lamination type specimens as shown in Table 2, have the mentioned orientation angles and stack sequences. The orientation angle for the tissue mats and chopped strand mats 450 is 0°. The woven roving mat 600 has a different angle orientation every specimen which are + 45° and 45°.

All specimens are impregnated with the mixture of an unsaturated polyester resin and 1% methyl ethyl ketone peroxide (MEKP) as catalyst. The ratio of resin to fibre is fixed at approximately 2:1 [12]. The laminated composite was cured at room temperature for 24 h. After the specimen has been cured, the laminates were cut according to the specimen size using a Tenoning and Squaring machine. The specimen size was cut to 250 mm × 25 mm. Table 2 shows the specimen label, with the

Fig. 2 Side and front views of the specimen size (unit in mm)

Table 2 Specimen labelling according to approaches

Specimen label	Approach
A + 45°	Experiment
B − 45°	Experiment
AA + 45°	Finite Element Analysis
BB − 45°	Finite Element Analysis

Table 3 Results for the tensile test for both approaches

Approaches	Specimen label	Load (kN)	Tensile strength (MPa)	Strain	Young's modulus (MPa)
Experiment	A + 45°	25.422	130.802	0.15	872.013
	B − 45°	17.384	80.28	0.11	729.818
FEA	AA + 45°	30	150	0.20	750.000
	BB − 45°	30	84.54	0.12	704.500

single alphabet representing *Experiment* and the double alphabet representing *Finite Element Analysis* (FEA).

2.3 Tensile Test – Experimental

The tensile test was performed by using a universal testing machine (UTM) which is a Galdabini Quasar 100 kN. The UTM has a two-column rigid system with 100 kN maximum capacity and is suitable for metals, plastics, composites, and other materials. This machine reveals a flexible and modular design for easy future expansion and the key technical advantages include extremely high resolution of load and stroke readings, as well as minimum test speed of 2 mm/min, for the high performance and most accurate results. Before the experiment was carried out, the specimen thickness and gauge length were measured and numbered for identification. The tensile testing starts after setting up the specimen on the gauge. Once the specimen fails, the machine stops the loading and saves the results. The result is shown in Fig. 3.

2.4 Tensile Test – Finite Element Analysis

The simulation of FEA uses the ANSYS software to create the specimen model. The specimens AA + 45° and specimens BB-45° are FRP flat bars as a specimen (250 mm × 25 mm × 8 mm). The brief procedure to design specimens in the Workbench ANSYS incorporates several steps. The first step is to design the tissue mat, CSM 450, and woven roving 600 with measurements 250 mm × 25 mm × 8 mm. Then,

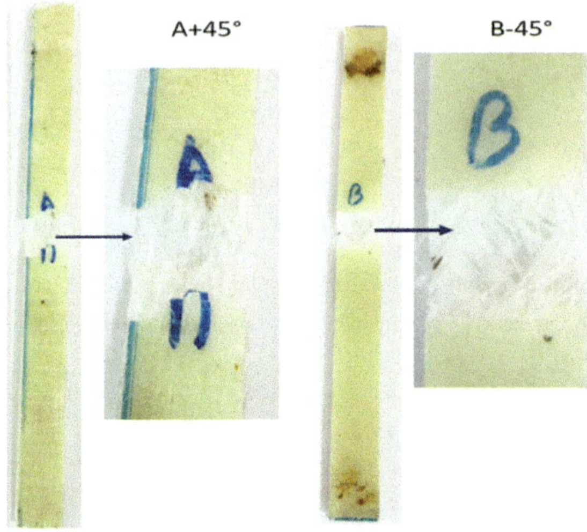

Fig. 3 Sample result of tensile test (experiment)

assemble all materials based on the sequence layers as Table 1. The specimens AA + 45° and specimens BB-45° are assemble the layers. Then, define the size of the 4 mm mesh element with a growth of 1.05 and the mesh metric skewness. The mesh contained 3402 elements and 26,946 nodes. Next, set up the material specification details such as density, modulus of elasticity, and shear modulus. The force at the UTM machine is set as tension-tension. In addition, the force is constant with 30 kN according to the tensile test. Therefore, insert the solution for the equivalent stress, the equivalent elastic strain, and the total deformation to be solved as shown in Fig. 4.

3 Results and Discussion

The specimen size of 250 mm × 25 mm and thickness of 8 mm was prepared according to ASTM 3039D and the ABS rules. The specimen has 2 groups (A + 45°and B − 45°), and each group has five samples. It is measured by the UTM, 100 kN for the experimental approach, and ANSYS Workbench software for the FEA approach. Figure 5 shows the comparison of the tensile strength results. In both approaches, the specimen A + 45° results were the highest, in which the UTM is 130.802 MPa, and the FEA is 150 MPa, with a slight difference being ± 13%. The specimen B + 45° results were the highest, in which the UTM is 80.28 MPa and the FEA is 84.543 MPa, with an insignificant difference of ± 5%.

The result of strain for specimen A + 45° in experiment was 0.15 and the FEA was 0.20. The modulus of elasticity can directly be calculated by using the formula as shown in Eq. (1). The specimen A + 45° has the highest elastic modulus 872.013 MPa and the specimen B-45° has the 729.818 MPa as shown in Table 2.

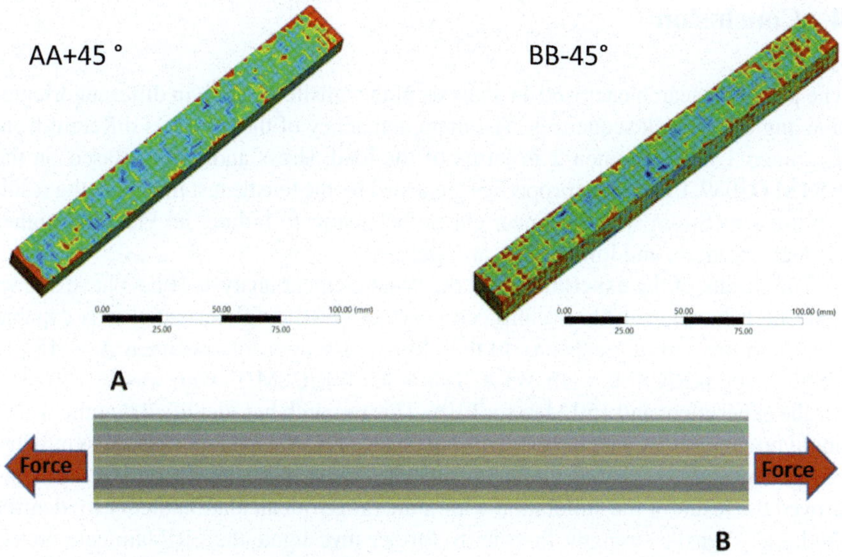

Fig. 4 Sample result of tensile test (FEA)

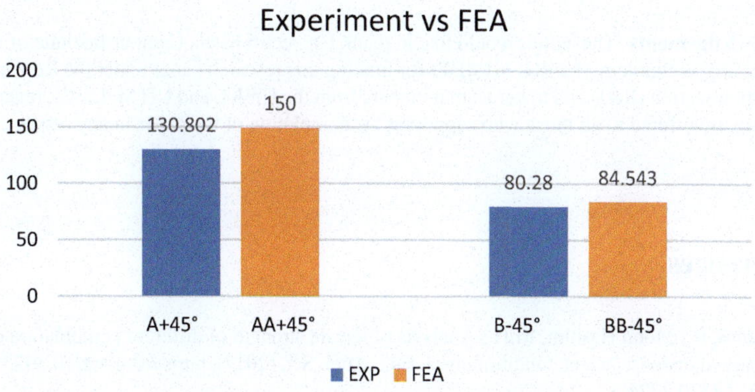

Fig. 5 Result of tensile strength universal testing machine (UTM) versus finite element analysis (FEA)

$$E = \frac{\sigma}{\varepsilon} \tag{1}$$

4 Conclusion

The present research enclosed an analysis of the tensile strength in different orientation angle and stack sequences. The main deficiency of the results is different from 1pecimen 1 and specimen 2 in terms of the load, stress and strain. Based on the ASTM D3039, FRP fabric properties are tested by the tensile test in which, the result for stress vs Strain shows a linear elastic behaviour to failure, no yielding, higher ultimate strength, and lower strain at failure.

The results of the experiment and the finite element analysis (FEA) have shown that, the behaviour of fibre reinforced polymer (FRP) angle orientation is capable to sustain the carrying capacity to the ultimate strength for specimen A + 45°(T/CSM/ + 45°WR/CSM/ + 45°WR/CSM/ + 45°WR/CSM/T) with was 130.80 MPa for the experiment and 150 MPa for FEA. This research has identified that the + 45° angle orientation, at which the stack sequences is isotropic and balanced symmetry of Specimens A + 45° has the highest tensile strength. The experimental work has proved the result of the simulation within the range of calculation theory to identify the highest tensile strength. In order to further investigate the FRP laminate on the mechanical properties and microstructure, it is scanning electron microscope (SEM). Therefore, it is recommended to investigate the details of the composite through interface debonding, tight binding, and delamination.

Acknowledgements The author would like to thank Universiti Kuala Lumpur Institute of Marine Engineering Technology (UniKL MIMET) for funding this research paper under the Research and Innovation section (R&I) and to her team members from the UniKL and UiTM. Lastly, regards and blessings are offered to all those who supported the completion of the paper in any respect.

References

1. Ramli, R., Mohd Hashim, M.H.: Analysis of tensile strength of different combination of FRP material under seawater conditioning. J. Eng. Appl. Sci. (2017). https://doi.org/10.3923/jeasci.2017.4320.4324
2. Rahman, R., Zhafer Firdaus Syed Putra, S.: Tensile properties of natural and synthetic fiber-reinforced polymer composites. In: Mechanical and Physical Testing of Biocomposites, Fibre-Reinforced Composites and Hybrid Composites (2019)
3. Masuelli, M.A.: Introduction of Fibre-Reinforced Polymers—Polymers and Composites : Concepts , Properties and Processes. In: Fiber Reinforced Polymers—The Technology Applied for Concrete Repair (2013)
4. Halim, N.H.F.A., Alih, S.C., Vafaei, M., Baniahmadi, M., Fallah, A.: Durability of fibre reinforced polymer under aggressive environment and severe loading: a review (2017)
5. Mallick, P.K.: Fiber-Reinforced Composites: Materials, Manufacturing, and Design, 3rd edn. (2007)
6. Tamon, U.: FRP for Construction in Japan. Jt. Semin. Concr. Eng. (2005)
7. Hossain, R.M., Islam, A., Van Vuure, A.W., Ignaas, V.: Effect of fiber orientation on the tensile properties of jute epoxy laminated composite. J. Sci. Res. (2012). https://doi.org/10.3329/jsr.v5i1.10519

8. Hudisteanu, I., Taranu, N., Isopescu, D.N., Entuc, I.S., Oprisan, G., Ungureanu, D.: Numerical analysis of intralaminar damage evolution on various composite laminates. In: IOP Conference Series: Materials Science and Engineering (2018)
9. Rubino, F., Nisticò, A., Tucci, F., Carlone, P.: Marine application of fiber reinforced composites: a review (2020)
10. Yuhazri, M.Y., Jennise, T.T.T., Sihombing, H., Mohamad, N., Yahaya, S.H., Zalkis, M.Y.A.: Water absorption and thickness swelling of laminated composite after cured at different angle. Appl. Mech. Mater. (2014)
11. ASTM: D3039/D3039M Standard Test Method for Tensile Properties of Polymer Matrix Composite Materials. Annu. B. ASTM Stand. (2014)
12. Hamid, N.A.: An experimental study of the influence of fiber architecture on the strength of polymer composite material. J. Mech. Eng. Technol. **2**, 1–16 (2010)

Characterisation of Oil Palm Fibres as a Strengthening Material in Ceramic Membranes

Maisarah Mohamed Bazin and Muhammad Akmal Mohd Ghazali

Abstract Ceramic membranes can be fabricated with low-cost materials such as Sayong ball clay, which is cheap and commonly available in Malaysia. Readily commercialised ceramic membranes available in the market use advanced ceramics such as alumina, which is quite expensive. Therefore, low-cost ceramic membranes are highly desirable. However, the mechanical properties of low-cost ceramic membranes are lower. Due to this, this work studied oil palm fibre (OPF) as a strengthening material for ceramic membranes through the gel casting technique. The effect of amount of OPF content on the properties of the ceramic membrane was studied. The process of gel casting involves the mixing of Sayong ball clay powder with the monomer, dispersant, solvent, initiator, catalyst, and pore former. Rectangular samples of ceramic membrane were fabricated with varied OPF content (0, 5, and 10 wt% of OPF). The physical characterisation (shrinkage, apparent porosity, and bulk density), microstructure analysis, and mechanical strength analysis were carried out on the ceramic membrane sintered at a temperature of 1,050 °C. The apparent porosity and bulk density of the samples were calculated according to the Archimedes method. The effect of adding OPF on the strength of the membrane was examined through the three-point bending test. The membranes with the addition of OPF showed varying bulk densities from 1.863 to 1.953 g/cm^3 and apparent porosities ranging from 32.93 to 30.17% at different OPF contents. In addition, the membrane exhibited an average pore size of 7.56 nm. The flexural strength of the membranes increased from 5.684 to 8.337 MPa as the wt% of OPF increased. As a conclusion, the OPF had successfully increased the strength of the membrane.

Keywords Ceramic membrane · Oil palm fibre · Sayong ball clay · Gel casting · Mechanical strength

M. M. Bazin (✉) · M. A. M. Ghazali
Mechanical Engineering Section, Malaysia France Institute, Universiti Kuala Lumpur, Section 14, Jalan Teras Jernang, 43650 Bandar Baru Bangi, Selangor, Malaysia
e-mail: maisarahmb@unikl.edu.my

M. A. M. Ghazali
e-mail: makmal.ghazali@unikl.edu.my

A. Ismail et al. (eds.), *Engineering Frontiers*, Advanced Structured Materials 208,
https://doi.org/10.1007/978-3-031-56844-2_5

1 Introduction

Ceramic membranes are known to have various advantages, such as high thermal stability, high resistance to pressure, and high separation efficiency as compared to polymeric membranes. It has been used in various industries, such as water treatment, chemical, pharmaceutical, and other industries that involve separation processes. It has been extensively used in microfiltration, ultrafiltration, and nanofiltration processes due to its desirable properties of being able to resist extreme levels of temperature, chemical, and pH [1]. The mechanical, chemical and thermal properties of clay mineral as a membrane material are comparable to those of advanced ceramics and are superior than that of polymers. These materials cost less to handle in terms of the sintering temperature and cost of raw materials. As compared to advanced engineering ceramics, clay minerals have been used to fabricate low-cost porous membranes in gas sensors, molten metal filters, and gas filters with reduced costs. Clay minerals can be a source of ceramic materials, such as nepheline syenite, cristobalite, and mullite [2]. Sayong ball clay has been introduced as a major material in the fabrication of low-cost ceramic membranes [3–5].

Oil palm fibre (OPF) is an agriculture waste commonly available in Malaysia as one of the major producers of the oil palm industry in the world. Numerous studies have been done on the use OPF as reinforcement in composites which have shown improvement in toughness, flexural and bending strengths [6, 7]. Amartey [8] found that the addition of OPF from 0 to 0.25% into concrete had successfully increased the compressive and tensile strengths of the concrete. On the other hand, a study by Faezatul [9] used OPF as a pore forming agent for ceramic membranes. It was found that the addition of OPF had increased the ceramic membrane porosity but decreased the compressive strength and density.

The main objective of this work was to study the effect of the OPF addition fabricated by the gel casting method on the strength of ceramic membranes. For this study, Sayong ball clay, which can be acquired easily in Malaysia, was used as the starting raw material, while OPF, which is an inexpensive material, was used as a strengthening material. The performance of OPF in a porous ceramic membrane were characterised.

2 Methodology

2.1 Materials

In this research, the raw materials that were used were Sayong clay powder and oil palm fibre. The materials used for gel casting were methacrylamide (MAM) solution, bisacrylamide (MAD), ammonium persulfate, tetra-methylethylene-diamine (TEMED), and polyethylene glycol (PEG). Distilled water was used as the solvent

Table 1 Raw materials used for gel casting process

Raw materials	Function/details
1. Sayong ball clay	Ceramic powder (70 μm)
2. Oil palm fibre	Strengthening material
3. Corn starch	Pore former
4. Methacrylamide (MAM)	Monomer
5. Bisacrylamide (MAD)	Cross-linker
6. Ammonium persulfate (APS)	Initiator
7. Tetramethylethylenediamine (TEMED)	Catalyst
8. Polyethylene glycol (PEG)	Dispersant

while corn starch was used as the pores former. Table 1 summarises the materials used and their respective functions.

2.2 *Fabrication of Membranes*

The fabrication technique used in this study was the gel casting method. The ceramic slurry was prepared by mixing 5% MAM solution with MAD powder. Then, Sayong ball clay powder (70 wt%) was added to the mixture with fine grinded oil palm fibres (0 wt%, 5 wt%, 10 wt%). After mixing for 30 min, 0.1% of PEG, 0.1% of APS solution (10 wt%), and 0.05% of TEMED were added and mixed for another 30 min. The mixing was done by using a IKA RW20 digital overhead stirrer at 1300 rpm. The slurry was casted into a prepared mould with dimensions of 80 mm × 30 mm × 2.5 mm and left to dry at room temperature for a day, and then dried in a muffle furnace (Daihan) at 100 °C for 30 min. Then, the casted slurry was unmoulded and the obtained green body was then pre-sintered at 500 °C for 1 h to remove all the binder, followed by sintering at 1050 °C. The sintering process was performed by using a LT FURNACE L12-1300.

In this study, the porosity was created through the burning out of the monomer and corn starch during the sintering process. Thus, the membranes were heated from room temperature until up to 500 °C at a heating rate of 5 °C per min, and holding at 250 °C for 30 min to evaporate any moisture present to prevent cracking during the heating process. Then, the temperature was raised to 500 °C and held for 1 h to eliminate the monomer and corn starch. Finally, the final sintering temperature was set to 1050 °C and held for 2 h, then cooled in the oven at room temperature for 3 h. Figure 1 shows the sintering profile used in this study.

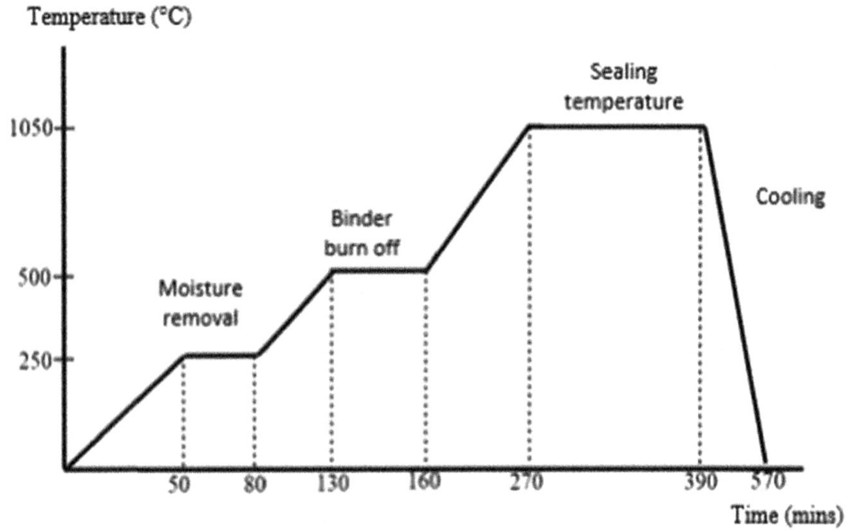

Fig. 1 Sintering profile of Sayong ball clay membrane

2.3 *Characterization Techniques*

Shrinkage measurements was done by weighing the membranes before and after the sintering process. The initial and final dimensions were taken to determine the shrinkage percentage.

Another important physical property is the bulk density and apparent porosity, following the Archimedes method according to ASTM C733-88 standard. The membrane weight was taken before and after soaking in distilled water. To ensure the pores were filled with water, the soaking lasted for 30 min. The suspended average weight of the membranes was also taken. The bulk density and apparent porosity of the membranes were calculated by using Eqs. (1) and (2), respectively.

$$BulkDensity = \frac{Drymass}{Soakedmass - Suspendedmass} \tag{1}$$

$$ApparentPorosity = \frac{Soakedmass - Massinair}{Soakedmass - Suspendedmass} \tag{2}$$

Scanning electron microscopy (SEM) was used to analyse the microstructure of the membranes. In addition, the three-point bending test was done to measure the flexural strength of the membranes. For every membrane with different OPF content (0, 5, and 10 wt%) sintered at 1050 °C, at least three samples were used in order to get the average number. This test was done by using the universal testing machine, Victor VEW 2302. The three-point bending test was performed with a span of 40 mm and a crosshead speed of 0.5 mm/min, according to the ASTM C-1161-02c standard.

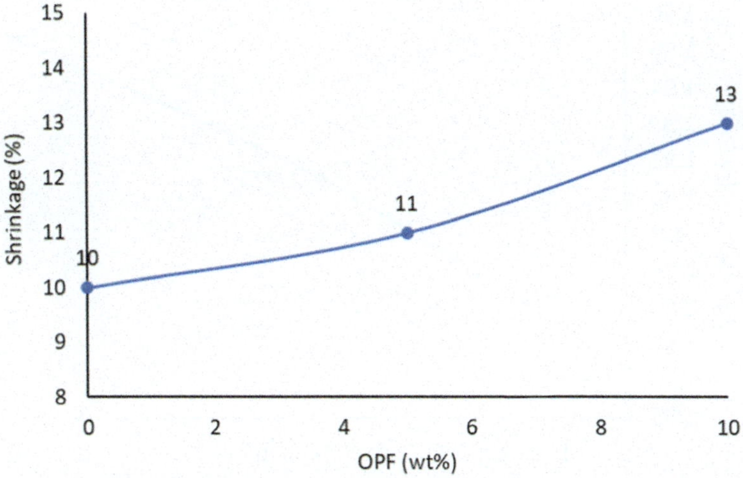

Fig. 2 Shrinkage of membranes with different OPF wt% at 1050 °C

3 Results and Discussion

3.1 Shrinkage

Figure 2 shows the results of membrane shrinkages at different OPF contents. From the graph, the shrinkage values increase as the OPF content increased. The shrinkage occurred due to the dimensional change caused by the physical and chemical process [3], and also due to the collapse of three-dimensional network structures of the cross-linked polymer gels [10]. Corn starch was used in this study as a pore forming agent, thus, adding corn starch affected the shrinkage as it created pores during the burn off of the membranes [11].

3.2 Bulk Density

Figure 3 shows that the bulk density of the membranes increased from 1.863 to 1.953 g/cm^3 as the OPF content increased. The densification happened after the burn-out of the monomer and pore former, which resulted in reduced pore size and porosity. This result was in good agreement with a report from Muazu et al. [12], whereby increasing the solid content made the membrane denser. From a previous study, Ahmad et al. [4] fabricated a Sayong clay membrane by gel casting with adjusted monomer (5, 10, and 15 wt%) at 1300 °C. The membrane exhibited a measured density ranging from 1.36 to 1.87 g/cm^3 as the monomer content increased. In another study, Bazin et al. [13] achieved a density of 1.6 g/cm^3 from Sayong clay

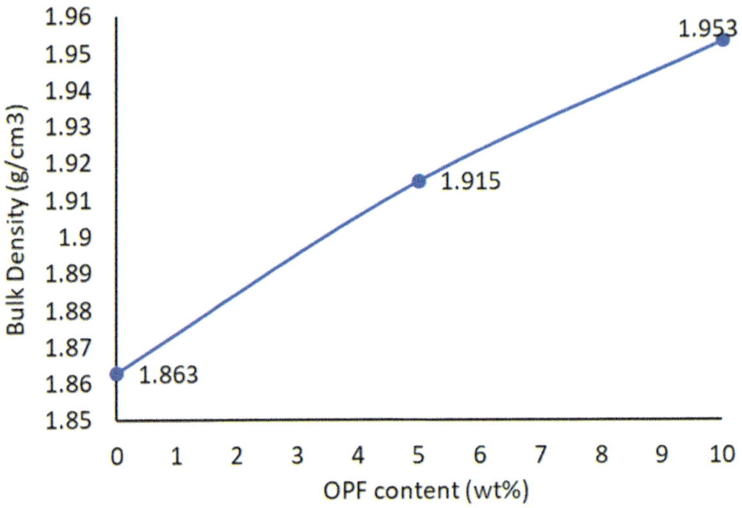

Fig. 3 Bulk density of membranes with different OPF wt% at 1050 °C

membrane made by compaction method with a sintering temperature of 1050 °C, making the membrane suitable for nanofiltration. In conclusion, adding OPF can increase the membrane densification at a lower sintering temperature of 1050 °C.

3.3 Apparent Porosity

The results of apparent porosity are shown in Fig. 4. From the result, the apparent porosity reduced with increasing OPF content. This may occur because as the OPF content increases, the OPF did not completely burn-out during the sintering process and made the membranes become denser. Therefore, the apparent porosity is inversely proportional to density [12].

3.4 Flexural Strength

The three-point bending test was conducted to determine the membranes flexural strength with the addition of OPF. Figure 5 shows the data for flexural strength of the ceramic membrane with different OPF wt% sintered at 1050 °C. The results showed that the flexural strength increased (from 5.7 to 8.3 MPa) with increased OPF content. This result was in good agreement with Amartey [8] and showed that OPF is a good strengthening material for Sayong ball clay ceramic membrane. In another study on Sayong ball clay membrane, Ahmad et al. [14] achieved a flexural strength of

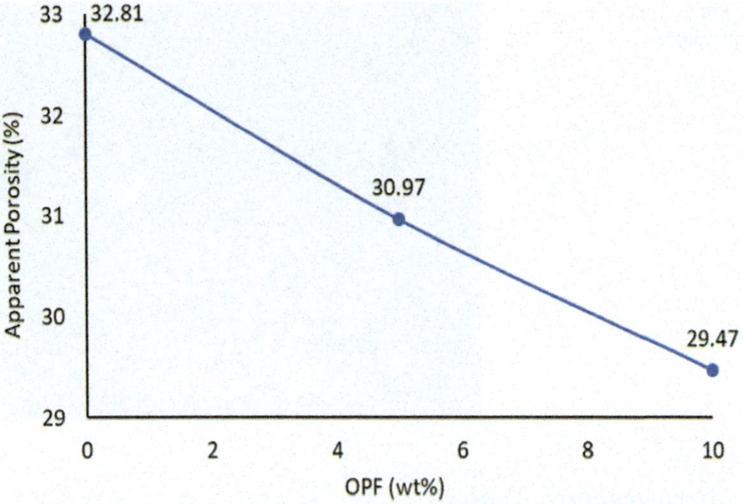

Fig. 4 Apparent porosity of membranes with different OPF wt% at 1050 °C

5 MPa for membrane with 5 wt% of starch content sintered at 1000 °C. Similarly, Bazin et al. [13] achieved a flexural strength of 6 MPa at 1050 °C for Sayong clay membrane prepared by compaction method. These results are lower as compared to the strength of membrane with addition of OPF.

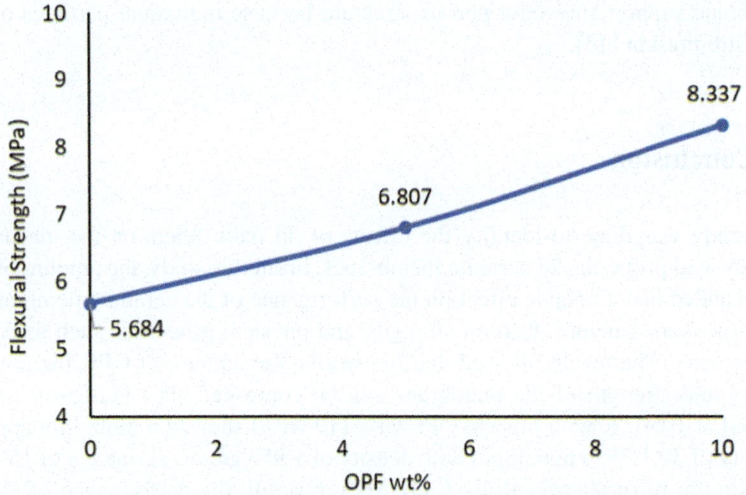

Fig. 5 Flexural strength of membranes with different OPF wt% at 1050 °C

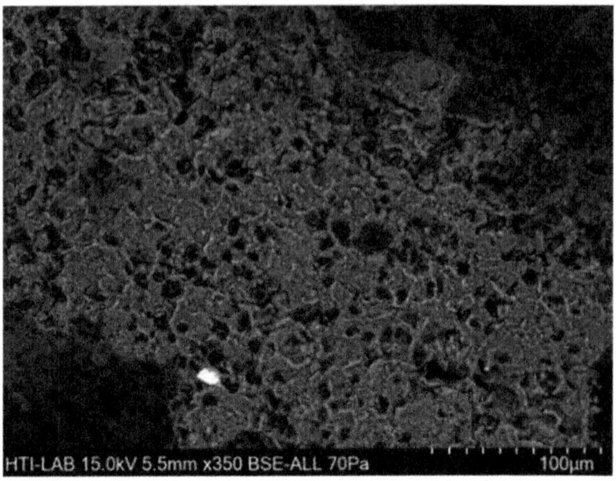

Fig. 6 Microstructure image of sample surface

3.5 Microstructure Analysis

The microstructure of the samples was observed by FESEM. The structure and pore size were analysed. The membrane sample with 10 wt% OPF sintered at 1050 °C was used. Figure 6 shows the microstructure of the membrane surface. From the figure, a rough surface and dot-like pores were observed.

In a study by Muazu et al. [12] the microstructure of porous Sayong membrane sintered at 1000 °C showed similar a image, whereby the membrane was rough and dot-like pores were observed. At sintering temperature range of 900–1150 °C, the membrane support showed a porous structure because individual particles of clay were still present [15].

4 Conclusion

This study was done to identify the effects of oil palm fibers on the mechanical and physical properties of ceramic membranes. From this study, the amount of OPF (wt%) added had a notable effect on the performance of the ceramic membranes in terms of microstructure, flexural strength, and physical properties such as density and porosity. The results showed that by varying the amount of OPF, the porosity, density and strength of the membrane can be controlled. The gel-casted sample sintered at 1050 °C with highest OPF wt% (10 wt%) showed a minimum apparent porosity of 30.17%, a maximum bulk density of 1.953 g/cm^3, shrinkage of 15% and a maximum flexural strength of 8.337 MPa. Overall, the performance of OPF in Sayong clay membrane was studied and it can be concluded that the addition of OPF had successfully enhanced the properties of the membrane.

Acknowledgements This work was supported by research university grant (STR19021) funded by Universiti Kuala Lumpur.

References

1. Usman, J., Othman, M.H.D., Ismail, A.F., Rahman, M.A., Jaafar, J., Raji, Y.O., Gbadamosi, A.O., El Badawy, T.H., Said, K.A.M.: An overview of superhydrophobic ceramic membrane surface modification for oil-water separation. J. Market. Res. **12**, 643–667 (2021). https://doi.org/10.1016/j.jmrt.2021.02.068
2. Jana, S., Purkait, M.K., Mohanty, K.: Preparation and characterizations of ceramic microfiltration membrane: effect of inorganic precursors on membrane morphology. Sep. Sci. Technol. **46**(1), 33–45 (2011). https://doi.org/10.1080/01496395.2010.503669
3. Ahmad, N., Binti, N., Zaidan, M.: Effect of sintering temperature on membrane properties of Sayong ball clay. **315**, 349–353 (2013). https://doi.org/10.4028/www.scientific.net/AMM.315.349
4. Ahmad, N., Binti, N., Zaidan, M., Bazin, M.M.: Fabrication and characterization of ceramic membrane by gel cast technique for water filtration. **686**, 280–284 (2013). https://doi.org/10.4028/www.scientific.net/AMR.686.280
5. Mohd Noor, S.F., Ahmad, N., Khattak, M.A., Khan, M.S., Anique Mukhtar, Kazi, S., Badshah, S., Khan, R.: Application of Sayong ball clay membrane filtration for Ni (II) removal from industrial wastewater. J. Taibah Univ. Sci. **11**(6), 949–954 (2017). https://doi.org/10.1016/j.jtusci.2016.11.005.
6. Bilba, K., Arsene, M.A., Ouensanga, A.: Sugar cane bagasse fibre reinforced cement composites. Part I. Influence of the botanical components of bagasse on the setting of bagasse/cement composite. Cement Concr. Compos. **25**(1), 91–96 (2003). https://doi.org/10.1016/s0958-9465(02)00003-3
7. Tolêdo Romildo, F.D., Ghavami, K., England, G.L., Scrivener, K.: Development of vegetable fibre-mortar composites of improved durability. Cement Concr. Compos. **25**(2), 185–196 (2003). https://doi.org/10.1016/S0958-9465(02)00018-5
8. Amartey, B.H.S., Kumator, T.J., Amartey, Y.D., Ali, A.: The use of oil palm fibre as an additive in concrete. Mater. Today: Proc. **86**, 111–115 (2023). https://doi.org/10.1016/j.matpr.2023.03.820
9. Faezatul, A.M.R., Mohamad, Z.N., Mohd Warikh, A.R., Julie, J.M., Mohd Al Amin, M.N.: Preparation and characterization of ceramic membrane by using palm fibers as pore forming agent. AIP Conf. Proc. **2068**, 020056 (2019) https://doi.org/10.1063/1.5089355
10. Tobita, H., Hamielec, A.E.: Crosslinking kinetics in polyacrylamide networks. Polymer **31**(8), 1546–1552 (1990). https://doi.org/10.1016/0032-3861(90)90163-S
11. Nie, Z., Lin, Y.: Fabrication of porous alumina ceramics with corn starch in an easy and low-cost way. Ceramics – Silikáty **59**(4), 348–352 (2015)
12. Bazin, M.M., Muazu, A.B., Norhayati, A.: A comparative study of porous support from Sayong and Kankara Clay. In: 3rd International Conference on Water Resources (ICWR 2015), Langkawi, Malaysia, 24–25 Nov 2015
13. Bazin, M.M., Ahmad, N., Nakamura, Y.: Preparation of porous ceramic membranes from Sayong ball clay. J. Asian Ceram. Soc. **00**(00), 1–9 (2019). https://doi.org/10.1080/21870764.2019.1658339
14. Bazin, M.M., Ahmat, M.A., Zaidan, N., Ismail, A.F., Ahmad, N.: Effect of starch addition on the microstructure and strength of the ball clay membranes. Jurnal Teknologi (Sci. Eng.). **69**(9), 117–120 (2014)
15. Johari, I., Abu, B.H., Bakar, B.H.A., Ahmad, Z.A.: Effect of the change of firing temperature on microstructure and physical properties of clay bricks from Beruas. Sci. Sinter. **42**, 245–254 (2010). https://doi.org/10.2298/SOS1002245J

Review of Fiber Bragg Grating Sensors for Marine Applications

Zulzamri Salleh, Muhammad Akmal Afiq Zullastri, and Septia Hardy Sujiatanti

Abstract Fiber Bragg grating (FBG) sensors have gained significant attention in recent years due to their numerous advantages in the field of marine applications. This review paper provides an overview of FBG sensors and their utilization in various marine applications, including structural health monitoring, underwater acoustic sensing, oceanography, and offshore oil and gas exploration. The paper discusses the working principles of FBG sensors, their fabrication techniques, and the challenges associated with their implementation in marine environments. Additionally, it highlights the benefits of FBG sensors, such as their compact size, immunity to electromagnetic interference, and ability to measure multiple parameters simultaneously. The review concludes with an assessment of the current state-of-the-art, identifying potential future directions and emerging trends in FBG sensor technology for marine applications.

Keywords Intensity · Sensor · Marine application · Wavelength

1 Introduction

The marine environment plays a vital role in various aspects of human life, including transportation, commerce, energy production, and environmental monitoring. Therefore, ensuring the safety, efficiency, and sustainability of marine activities is of utmost importance. Reliable sensing technologies are essential for monitoring and assessing

Z. Salleh (✉) · M. A. A. Zullastri
Universiti Kuala Lumpur, Malaysian Institute of Marine Engineering Technology, Dataran Industri Teknologi Kejuruteraan Marin Bandar Teknologi Maritim Jalan Pantai Remis, 32200 Lumut, Perak, Malaysia
e-mail: zulzamri@unikl.edu.my

M. A. A. Zullastri
e-mail: makmal.zullastri@s.unikl.edu.my

S. H. Sujiatanti
Department of Naval Architecture, Faculty of Marine Technology, Institut Teknologi Sepuluh Nopember, Surabaya, Indonesia

A. Ismail et al. (eds.), *Engineering Frontiers*, Advanced Structured Materials 208,
https://doi.org/10.1007/978-3-031-56844-2_6

the condition of marine structures, tracking environmental parameters, detecting potential hazards, and optimizing operations.

Marine infrastructures, such as ships, offshore platforms, subsea pipelines, and deep water infrastructure, are subjected to harsh environmental conditions, including corrosive saltwater, extreme temperatures, and high-pressure depths. These structures are susceptible to degradation, fatigue, and structural failures over time. To ensure their integrity and prevent catastrophic incidents, continuous monitoring and early detection of structural defects, such as cracks, deformations, and corrosion, are crucial [1]. Reliable sensing technologies provide real-time data on structural health, allowing for proactive maintenance and timely interventions, thus reducing the risk of accidents and enhancing operational efficiency.

Monitoring the marine environment is essential for understanding and mitigating the impacts of human activities on marine ecosystems. Environmental parameters such as temperature, pressure, salinity, dissolved oxygen levels, and pollutant concentrations directly affect marine life and ecological balance. Reliable sensing technologies enable scientists and researchers to gather accurate and high-resolution data on these parameters, facilitating the assessment of marine ecosystems, climate patterns, and the effects of pollution [2]. This information is vital for sustainable management practices, conservation efforts, and the development of mitigation strategies.

In addition, reliable sensing technologies are crucial for marine exploration and resource extraction, particularly in the offshore oil and gas industry. Monitoring the integrity of wells, pipelines, and subsea equipment ensures safe and efficient extraction operations. Leakage detection systems and real-time monitoring of flow rates contribute to environmental protection by minimizing the risk of oil spills and preventing costly damage to marine ecosystems. Moreover, the use of reliable sensors enables efficient reservoir management, leading to optimized production and reduced operational costs.

Given the challenging and dynamic nature of the marine environment, traditional sensing technologies often face limitations, such as susceptibility to corrosion, electrical interference, and limited sensing capabilities. Fiber Bragg grating (FBG) sensors offer a promising solution for marine applications due to their unique characteristics, including compact size, high sensitivity, immunity to electromagnetic interference, and multiplexing capabilities. These sensors can monitor various parameters, such as strain, temperature, pressure, and vibration, simultaneously and with high accuracy. The development and deployment of reliable sensing technologies, such as FBG sensors, significantly contribute to the safety, efficiency, and sustainability of marine activities, making them essential for marine applications.

Figure 1 shows the FBG installation setup at the laboratory before starting the experiments. Fiber Bragg grating sensors are optical devices that utilize the principle of wavelength-selective reflection in optical fibers to measure various physical parameters. FBGs consist of periodic variations in the refractive index along a section of an optical fiber, creating a wavelength-specific reflection or "Bragg wavelength." When light is propagated through the fiber, a specific wavelength of light is reflected back due to the grating structure. There are many advantages of FBG sensors widely used in many areas and most of them not only in the engineering fields but also in the medical industry. FBG sensors offer several unique characteristics that make them well-suited for a wide range of applications, including marine environments. FBG sensors are small and lightweight, allowing for their integration into structures or equipment without significant additional weight or space requirements. Multiple FBG sensors can be multiplexed along a single fiber, enabling the simultaneous measurement of various parameters, such as strain, temperature, pressure, and vibration, at different locations. FBG sensors also have high sensitivity, allowing them to detect small changes in the measured parameters. They offer excellent accuracy and repeatability, providing reliable and precise measurements. They are immune to electromagnetic interference since they operate based on optical principles, making them suitable for applications in electrically noisy environments. Usage for engineering characteristics, it is widely used to measure related to the physical properties such as strain, temperature, pressure, acceleration, and more, making them versatile for various applications. In the reliability characteristic behavior, it exhibits excellent long-term stability and does not suffer from drift or degradation over time, ensuring reliable and consistent measurements.

The purpose of this review paper is to provide a wide-ranging overview of FBG sensors' utilization in marine applications. The review aims to cover the working principles of FBG sensors, applications in marine environment and challenges using FBG sensors. The review explains the basic principles of FBG sensors, including their fabrication techniques and spectral interrogation methods. It also discussed various marine applications where FBG sensors have been employed, such as for structural health monitoring, underwater acoustic sensing, oceanography, and offshore oil and gas exploration. The review addresses the challenges associated with implementing FBG sensors in marine environments, including issues like corrosion, biofouling, high-pressure and deep-sea deployment, and compensation techniques for temperature and strain. The review highlights the unique benefits and advantages

Fig. 1 FBG sensor installation setup

of FBG sensors, such as their compact size, immunity to electromagnetic interference, multiplexing capabilities, high sensitivity, and potential cost-effectiveness including identify emerging trends and potential future directions in the field.

By encompassing these key aspects, the review aims to provide researchers, engineers, and practitioners in the field of marine applications with valuable insights into the capabilities, challenges, and opportunities associated with the FBG sensor technology.

2 Methodology

2.1 *Working Principle of FBG Sensors*

Fiber Bragg grating sensors have been operated based on the fundamental of wavelength-selective reflection in optical fibers. The core of a FBG sensor consists of a section of optical fiber that contains periodic variations in the refractive index [3]. These variations create a grating structure, which exhibit as a wavelength-specific reflector.

When light is launched into the fiber, it propagates along the fiber until it reaches the FBG. At the grating region, a definite wavelength of light, known as the Bragg wavelength (λ), matches the period of the grating structure [4]. This specific wavelength is reflected back in the opposite direction, while other wavelengths continue to propagate. The schematic diagram of FBG sensor shown as in Fig. 2, for details see [5]. The wavelength of the light source is related to the Bragg grating period do through the equations below:

$$\lambda_0 = 2n_0 d_0 \tag{1}$$

$$\Delta\lambda = 2n_0 \Delta d + 2d_0 \Delta n \tag{2}$$

where λ is the Bragg wavelength and n is the effective refractive index.

The Bragg wavelength's sensitivity to strain results from the sensor's physical expansion or contraction and the alteration of the refractive index caused by the photoelastic effects. Equation (3) allows us to calculate the wavelength shift resulting from axial strain.

$$\Delta\lambda = (1 - p_e)\lambda_0 \varepsilon \tag{3}$$

where ε is the applied strain, and p_e is the photoelastic coefficient term given by

$$p_e = \left(\frac{n_0^2}{2}\right)[p_{12} - \nu(p_{11} + p_{12})] \tag{4}$$

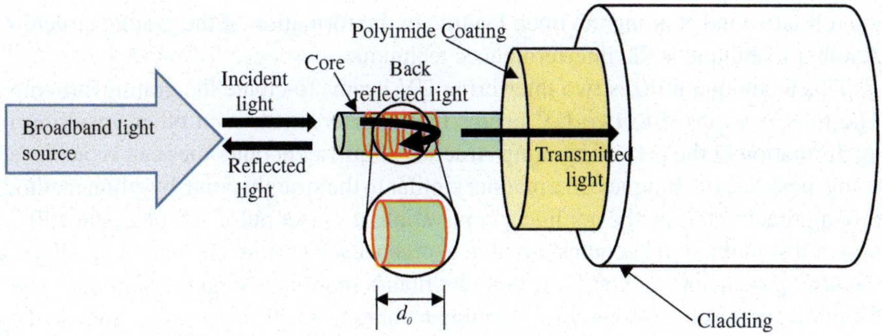

Fig. 2 Schematic diagram of fiber Bragg gratings [5]

where p_{ij} coefficients are the Pockel's coefficients of the strain-optic tensor and ν is the Poisson's ratio.

By measuring the reflected or transmitted wavelength, the physical parameter being monitored, such as strain or temperature, can be determined shows in Fig. 3. Changes in the monitored parameter cause shifts in the Bragg wavelength, allowing for the measurement of the corresponding parameter [6].

There are several methods for fabricating FBGs in optical fibers. The most common techniques include the segment facade technique. In this method, a segment facade with a periodic pattern is placed in contact with the optical fiber during the UV exposure [7]. The interference pattern created by the phase mask causes variations in

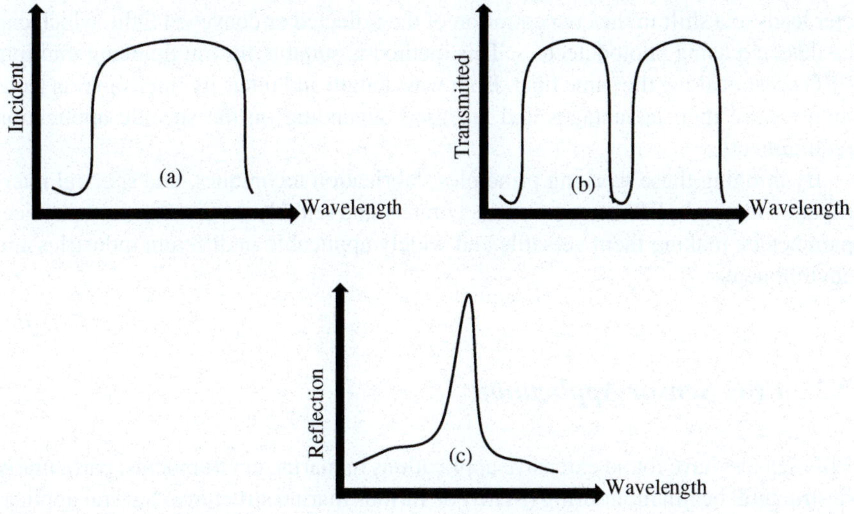

Fig. 3 Profile of the spectrum FBG index shows the **a** incident, **b** transmitted and **c** reflection beams

the refractive index along the fiber, leading to the formation of the grating structure. Another technique is the interferometric technique.

This technique utilizes two interfering UV beams to create the grating structure. The fiber is exposed to two UV beams that interfere with each other, resulting in the formation of the periodic grating structure. Equivalent outcomes can be achieved using mask-less techniques. In a manner similar to the straight point-by-point method, this approach relies on the nonlinear captivation of a laser pulse. Through the utilization of this method, it becomes possible to create each grating element by modifying the laser parameters and orchestrating the fiber's motion using a translational stage. Similarly, in the continuous core-scanning technique, inscription is accomplished by controlling the movement parameters of the translational frame to which the fiber is anchored. Various types of index modulations exist. In the case of a simple Fiber Bragg grating sensors, as depicted in Fig. 3, the disturbance occurs at regular intervals, repeating in a predictable manner, and consistently results in a positive change in modulation strength. As a result, the refractive index within the grating adopts a square wave pattern with a duty cycle equal to half of the period.

The point-by-point technique is the method using the grating structure is created by directly exposing the fiber to a focused UV laser beam. The laser beam is scanned along the fiber, inducing localized refractive index changes, and forming the grating structure. FBG sensors can be interrogated using two main spectral interrogation techniques. In this technique, the reflected or transmitted wavelength of the FBG is measured using a broadband light source and a spectrometer. The spectral shift of the Bragg wavelength corresponds to the change in the monitored parameter, such as strain or temperature. Intensity interrogation is also technique where the reflected or transmitted intensity of the FBG is measured. The change in the monitored parameter leads to a shift in the concentration of the reflected or conveyed light, which can be detected using photodetectors. This method is suitable for multiplexing multiple FBG sensors along the same fiber. Both wavelength and intensity interrogation techniques have their advantages and are used depending on the specific application requirements.

By utilizing these working principles, fabrication techniques, and spectral interrogation methods, FBG sensors can accurately and reliably measure various physical parameters, making them versatile and widely applicable in different industries and environments.

2.2 FBG Sensor Application

FBG sensors have found extensive applications in marine environments, particularly in structural health monitoring (SHM) of various marine structures. Several applications of FBG devices in marine SHM including monitoring of ship structures. FBG sensors are utilized to monitor critical components of ships, such as hulls, decks, and bulkheads. They enable the detection of structural defects, such as cracks, deformations, and corrosion, providing early warnings for maintenance and preventing catastrophic failures [8]. FBG sensors can also measure strain and fatigue in ship structures, ensuring their structural integrity and safety. While in the oil and gas industry such as offshore platforms and subsea pipelines. FBG sensors are employed for continuous monitoring of offshore platforms and subsea pipelines. They enable real-time detection and localization of structural defects, including cracks, weld quality, and pipeline integrity [9]. By monitoring strain and stress levels, FBG sensors contribute to the assessment of structural performance and enable proactive maintenance, minimizing downtime and reducing the risk of accidents. The structural integrity particularly for offshore platforms is related to the exposure and creating the structural defects: FBG sensors offer the capability to identify and concentrate structural defects in marine structures. By measuring strain and deformation along the structure, they can identify areas of concern and provide valuable information for repair and maintenance activities [10]. This helps to prevent structural failures and extends the lifespan of marine assets. The catastrophic failure such as installation of pipelines industry is related to the welded joints, which require strain and fatigue monitoring. FBG sensors are used to measure strain and fatigue in marine structures subjected to dynamic loading conditions. They provide valuable data for assessing the structural health and performance of offshore platforms, ship hulls, and subsea pipelines [11]. By monitoring strain and fatigue levels, FBG sensors assist in optimizing maintenance schedules and ensuring the longevity of marine structures. Any mechanical instrument and machine such as a pumping system, propeller system in a ship will be exposed to vibration failure. FBG sensors are employed to monitor vibration and impact events in marine structures. They can detect and measure vibrations caused by wave action, machinery operation, or external forces. This data helps to evaluate the structural response and integrity of marine structures, enabling necessary adjustments and maintenance.

Meanwhile the size is also concerned in shipyard companies and definitely FBG sensors offer several advantages in marine SHM applications, including their compact area, very sensitive to environment, susceptibility to EM interference, and ability to measure multiple parameters simultaneously. These characteristics make FBG sensors well-suited for monitoring and ensuring the structural integrity and safety of marine structures, contributing to the overall efficiency and sustainability of marine operations.

Underwater acoustic sensing involves the use of sensors to detect and analyze sound waves in marine environments. It has numerous applications, including hydrophone arrays and distributed acoustic sensing, detection of underwater noise, marine mammals, and seismic activity, as well as sonar applications and underwater communication systems [12–15]. Oceanography involves the study of the physical, chemical, biological, and geological aspects of the ocean. It encompasses a wide range of research areas, including the measurement of temperature, pressure, and salinity, current profiling and wave measurement, as well as the monitoring of tsunamis, tides, and other oceanic phenomena [16–19]. Complexity of heavy industry such as oil and gas exploration involves the crude oil exploration, production, and extraction of oil and gas reserves located beneath the seabed. It requires advanced technologies and techniques to ensure efficient and safe operations [20–23].

3 Results and Discussion

Based on the analysis of existing literature, the attention in FBG sensor applications is directed towards aspects such as the specific type of single-point sensing elements, interrogation methods, the detection threshold, and/or sensitivity. The details from the evaluation of journals have been summarized in Table 1 [24]. In the literature, two primary research approaches are commonly pursued. The first approach aims to enhance the credibility of the sensing elements, while the second approach focuses on enhancing the read-out systems. The summary mentioned that a single-point sensor configuration is described, using an infrared femtosecond laser and a near-field phase mask, an all-solid FBG is embedded within an optical fiber [25]. This structure's cross-section is defined by concentric circles that showcase refractive indices alternatingly. To interrogate the sensor, a laser beam is used while the FBG is placed on a mechanical elongating stage. Strain measurement is achieved by observing the shift in the transmission peak of the FBG, with a sensitivity of 1.10 pm/με and a detection limit of 26 με. On the other hand, the proposed FBG sensor is positioned between two extensive materials with different expansion coefficients (one positive and one negative) [26]. This configuration enables strain measurement by observing the change in the Bragg wavelength. The sensor exhibits a sensitivity of 1.5 pm/με when strain is measured based on the Bragg wavelength change. In a different approach, a FBG is created using a stressed polydimethylsiloxane (PDMS) micro-diffraction grating. This grating is formed by patterning a layer of the biocompatible soft polymer, PDMS, using soft-lithography techniques. The experimental setup involves placing the grating between two glass plates that are fixed onto interpreting stages. The reported results from the experiments demonstrate a resolution of 0.005 ε.

Nevertheless, a previous has been reported, consisting of both impassive and active configurations [25]. In the passive configuration, the strain variation is analyzed using the change in Bragg wavelength and the change in full width half maximum, resulting in sensitivities of 104.1 pm/με and 61.6 pm/με, respectively. By the way, in the active configuration, a FBG strainer is integrated into a ring laser cavity and achieves a resolution of 7.0×10^2 nε and a sensitivity of 74.5 pm/με.

In reference [26], the sensor configuration is referring to a pair of FBGs with different Bragg wavelengths. One of the FBGs is stamped, while the other exhibits an un-stamped constituency of polymer fiber. Strain is useful using a translation stage to calculate the sensitivity. The un-stamped FBG demonstrates a sensitivity of 1.24 pm/με, while the stamped FBG shows a sensitivity of 1.65 pm/με. By carefully calibrating the characterization matrix coefficients, it becomes possible to measure both temperature and strain simultaneously.

Table 2 presents an overview of the primary sectors that make up the global market for fibre-optic strain sensors applied in maritime application, comparing them to previous strain sensing technologies. It is important to note that the summary provided does not comprehensive and it is only the major role from industry currently utilizing fibre-optic strain sensors. However, as the technology becomes more accessible and

Table 1 List of interrogation/read-out technique from previous works [24]

Interrogation technique	Punctual optical sensing element	Sensitivity	Detection limit	References
Laser, and mechanical stretching	All solid FBG	1.10 pm/με	26 με	[27]
Laser beam, optical analyser	FBG embedded with expansive material	1.5 pm/με	–	[28]
Laser	Stretched polydimetilsiloxane (PDMS) micro-diffraction grating	–	0.005	[29]
A broadband optical basis OSA to perform readings	3 mm long FBG written in photosensitive SMF	Wavelength shift and FWHM of 104.1 pm/με and 61.6 pm/με respectively	700 nε (active)	[25]
Spectroscopic techniques OSA	Bragg grating pair, with one grating in the stamped and the other in un-stamped area of polymer fibre strain applied using a translation stage setup	Un-stamped fibre Bragg grating 1.24 pm/με, Stamped FBG 1.65 pm/με	–	[26]
Wavelength shift through OSA	Dual-FBG strain sensor in PMMA single-mode microstructured POF	0.73 pm/με	–	[30]

Table 2 List of application using FBG sensors in the market [24]

Oil and gas industry	Energy power generation	Military defense
Pipelines work [31–33]	Wind turbines [34]	Hydrophones [35, 36]
Seismic data [37, 38]	Geothermal [39]	Shipboard [40]
In-well [41]	Smart grids [42]	Intrusion detection [43, 44]

cost-effective, it is anticipated that the number of sectors in this list will expand in the future, potentially leading to its adoption in other market segments.

4 Conclusion

This review presents an overview of FBG strain sensing technologies including it applications. It discusses the fundamental properties and functional involved, interrogation/readout techniques, and important parameters for assessing the performance of FBG strain sensors. The paper also delves into recent advancements in highly-sensitive FBG heads and their specific applications. Additionally, novel configurations such as mode splitting-based Bragg grating strain sensors are explored. The operational principle, performance, and advantages of FBG strain sensors are emphasized and compared to provide insights into the current state of the art. Furthermore, the review investigates the marketable and major players in the international global market.

Acknowledgements The first author extends their gratitude to the Ministry of Higher Education (MoHE), Malaysia, for their grant assistance (FRGS/1/2021/TK0/UNIKL/02/20), which played a pivotal role in carrying out and finishing this research.

References

1. Taylor, D.A.: Introduction to Marine Engineering. CRC Press, Boca Raton, FL (2013)
2. Fanellia, P., Mercuria, A., Trupiano, S., Vivio, F., Falcucci, G., Jannelli, E.: Live reconstruction of global loads on a powerboat using local strain FBG measurements. Paper presented at International Conference on Stress Analysis, Assisi, Italy 4–7 Sept 2019
3. Hisham, K.H.: Fiber Bragg Grating Sensors: Principles and Applications. CRC Press, Boca Raton, FL (2019)
4. Eric, U., William, B.S.: Fiber Optic Sensing: An Introduction for Engineers and Scientists. Wiley (2018)
5. Ramly, R., Kuntjoro, W., Rahman, M.K.A.: Using embedded fiber Bragg grating (FBG) sensors in smart aircraft structure materials. Procedia Eng. **41**, 600–606 (2012). https://doi.org/10.1016/j.proeng.2012.07.218
6. Raman, K.: Fiber Bragg Gratings. Academic Press (2010)
7. Andreas, O., Kyriacos, K., Glenn, E.K.: Fiber Bragg gratings: fundamentals and applications in telecommunications and sensing. Phys. Today (2000). https://doi.org/10.1063/1.883086
8. Magdalena, M., Katarzyna, M., Wieslaw, O.: Application of embedded fibre Bragg grating sensors for structural health monitoring of complex composite structures for marine applications. J. Mar. Struc. **76**, 102–103 (2021)
9. Wang, Y., Zhao, S.M.: Structural health monitoring of offshore structures using fiber Bragg grating sensors: a review. Sensors **17**, 10 (2017)
10. Kim, C.S., Kang, H.K., Cho, Y.P.: Structural health monitoring of subsea pipelines using fiber Bragg grating sensors. J. Mec. Sci. and Tec. **31**, 5415–5424 (2017)
11. Nair, R., Asokan, A.G.: Fiber Bragg grating sensors for vibration monitoring: a review. J. Rei. Plas. and Com. **38**, 103–120 (2019)
12. Etter, P.C.: Underwater Acoustic Sensor Networks. CRC Press (2018)
13. Holiday, C.W.: Underwater Acoustics: Analysis, Design and Performance of Sonar. John Wiley & Sons (2012)
14. Ainslie, M.A.: Principles of Sonar Performance Modeling. Springer (2010)
15. Papadakis, J.S.: Underwater Acoustic Digital Signal Processing and Communication Systems. CRC Press (2001)
16. Dickey, T.D.: Introduction to Physical Oceanography. Texas A&M University (2019)
17. Knauss, J.A.: Introduction to Physical Oceanography. Prentice Hall (1997)
18. Riley, J.P., Chester, R.: Introduction to Marine Chemistry. Academic Press (2005)
19. Robinson, A.R., McCarthy, J.J., Rothschild, B.J.: The Sea. Tsunamis, 10, John Wiley & Sons (2005)
20. Charpentier, R.R., Klett, T.R.: Assessment of undiscovered oil and gas resources of the nation's outer continental shelf. U.S. Geological Survey Fact Sheet, 2016–3032 (2016)
21. Chakrabarti, S.: Offshore Structure Modeling: Advanced Series on Ocean Engineering. World Scientific Publishing (2013)

22. Liu, H., Bai, Y.: Pipeline Leak Detection Handbook. Gulf Professional Publishing (2016)
23. Bai, Y.: Subsea Pipelines and Risers. Elsevier (2005)
24. Campanella, C.E., Cuccovillo, A., Campanella, C., Urt, A., Passaro, V.M.N.: Fibre Bragg grating based strain sensors: review of technology and applications. Sensors **18**, 3115 (2018). https://doi.org/10.3390/s18093115
25. Ferreira, M., Bierlich, J., Becker, M., Schuster, K., Santos, J., Frazão, O.: Ultra-high sensitive strain sensor based on post-processed optical fiber Bragg grating. Fibers **2**, 142–149 (2014)
26. Bhowmik, K., Peng, G.D., Luo, Y., Ambikairajah, E., Lovric, V., Walsh, W.R., Rajan, G.: High intrinsic sensitivity etched polymer fiber Bragg grating pair for simultaneous strain and temperature measurements. IEEE Sens. J. **16**, 2453–2459 (2016)
27. Liu, N., Li, Y., Wang, Y., Wang, H., Liang, W., Lu, P.: Bending insensitive sensors for strain and temperature measurements with Bragg gratings in Bragg fibers. Opt. Express **19**, 13880 (2011)
28. Li, J., Du, Y., Liu, C.: FBG strain sensor based on thermal stress mechanism. In: Proceedings of the 2008 Second International Symposium on Intelligent Information Technology Application, Shanghai, China, 20–22 Dec 2008
29. Moscato, M., Schena, E., Saccomandi, P., Francomano, M., Accoto, D., Guglielmelli, E., Silvestri, S.A.: Micromachined intensity-modulated fiber optic sensor for strain measurements: working principle and static calibration. In: Proceedings of the 2012 Annual International Conference of the IEEE Engineering in Medicine and Biology Society, San Diego, CA, USA, 28 Aug–1 Sept 2012; pp. 5790–5793
30. Yuan, W., Stefani, A., Bang, O.: Tunable polymer fiber Bragg grating (FBG) inscription: fabrication of dual-FBG temperature compensated polymer optical fiber strain sensors. IEEE Photonics Technol. Lett. **24**, 401–403 (2012)
31. Qingmin, H., Wenling, J., Shuhui, Z., Liang, R., Ziguang, J.: Natural gas pipeline leakage detection based on FBG strain sensor. In: Proceedings of the 2013 Fifth International Conference on Measuring Technology and Mechatronics Automation, Hong Kong, China, 16–17 Jan 2013
32. Han, B., Li, L., Wu, Z., Jing, H.: Applications of FBG and ZigBee in telemetering of vortex-induced vibration for pipelines. In: Proceedings of the 2013 22nd Wireless and Optical Communication Conference, Chongqing, China, 16–18 May 2013.
33. Wang, J., Zhao, L., Liu, T., Li, Z., Sun, T., Grattan, K.T.V.: Novel negative pressure wave-based pipeline leak detection system using fiber Bragg grating-based pressure sensors. J. Ligh. Tech. **35**, 3366–3373 (2017)
34. Kim, C.H., Paek, I., Yoo, N. Monitoring of small wind turbine blade using FBG sensors. In: Proceedings of the 2010 International Conference on Control, Automation and Systems (ICCAS 2010), Gyeonggi-do, Korea, 27 October 2010
35. Campopiano, S., Cutolo, A., Cusano, A., Giordano, M., Parente, G., Lanza, G., Laudati, A.: Underwater acoustic sensors based on fiber Bragg gratings. Sensors **9**, 4446–4454 (2009)
36. Zhang, W., Li, F., Liu, Y., Xiao, H.: FBG hydrophone: theory and experiment. In: Proceedings of the 2008 1st Asia-Pacific Optical Fiber Sensors Conference, Chengdu, China, 7–9 Nov 2008
37. Zhang, W., Huang, W., Li, L., Liu, W., Li, F.: High resolution FBG sensor and its applications in geophysics. In: Proceedings of the 2017 16th International Conference on Optical Communications and Networks (ICOCN), Wuzhen, China, 7–10 Aug 2017
38. Hu, B., Song, G., Zhang, H., Liu, T.: A precision edge filter interrogation system for mine Fiber Bragg Grating micro-seismic sensors. In: Proceedings of the 2017 16th International Conference on Optical Communications and Networks (ICOCN), Wuzhen, China, 7–10 Aug 2017
39. Bremer, K., Lewis, E., Leen, G., Moss, B., Lochmann, S., Mueller, I., Reinsch, T., Schroetter, J.: Fibre optic pressure and temperature sensor for geothermal wells. In: Proceedings of the 2010 IEEE Sensors, Kona, HI, USA, 1–4 Nov 2010.
40. Wang, G., Pran, K., Sagvolden, G., Havsgård, G.B., Jensen, A.E., Johnson, G.A., Vohra, S.T.: Ship hull structure monitoring using fibre optic sensors. Smart Mater. Struct. **10**, 472–478 (2001)

41. Kamikawachi, R.C., Abe, I., Kalinowski, H.J., Fabris, J.L., Pinto, J.L.: Thermal characterization of etched FBG for applications in oil and gas sector. In: Proceedings of the 2007 SBMO/IEEE MTT-S International Microwave and Optoelectronics Conference, Brazil, Brazil, 29 Oct–1 Nov 2007
42. Huang, Q., Zhang, C., Liu, Q., Ning, Y., Cao, Y.: New type of fiber optic sensor network for smart grid interface of transmission system. In: Proceedings of the IEEE PES General Meeting, Providence, RI, USA, 25–29 July 2010
43. Allwood, G., Hinckley, S., Wild, G.: Optical fiber Bragg grating based intrusion detection systems for homeland security. In: Proceedings of the 2013 IEEE Sensors Applications Symposium Proceedings, Galveston, TX, USA, 19–21 Feb 2013
44. Catalano, A., Bruno, F., Pisco, M., Cutolo, A., Cusano, A.: An Intrusion detection system for the protection of railway assets using fiber Bragg grating sensors. Sensors **14**, 18268–18285 (2014)

Numerical Simulation of a Fixed Oscillating Water Column Wave Energy Converter Device

Rohaizad Hafidz Rozali, Muhammad Adli Mustapa, Mohd Amin Hakim Ramli, Muhammad Nasuha Mansor, Iwan Zamil Mustafa Kamal, and Engku Ahmad Azrulhisham Engku Mohamed

Abstract Wave energy converters are devices designed to harness the energy from ocean waves and convert it into usable electricity. A prototype of the oscillating water column (OWC) device was developed since 1983 and showed unsuitable output for low to mild wave conditions. Thus, several studies have been conducted to increase the pneumatic performance of the OWC, particularly by improving their structural geometry and integration with breakwaters. Thus, the primary objectives of this study are to investigate the optimum slope angle for the back wall and to find out the performance of the proposed slope in Malaysia's coastline. Computational fluid dynamics (CFD) method was used in this study. The design performance is analysed by using two indication variables, i.e., pressure and velocity. The pressure would accelerate the motion and increase the output energy. The contribution of this study is the investigation on the variation of slope angle that can match with Malaysia low wave conditions. Based on the numerical simulation, a design with 19° angle has a good performance at calm wave with 9.88 kJ/kg energy estimation. However, at high wave amplitude, a design with 10° angle has a better performance with 10.52 kJ /

R. H. Rozali (✉) · M. A. Mustapa · M. A. H. Ramli · M. N. Mansor · I. Z. M. Kamal
Maritime Engineering Technology Section, Universiti Kuala Lumpur, Malaysian Institute of Marine Engineering Technology 32200, Lumut, Perak, Malaysia
e-mail: rohaizad@unikl.edu.my

M. A. Mustapa
e-mail: adli@unikl.edu.my

M. A. H. Ramli
e-mail: mohdamin@unikl.edu.my

M. N. Mansor
e-mail: mnasuha@unikl.edu.my

I. Z. M. Kamal
e-mail: iwanzamil@unikl.edu.my

E. A. A. E. Mohamed
Water & Energy Section, Universiti Kuala Lumpur, Malaysian France Institute 43650, Bangi, Selangor, Malaysia
e-mail: engku@unikl.edu.my

A. Ismail et al. (eds.), *Engineering Frontiers*, Advanced Structured Materials 208,
https://doi.org/10.1007/978-3-031-56844-2_7

kg energy estimation.Please check and confirm if the authors and their respective affiliations have been correctly identified. Amend if necessary.Only one amendment for co-authors; Azruhisham to Azrulhisham

Keywords Wave energy device · Fixed oscillating water column · Slope angle low wave energy · Computational fluid dynamics

1 Introduction

The demand of renewable energy has increased due to the climate change and increase of carbon dioxide emission. The popular types of renewable energy are solar, hydro, wind biomass, geothermal and wave [1]. The ocean contains enormous amounts of energy that is dissipated along the world's coastlines. To harvest the energy from the wave, the wave energy converter (WEC) device is needed. There are many WEC devices that can extract energy from the waves, including attenuators, point absorbers, overtopping, oscillating water columns (OWC) and other new technologies [2]. An OWC is the most widely used wave energy converting technology for onshore structures [3]. The technology has the advantages of a simple mechanisms, low cost of manufacturing and transportation, no moving parts in the water, and low effect on marine life [4]. The components of an OWC consist of a water column region, an air chamber, and a duct. Airflow in the air chamber spins an air turbine, and the turbine generates electricity finally [5, 6]. An OWC can be categorized as fixed and floating. A fixed OWC can be either mounted to a cliff or seabed and floating can be restrained by moorings [7]. In order to develop an efficient OWC design, good knowledge of the OWC geometry important components is necessary. Opening slope, depth at opening mouth, shape of the chamber, chamber diameter/width are the geometric parameters that can influence the OWC performance [8]. By focusing on one feature of OWC, the design could be optimized and operates efficiently in Malaysia's coastline. This research is carried out to find the optimum slope angle of the OWC device. This present study will concentrate on the back wall of the slope angles to improve the OWC device performance. The proposed slope angle was simulated numerically using a computational fluid dynamics (CFD) software. Thus, this paper will show the average pressure, average velocity and energy estimation.

Table 1 Wave condition in Kapas Island

Wave condition	Model scale		Full scale	
	Wave height, Hs (m)	Period, Tp (s)	Wave height, Hs (m)	Period, Tp (s)
1	0.0167	0.9037	0.2505	3.5
2	0.05	1.4201	0.75	5.5
3	0.15	2.1947	2.25	8.5

2 Methodology

2.1 Case Study Location and Wave Data

Kapas Island Terengganu that is facing the South China Sea was chosen as the case study location. Table 1 shows the summary of the wave condition at Kapas Island, Terengganu. Condition 1, 2 and 3 represent calm waves, medium waves and strong waves, respectively. The wave condition was selected according to the probability of occurrence [9].

2.2 Model Dimension, Wave Tank Set-Up and Wave Boundary, Location of Probe and Grid-Dependent Study

By using the Froude scaling method, the research was carried using a downscale model at 1:15 ratio. Due to this modification, certain constraints would eventually change too. The length of the back wall increases along with the increment of back wall slope angle. Plus, the opening air duct had become smaller as the slope angle increased. Other than that, the dimensions remain unchanged. The basis and proposed designs are show in Fig. 1. The wave tank sidewall is approximately 1.38 m away from the model sidewall. The main purpose is to prevent any undesirable contact that can result in wave reflection in front of the test model. The wave tank was designed in Autodesk Inventor. The tank was built with the slope at the back which helps the wave remain stable after hitting the model. It is also built with the sidewall on the side of the tank. Three models (design 1, design 2 and design 3) were placed in the same location that is 17.55 m from the incoming wave as shown in Fig. 2. The model is located so that the wave has enough time to be fully developed before reaching the model. The boundary condition is defined in Fig. 3. The incoming wave location is set as wave boundary type represented as WV. The sidewalls of the boundary are set as symmetry (S). The bottom floor of the boundary is set as wall (W). Meanwhile, the top side of the boundary is set as specified pressure, represented as P. Lastly, the end side of the boundary is set as outflow (O) with restriction one-way fluid direction so that the water does not return to the model. This simulation only uses one gridline configuration. More than one gridline configuration could be used to obtain more precise results.

In order to obtain results at various parts of the model, the probe configuration needs to be set-up. All models included the basic design and should have the same probe position as shown in Fig. 4. The smaller the grid cell size, the more precise the simulation results. However, small cell size requires a longer time to complete the simulation as the total cell increases when the size cell decreases. Thus, it is important to perform a grid-dependent study before starting a full simulation. Through the grid-dependent study, the optimum grid cell size for each design is obtained. The optimum mesh size is taken when the graph line starts to be constant. The optimum cell size

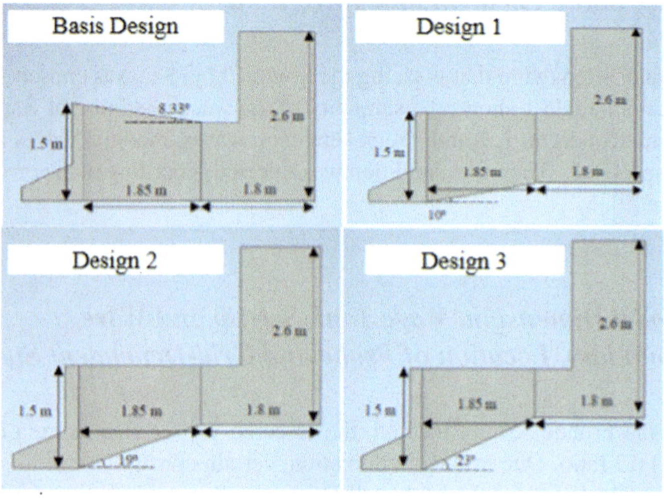

Fig. 1 OWC basic and proposed design with different angles

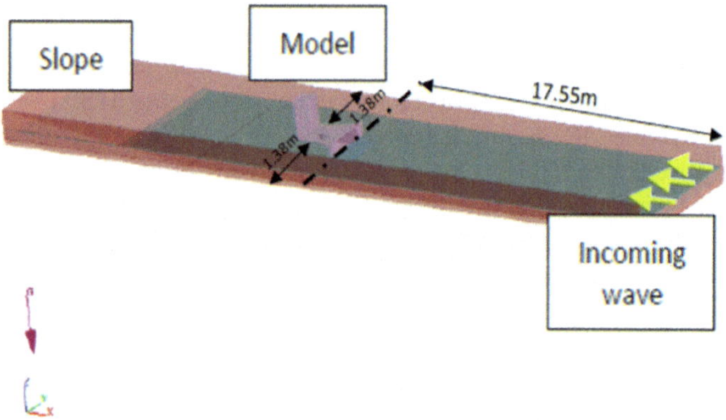

Fig. 2 Wave tank arrangement

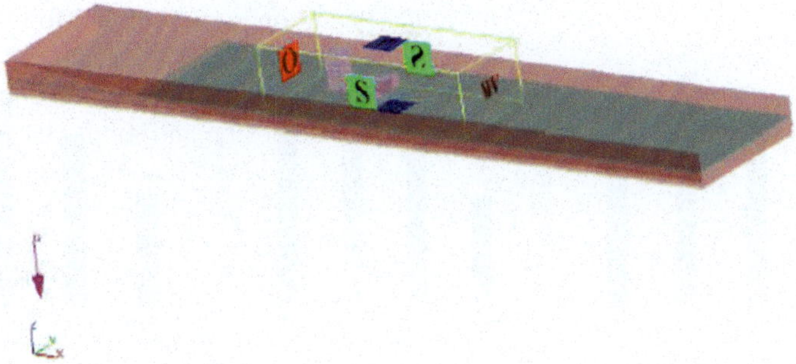

Fig. 3 Boundary condition of simulation

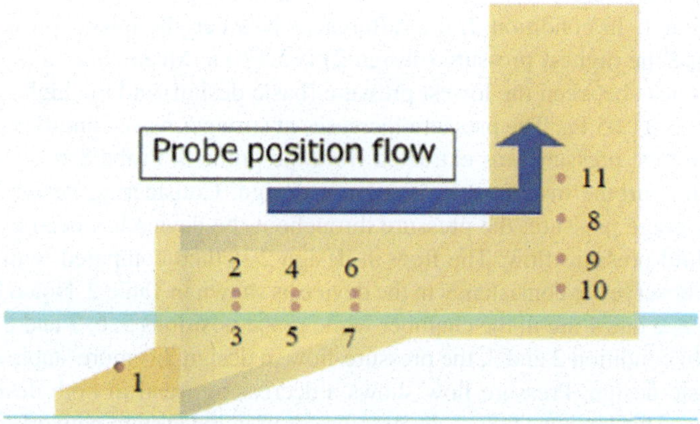

Fig. 4 Probe configurations

for design 1, 2 and 3 is 0.055 m, 0.055 m and 0.05 m respectively. These optimum cell size are used to run a full simulation.

3 Results and Discussion

3.1 Average Pressure, Fluid Pressure Flow and Average Velocity Magnitude

From Fig. 5, the graph shows the average pressure for condition 1. The basic design has the lowest pressure among all designs which is 11.55 Pa. The highest pressure belongs to design 2 which is 21.49 Pa. Design 3 has 20.86 Pa average pressure while design 1 has 16.86 Pa average pressure. The gap between the highest pressure and lowest pressure is 9.93 Pa and considering this is model size, the differences

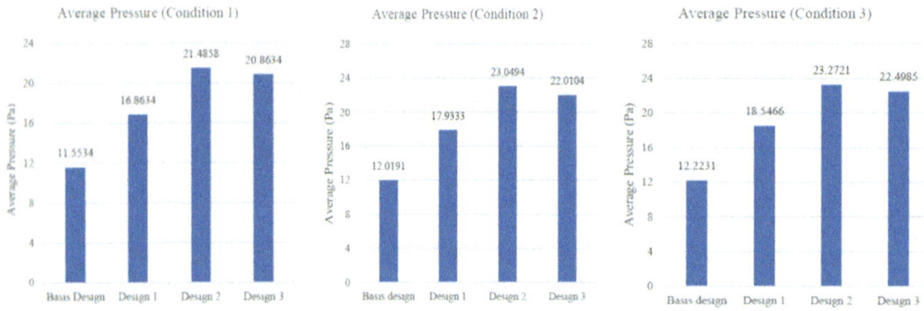

Fig. 5 Pressure comparison

are significant. Based on Fig. 5, the trend in condition 2 and 3 are quite similar to condition 1. In condition 2, the differences between the lowest pressure (basic design) and the highest pressure (design 2) is 5.12 Pa. Meanwhile, in condition 3, the differences between the lowest pressure (basic design) and the highest pressure (design 2) is 11.05 Pa. The pressure increases at stronger wave conditions.

The average pressure was extracted from the probe 8. Probe 8 is the suggested location for turbine instalment in the basic design. Considering design 2 has the highest average pressure, the pressure throughout the device has been extracted to see the fluid pressure flow. The flow in design 2 is then compared with the basic design to observe the consistency in the device as shown in Table 2. Noted that probe positions 2, 3 and 4 are in the chamber while probe positions 5, 6, 7 and 8 are in the air duct. In condition 2 and 3, the pressure-flow in design 2 is more stable compared to the basic design. Pressure flow shows a decreasing trend in both designs at all conditions. This happens due to the pressure drop in the system network.

As shown in Fig. 6, condition 1, design 1 has the highest velocity of air which is 0.0637 m/s while the basic design has the lowest velocity of air which is 0.002 m/s. The differences between the highest velocity and lowest velocity are significant which is 0.0617 m/s. In condition 2, design 2 still has the highest average velocity which is 0.0849 m/s. Design 1 has 0.0659 m/s while design 3 and the basic design

Table 2 Average pressure flow from chamber to air duct

		Chamber probe			Air duct probe			
	Probe	2	4	6	10	9	8	11
	Position	2	3	4	5	6	7	8
Basis design	Condition 1	16.25	16.24	16.21	15.76	13.96	11.55	9.15
	Condition 2	20.60	20.21	19.41	16.41	14.53	12.02	9.50
	Condition 3	24.84	24.75	24.39	16.88	14.85	12.22	9.72
Design 2	Condition 1	26.35	26.32	26.26	25.73	23.91	21.48	19.07
	Condition 2	31.38	30.94	29.91	27.51	25.62	22.98	20.32
	Condition 3	32.08	31.63	30.82	27.71	25.78	23.16	20.45

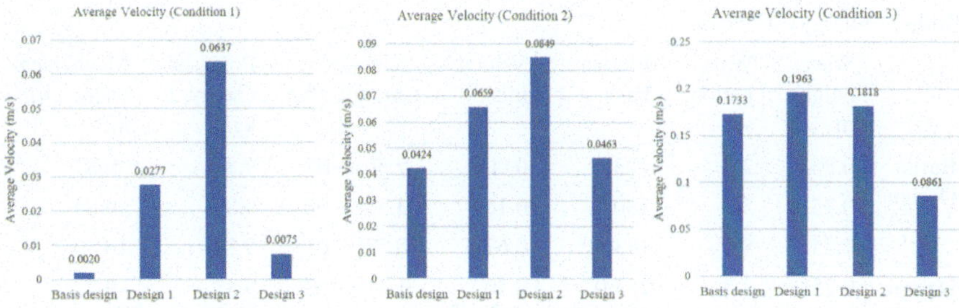

Fig. 6 Average velocity

have 0.0463 m/s and 0.0424 m/s respectively. The basic design also remains with the lowest average velocity of air. In condition 3, the differences in average velocity between design 1, design 2 and the basic design are trivial. Plus, the highest average velocity is produced by design 1 while design 3 has the lowest average velocity. The velocity magnitude is a vector quantity that has direction and theoretically if the pressure is high, the speed also high. Even so, many factors could diminish the velocity of air along the process such as the wave reflection. This phenomenon would decrease the performance of the device eventually.

3.2 Energy Estimation

The energy is calculated at the suggested location for the turbine in full-scale measurement. Table 3 shows the estimation of mechanical energy for all conditions. The potential energy for all designs has no change as the probe 8 (turbine location) remains constant. Meanwhile, the kinetic energy is different in all cases due to a derivation of the energy from the velocity magnitude. At condition 1, the highest mechanical energy is 9.88 kJ/kg which is calculated for design 2. The estimated energy calculated at design 1, design 3 and the basic design is quite insignificant which is 9.82 kJ/kg, 9.81 kJ/kg and 9.81 kJ/kg respectively. At condition 2, the highest mechanical energy is 9.94 kJ/kg which is calculated for design 2. The lowest mechanical energy estimation is 9.84 kJ/kg at for the basic design. Lastly, at condition 3, the highest mechanical energy is 10.52 kJ /kg for design 1. Followed by design 2 with 10.42 kJ/kg estimation, the lowest mechanical energy is 9.95 kJ/kg which is calculated for design 3.

Table 3 Energy estimation

	Wave height, m	Wave period, sec	Pressure, Pa	Velocity, m/s	Kinetic energy, J/kg	Potential energy, J/kg	Mechanical energy, kJ/kg
Basis Design	0.25	3.5	173.3	0.008	7.50E−05	9.81	9.81
	0.75	5.5	180.3	0.164	3.31E−02	9.81	9.84
	2.25	8.5	183.3	0.671	5.52E−01	9.81	10.36
Design 1	0.25	3.5	253.0	0.107	1.41E−02	9.81	9.82
	0.75	5.5	269.0	0.255	7.98E−02	9.81	9.89
	2.25	8.5	278.2	0.760	7.08E−01	9.81	10.52
Design 2	0.25	3.5	322.3	0.247	7.46E−02	9.81	9.88
	0.75	5.5	345.7	0.329	1.32E−01	9.81	9.94
	2.25	8.5	349.1	0.704	6.07E−01	9.81	10.42
Design 3	0.25	3.5	313.0	0.029	1.03E−03	9.81	9.81
	0.75	5.5	330.2	0.179	3.94E−02	9.81	9.85
	2.25	8.5	337.5	0.333	1.36E−01	9.81	9.95

4 Conclusion

The design performance is analysed by using two indication variables, i.e., pressure and velocity. The pressure would accelerate the motion of the particle and increase the output energy. Design 1, 2 and 3 are the proposed designs that have been evaluated and compared with the basic design. From all aspects, the basic design is inefficient compared to the proposed designs. Based on the analysis, design 2 with 19° and 9.88 kJ/kg have good performance at calm waves compared to the basic design and other proposed designs. However, at high wave amplitude, design 1 (10° and 10.52 kJ /kg) has a better performance compared to design 2, design 3 and the basic design. The comparison of two cases shows that the increase in the shore slope angle means decrease in the output power in the air chamber of the OWC device. This observation was also discussed by the literature for the fixed OWC case [10]. Yet, the proposed design could be further improved to reduce the disturbance caused by the sloshing effect and wave reflection. The frictional loss and hydrostatic loss also can be reduced along the fluid channel from the chamber to the air duct to increase the performance of the air turbine.

Acknowledgements The authors would like to thank Universiti Kuala Lumpur (UniKL) for supporting this research under the Short Term Research Grant (str22024).

References

1. Kumar, S., Rathore, K.: Renewable energy for sustainable development goal of clean and affordable energy. Int. J. Mater. Manuf. Sustain. Technol. **2**(1), 1–15 (2023). https://doi.org/10.56896/IJMMST.2023.2.1.001
2. Wang, H., Falzarano, J.M.: Energy extraction from the motion of an oscillating water column. Ocean Syst. Eng. **3**(4), 327–348 (2013). https://doi.org/10.12989/OSE.2013.3.4.327
3. Cui, Y., Liu, Z., Zhang, X., Xu, C.: Review of CFD studies on axial-flow self-rectifying turbines for OWC wave energy conversion. Ocean Eng. **175**, 80–102 (2019). https://doi.org/10.1016/j.oceaneng.2019.01.040
4. Wu, B., Li, M., Wu, R., Zhang, Y., Peng, W.: Experimental study on primary efficiency of a new pentagonal backward bent duct buoy and assessment of prototypes. Renew. Energy **113**, 774–783 (2017). https://doi.org/10.1016/j.renene.2017.06.010
5. Ikoma, T., Masuda, K., Eto, H., Shibuya, S.: Basic characteristics of the primary conversion of an oscillating water column type wave energy converter installed on a wave-dissipating double caisson. J. Offshore Mech. Arct. Eng. **141**(6), 061902 (2019). https://doi.org/10.1115/1.4042943
6. Falcão, A.F.O., Henriques, J.C.C.: Oscillating-water-column wave energy converters and air turbines: a review. Renew. Energy **85**, 1391–1424 (2016). https://doi.org/10.1016/j.renene.2015.07.086
7. Bailey, H., Robertson, B.R.D., Buckham, B.J.: Wave-to-wire simulation of a floating oscillating water column wave energy converter. Ocean Eng. **125**, 248–260 (2016). https://doi.org/10.1016/j.oceaneng.2016.08.017
8. Yacob, D.H., Sarip, S., Kaidi, H.M., Ardila-Rey, J.A., Muhammad-Sukki, F.: Oscillating water column geometrical factors and system performance: a review. IEEE Access **10**, 32104–32122 (2022). https://doi.org/10.1109/ACCESS.2022.3160713
9. Yaakob, O., Hashim, F.E., Mohd Omar, K., Md Din, A.H., Koh, K.K.: Satellite-based wave data and wave energy resource assessment for South China Sea. Renew. Energy **88**, 359–371 (2016). https://doi.org/10.1016/j.renene.2015.11.039
10. Yamaç, H.İ., Koca, A.: Numerical wave tank analysis for energy harvesting with oscillating water column. In: Mechatronics 2017, Březina T., Jabłoński, R. (eds.) in Advances in Intelligent Systems and Computing, vol. 644. Cham: Springer International Publishing, pp. 726–734 (2018). https://doi.org/10.1007/978-3-319-65960-2_89

A Marine Flame Resistance Wall from Renewable Local Natural Fiber

Md Redzuan Zoolfakar, Khairul Akmal Azmi, and Wasis Dwi Aryawan

Abstract Natural fibers are widely used in industries such as automotive, building and marine due to their environmental benefits and lower cost compared to synthetic fibers. However, natural fibers are not naturally fire resistant. To address this concern, natural fiber composites are treated with flame retardant additives and other methods. In this research, a comparison of the fire performance of three different natural fibers was performed by manipulating the composition of the composite specimens. Factors such as the fiber's ability to withstand fire, the amount of smoke produced, the toxicity of the fiber, and the rate of heat release during combustion were evaluated. Standard fire tests are conducted to obtain reliable data. The aim of this study was to determine the best natural fibers in terms of fire resistance. In doing so, we can provide valuable insight for industries looking to replace synthetic fibers with more fire-resistant natural alternatives.

Keywords Natural fiber · Fire resistance · ASTM D635

1 Introduction

The composites business mostly favors synthetic fibers over natural fibers because of their better understood properties and benefits. Synthetic fibers such as Kevlar, carbon, and fiberglass are commonly used in composites because of their flame retardant properties. There is a lack of knowledge about the quality and uses of natural fibers, leading to the perception that composites with natural fibers are more flammable and less able to withstand intense heat. However, with increasing oil

M. R. Zoolfakar (✉) · K. A. Azmi
Maritime Engineering Technology Section, Universiti Kuala Lumpur, Malaysian Institute of Marine Engineering Technology, Lumut, Perak, Malaysia
e-mail: redzuan@unikl.edu.my

W. D. Aryawan
Department of Naval Architecture, Faculty of Marine Technology, Institut Teknologi Sepuluh Nopember, Surabaya, Indonesia
e-mail: wasis@na.its.ac.id

A. Ismail et al. (eds.), *Engineering Frontiers*, Advanced Structured Materials 208,
https://doi.org/10.1007/978-3-031-56844-2_8

prices and environmental concerns, there has been a resurgence in the use of natural fibers in industries such as textiles, construction, plastics, and automobiles [1]. These composites are gaining importance because of their non-carcinogenic and biodegradable properties. It can be concluded that natural fibers have advantages for marine applications with low density, low cost, and biodegradability. However, the main disadvantage of natural fibers and matrices is the high relative moisture absorption [2]. Therefore, chemical treatment is considered in modifying the surface properties of fibers [3]. This is driven by economic growth and the freedom provided by local agro-industrial markets and products, as well as the desire to reduce dependence on imported resources.

It is important to note that natural fibers have a lower flash point than synthetic fibers, making them more flammable [4, 5]. However, by combining natural fibers with suitable polymers, their flammability can be reduced [6]. For applications where cost and aesthetics are paramount, natural fibers may be preferred for structures such as ships. This study aims to collect data on local natural fiber composites regarding their flame resistance and identify the most sustainable natural fiber composites that can resist fire, according to the ASTM D-635 procedure [7]. By conducting this research, we are able to evaluate the fire characteristics of different composites reinforced with natural fibers after being exposed to fire [7]. Additionally, determining which natural fibers exhibit the highest fire performance data allow us to identify the most effective fire-retardant natural fibers.

2 Methodology

In this project, three different natural fibers (coconut coir, rice straw and kapok) were used to create a natural fiber reinforced composite. The first step involves preparing these fibers. Coir fiber is obtained from coconut which is used as fuel in traditional food. Kapok fibers are commonly used in pillows, clothing, and bedding. Rice straw fibers are widely used in the manufacture of archery targets. After obtaining natural fibers, the next step is to treat them by exposing them to sunlight to ensure they are dry and free from fungus. Then, a sensitivity analysis is conducted to examine how variations in the independent variable affect the dependent variable. This analytical analysis helps to understand the effects of uncertainty in mathematical models. Therefore, three batches, to simulate the outcome data pattern were created. By conducting a sensitivity analysis, researchers can gain an insight into the behaviour of natural fiber reinforced composites and better understand the variations in independent variables that affect the final product.

Next, the specimen is fabricated according to the ASTM D635 standard. In this case, the specimen for batch 1 was set for 1 layer with coir fiber and the type of resin was manipulated. Specimens for batch 2 were fixed for 2 layers with polyester resin and the fiber material was manipulated. Specimens for batch 3 were set to kapok fibers with epoxy resin and the fiber layers were manipulated. Each specimen consists

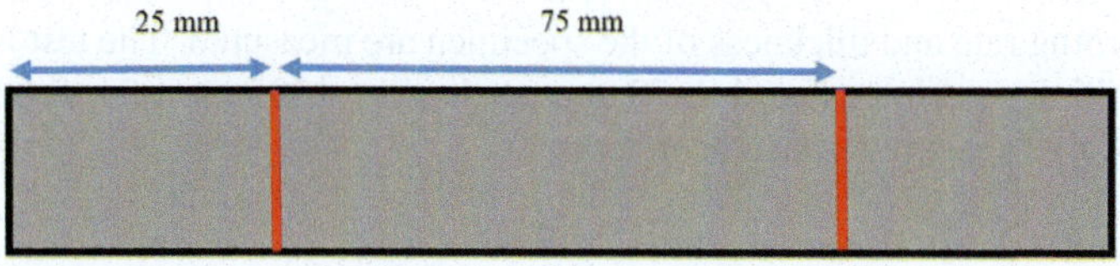

Fig. 1 Specimen marking

of a combination of natural fibers and obtained resin. Here are the instructions to run this project by following the ASTM D635 [8]:

1. Fabricate the specimens according to the ASTM D635 standard.
2. Cut and marking the specimens to the standard size as shown in Fig. 1.
3. Prepare the test apparatus.
4. Perform the fire test in a smoke-free environment.
5. Measure the burning rate and specimen thickness.
6. Repeat the test for multiple specimens.

Here are some additional details about each step:

- Fabricating specimens: Specimens are made using a wet lay-up process, where natural fibers are blended with resin. Wax is applied to the glass surface before placing the specimen to ensure easy removal after curing. The specimen is then cured at room temperature under pressure. The specimens are cut to the standard size of 125 mm length, 13.0 mm width, and with a minimum thickness of 3.0 mm. The maximum thickness should not exceed 13 mm.
- Prepare test apparatus: Test apparatus includes combustion chamber, wire gauze, Bunsen burner, ruler, and clamp retort stand as shown in Fig. 2.
- Conduct fire tests: Fire tests are conducted in a smoke-free environment, preferably in a combustion chamber. A test flame is applied to the specimen for 30 s, and

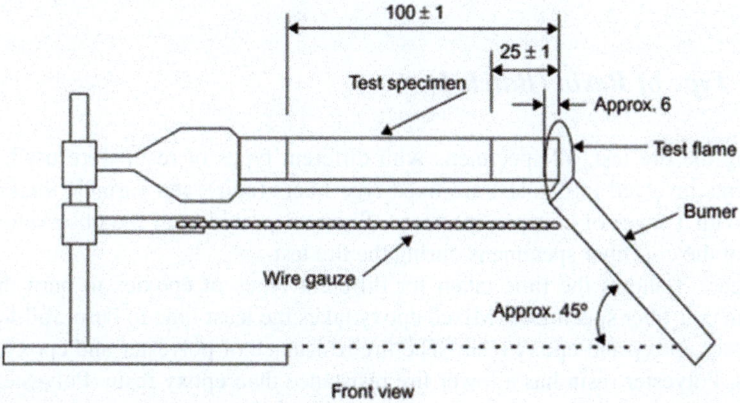

Fig. 2 Illustration of ASTM D635 standard

the burning rate and thickness of the specimen are measured. The test is repeated for multiple specimens until the flame reaches the 100 mm reference mark or a total of five specimens have been tested.

3 Results and Discussion

The composite network was created using three local natural fibers: coir fiber, rice straw fiber and kapok fiber. These fibers have been fabricated into different layers, with variations in the type and thickness of the resin. After all the fabrication materials are prepared, a laboratory scale fire test is conducted according to the ASTM D635 standard. This test is designed to determine the burning rate of a plastic in a horizontal position and is carried out in a confined burning chamber.

During the burning process, careful observations were made, and the time taken for the fibers to burn completely was recorded. These observations and measurements are used to calculate the burn rate for each specimen. The specimens were divided into three groups, and each group consisted of 12 specimens that were manipulated differently to account for factors that could potentially influence the test results.

3.1 Observation

Throughout the test, observation is an important action. It is done to see if anything unusual happens during the test. Additionally, observations were made to determine the exact time for the specimen to burn from the 25 mm marking to the 100 mm marking on the specimen. It is then recorded as elapsed time for use in the linear burn rate calculations. Finally, all observations are stored in a different table to distinguish them for each sample.

3.2 Type of Resin (Batch 1)

During the fire test, all specimens with different types of resin were used. Other elements involved in this test are fixed type fibers (coir), and variable thickness is used with 1 layer of coir per specimen. Figure 3 provides all the observations for each of the coir fiber specimens during the fire test.

Figure 3 shows the time taken for different types of epoxies to burn through 75 mm coir fiber specimens. Mixed epoxy takes the least time to burn, followed by polyester epoxy and epoxy resin. The fire resistance of polyester and epoxy resins varies. Polyester resin has a lower fire resistance than epoxy resin. Polyester resin can melt and burn when exposed to fire, producing harmful gases and smoke. As a result, they can be dangerous to be around in the event of a fire. Epoxy resin, on the

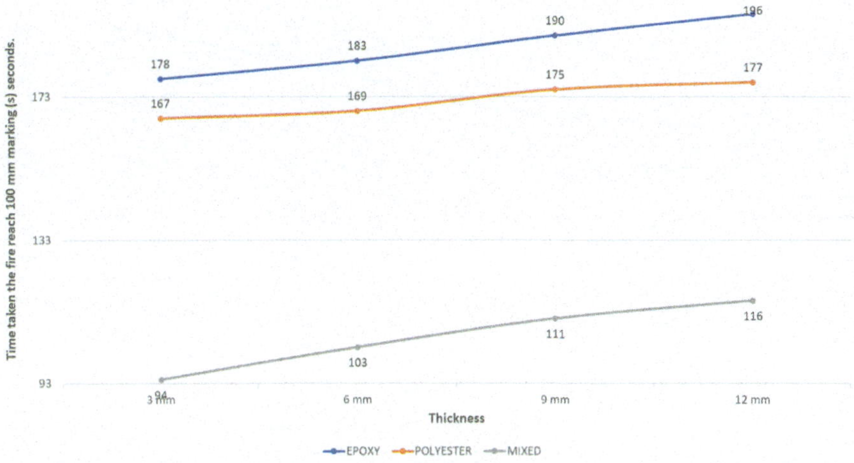

Fig. 3 Result on batch 1 specimen (resin manipulated)

other hand, is often more fire resistant than polyester resin. Epoxy resins are more resistant to heat and flame, and they are less prone to ignite or burn. Also, when exposed to fire, epoxy resin emits fewer harmful gases and fumes than polyester resin [3]. Epoxy resin is a thermoset polymer that cures through a chemical reaction, while polyester resin is a thermoset polymer that cures through a combination of heat and a catalyst [9].

3.3 *Type of Fiber Material (Batch 2)*

During the fire test, all the specimens with different types of fibers were used. Another component in this test is a set of resin types (polyester) with variable thickness and two fiber layers for each specimen. Figure 4 summarizes all the observations made during the fire test for each of the nine bagasse fiber specimens.

Figure 4 shows the time taken for polyester resin mixed with different types of fibers to burn from 25 to 100 mm. Kapok fiber and rice straw specimens burned in less than 3 min, while coconut coir fiber specimens took more than 3 min to burn. Burning time increases as the thickness of the specimen increases. Coir fiber is the most fire resistant among coir fiber, kapok, and rice straw because of its high ignition temperature and low heat release rate. The Kapok fiber has moderate fire resistance, but it can be highly flammable under certain conditions. The rice straw fiber is the least fire resistant due to its low ignition temperature and high heat release rate. However, the fire resistance of these fibers can vary depending on various factors such as thickness, ambient temperature and duration of ignition. Therefore, it is important to consider these factors when selecting fibers for applications where fire resistance is important.

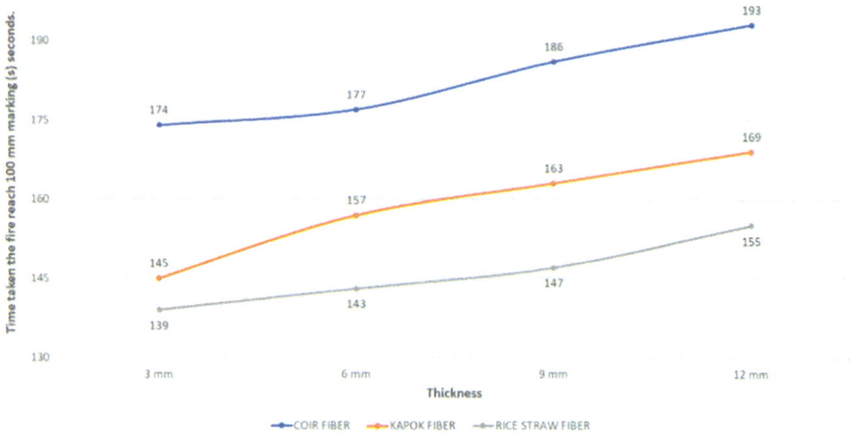

Fig. 4 Result on batch 2 specimen (fiber manipulated)

3.4 Layer of Fiber Apply (Batch 3)

In the fire test, the specimens with different fiber layers were used, while other test components, such as the type of resin (epoxy) and fiber used (coir), remained consistent. Specimen thickness varies. Figure 5 presents a summary of all observations recorded during the fire test for each bagasse fiber specimen.

Figure 5 shows the time taken for epoxy resins with different fiber layers to burn. The average time taken for most of the 75 mm shot specimens was long, except for the 3 mm thick single layer fiber specimen. Based on the fire test results with different fiber layers, it can be concluded that the fire performance of the specimen is affected by the number of fiber layers and the thickness of the specimen. The 3 mm-thick single-layer fiber specimen exhibited the shortest burning period, while the 3-layer

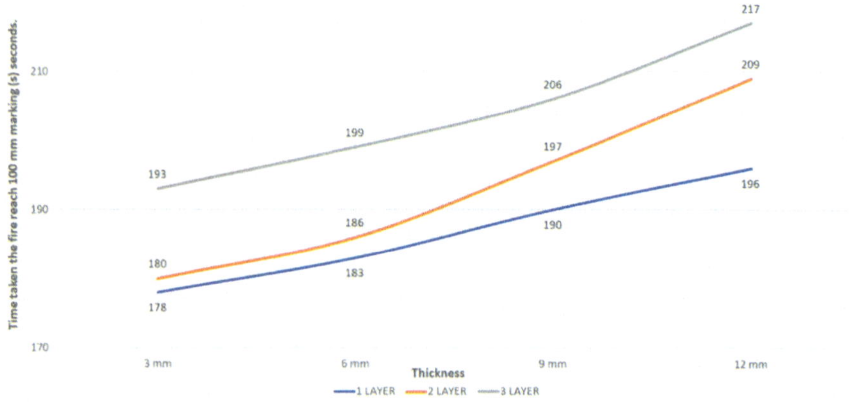

Fig. 5 Result on batch 3 specimen (layer manipulated)

fiber specimen took the longest to burn. This shows that the number of fiber layers has a significant effect on the fire performance of the specimen. Additionally, the duration required for the specimen to burn increases as its thickness increases, indicating that thicker specimens may be more susceptible to fire. Therefore, it is important to consider the number of fiber layers and thickness when designing materials for applications where fire resistance is important.

3.5 Linear Burning Rate

After all the specimens of the three different natural fibers (kapok, coir and rice straw fiber) have undergone a fire test according to the ASTM D635 standard, the next step is to calculate their linear burning rate. The linear burn rate is a quantitative measure of how fast a material burns in a straight line, expressed in millimeters per minute. This calculation involves determining the time required for the flame front to reach the 100 mm reference point on each specimen and using the appropriate formula.

$$LBR = 60\,L/t \tag{1}$$

where *LBR* is the linear burn rate in mm/min, *L* is the burn distance between reference marks in mm, and *t* is the time in seconds. According to the Eq. (1), the variable L represents the length of the burned specimen, starting from the 25 mm mark and up to the point where the flame front reaches the 100 mm mark. Since all specimens were fired up to the 100 mm mark, the shot length for each specimen was maintained at 75 mm. On the other hand, the change time (t) corresponds to the elapsed time required for each specimen to burn from the 25 mm mark up to the 100 mm mark. Following the objective of finding the fire resistance of various materials, the data of group 2 specimens was chosen (manipulated materials) to calculate the linear burning rate of the test. Figure 6 shows the results of the linear rate of combustion of the manipulated material. The slowest burning fibers are the safest in a fire. Coir fiber burns the slowest, at an average rate of 24.695 mm/s. Rice straw fibers burn the fastest, at an average rate of 30.775 mm/s. Kapok fibers burn in between, at an average rate of 28.335 mm/s. The average fiber linear burning rate for the specimens is shown in Fig. 7.

4 Conclusion

This study investigated the fire resistance of natural fibers, including coir, kapok, and rice straw, as an alternative to synthetic fibers in composite structures. The results of the study show that coir fiber has the highest fire resistance, followed by kapok fiber and rice straw fiber. Coir fiber has a low linear burning rate and minimal production of smoke and toxic fumes when exposed to fire. Rice straw fiber has the

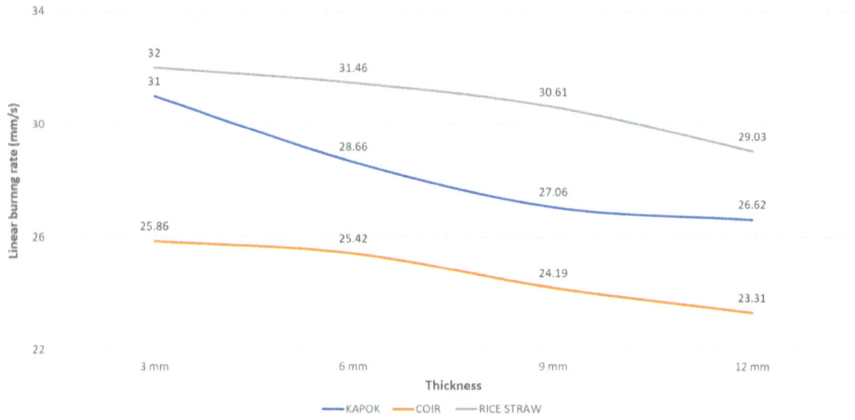

Fig. 6 Linear burning rate based on batch 2 specimen (fiber manipulated)

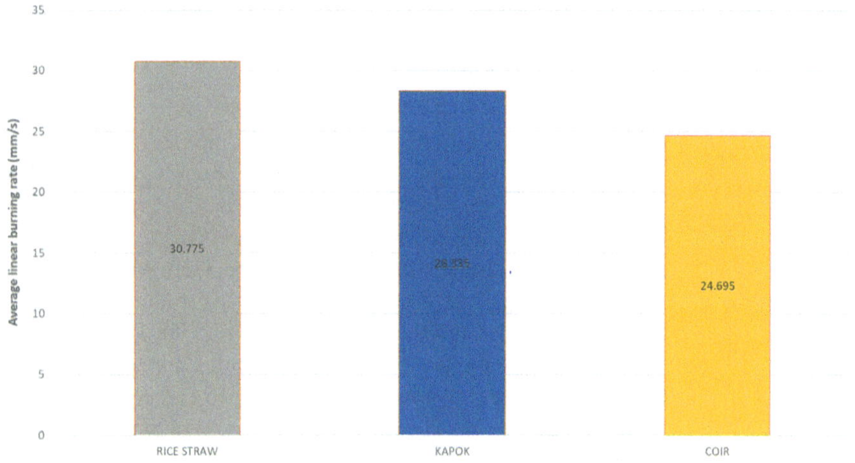

Fig. 7 Average linear burning rate based on batch 2 specimen (fiber manipulated)

lowest fire resistance, with a high heat release rate and significant smoke and toxic smoke emissions during combustion. Fire resistance is also affected by specimen thickness and layering, with thicker multi-layered specimens offering better fire resistance. Epoxy resins generally exhibit better fire resistance compared to polyester resins. Mixing epoxy and polyester resins is not recommended due to their chemical differences and potential adverse effects on curing, bond strength and mechanical properties. Overall, coir fiber was found to be the most fire resistant natural fiber, and it is a promising alternative to synthetic fibers in composite structures where fire resistance is important.

Acknowledgements The author would like to express his gratitude to the Institute of Marine Engineering Technology, Universiti Kuala Lumpur Malaysia for providing a constructive environment to present this research.

References

1. Thule, A., Shanks, R. (eds.): Natural Fibre Composites: Materials, Processes and Properties. Woodhead Publishing (2014)
2. Moudood, A., Rahman, A., Khanlou, H.M., Hall, W., Öchsner, A., Francucci, G.: Environmental effects on the durability and the mechanical performance of flax fiber/bio-epoxy composites. Compos. Part B-Eng. **171**, 284–293 (2019). https://doi.org/10.1016/j.compositesb.2019.05.032
3. Ramadan, N., Taha, M., La Rosa, A.D., Elsabbagh, A.: Towards selection charts for epoxy resin, unsaturated polyester resin and their fibre-fabric composites with flame retardants. Materials (Basel) **14**, 1–44 (2021). https://doi.org/10.3390/ma14051181
4. Madyaratri, E.W., Ridho, M.R., Aristri, M.A., Adly, M., Lubis, R., Iswanto, A.H., Nawawi, D.S., Antov, P., Kristak, L., Majlingov, A., Fatriasari, W.: Recent advances in the development of fire-resistant biocomposites—a review (2022)
5. Benedik, V.: Fire resistance of wooden materials and its improvement by means of protective agents (2017)
6. Bhattacharyya, D., Kim, N.K.: Flammability of natural fibre-reinforced polymeric composites. ICCM Int. Conf. Compos. Mater. (2017)
7. Hamid, M.R.Y., Ahmad, S.: Effect of flame retardants on wood plastic composites-HDPE based. Key Eng. Mater. **471–472**, 640–645 (2011). https://doi.org/10.4028/www.scientific.net/KEM.471-472.640
8. ASTM: Rate of Burning and/or Extent and Time of Burning of Plastics in a Horizontal Position 1. 1–7 (2010)
9. Pascault, J., Sautereau, H., National, I., Verdu, J., Williams, R.J.J., Plata, M.: Thermosetting polymers (2002)

Changes in Chemical Concentration of Silty Clay Added with Lime and Nanolime

Panbarasi Govindasamy and Mohd Raihan Taha

Abstract Stabilization of clayey soils with lime has shown that small amounts of lime considerably improve the workability whereas larger amounts of lime only improve the strength which requires several months or years and is thus considered the long-term treatment. Nanoparticles are excellent candidate materials for soil improvement. Due to their smaller dimensions, nanoparticles possess a very high specific surface and react more actively with other particles in the soil matrix. The existence of even a minute amount of these nanoparticles can result in extraordinary effects on the engineering properties of soil. Thus, this study was carried out to validate the chemical effectiveness of nanolime additives as soil stabilizer. Lime and nanolime were chosen as additive to investigate its reactivity on silty clayey soil. The results of pH test revealed that the pH increased proportionally, changing from 4.07 (acidic range) to about 7.85 and 12.47 for a dose of 1.0% of lime and nanolime, respectively. X-ray diffraction (XRD), energy-dispersive X-ray spectroscopy (EDX), X-ray fluorescence (XRF) and field emission scanning electron microscopy (FESEM) tests were run for original and treated soil samples to characterize the mechanisms of chemical improvement. Nanolime shows superiority in soil improvement compared with lime even at the dosage of 0.5%. The results indicate that nanoparticles have higher specific surface area and this will lead to a larger amount of water encompass the outer surface of particles. Due to higher ability of nanolime to flocculate and agglomerate, it could fill the micro and nano pores as well which dramatically enhances the chemical properties of the soil.

Keywords Lime · Nanolime · Chemical properties · Nanomaterial

---

P. Govindasamy (✉)
Politeknik Ungku Omar, Jln Raja Musa Mahadi, 31400 Ipoh, Perak, Malaysia
e-mail: panbarasi@puo.edu.my

M. R. Taha
Universiti Kebangsaan Malaysia, 43600 Bangi, Selangor, Malaysia

A. Ismail et al. (eds.), *Engineering Frontiers*, Advanced Structured Materials 208,
https://doi.org/10.1007/978-3-031-56844-2_9

1 Introduction

Clay soils in particular can present great problems in pavement design due to uncertainty associated with their performance. They are often unstable beneath a pavement and they are the most susceptible to problems from changes in moisture content. Clay soils tend to swell and become soft when they become wet and may shrink and become stiff when dried. Additives, including lime, fly ash, Portland cement, and more recently synthetics are available that will lessen these problems when mixed in the proper amounts with problem soils. These additives may be used separately or in combination and each has construction issues related to its performance.

Lime is generally used to treat clay soils, although not all clay soils are subject to significant changes in engineering properties when mixed with lime. The changes that can occur in a soil include soil modification and soil stabilization. Previous work in stabilization of clayey soils with lime has shown that small amounts of lime considerably improve the workability but contribute little to strength, whereas larger amounts of lime also improve the strength and bearing capacities of these soils. Addition of lime to clay soil provides an abundance of calcium ions (Ca^{2+}) and magnesium ions (Mg^{2+}). These ions tend to displace other common cations such as sodium (Na^{+}) or potassium (K^{+}), in a process known as cation exchange [1].

Replacement of sodium or potassium ions with calcium significantly reduces the plasticity index of the clay. A reduction of plasticity is usually accompanied by reduced potential for swelling. The addition of the lime increases the soil pH, which also increases the cation exchange capacity. The above reaction is accompanied by pozzolanic reaction, which is a slower or long-term reaction. The reactions are little understood, but apparently involve interaction between hydrated lime and siliceous and aluminous minerals in soils. The resulting gel cements the soil and may be similar to certain reaction products from the hydration of Portland cement. A major difference is that under normal curing conditions considerably more time is required for Pozzolanic—cementation reactions to contribute much strength.

The presence of nanoparticles, even at a small fraction, can dramatically influence the engineering properties of soil. At the nanoscale, a higher ratio of surface to volume and, in turn, a higher cation exchange capacity exists. Therefore, they interact very actively with other particles and solutions such that very minute amounts may lead to considerable effects on the physico-chemical behavior and engineering properties of soil [1].

2 Methodology

Samples prepared for chemical tests are original soil, soil mixed with lime and soil mixed with nanolime. The percentages of nanolime mixed with soil were 0.2, 0.3, 0.5, 0.8 and 1% each. 5% lime was added to soil. This amount was the value of initial lime consumption (ILC) for the studied soil which was determined based on the pH test. The dry original soil was mixed with lime or nanolime percentages and then the water was added to the mixture. In this method, mixing is carried out into two stages. Initially premix or hand mixed [2], the quantity of soil was divided into ten layers and each layer was sprayed with the required amount of additives. Each layer was mixed alone then put in the pot and then the mixture is mixed again by a horizontal cylindrical mixer for at least 3 h [3]. The dry mixtures in the pot were sprayed with the required amount of water during the mixing process. The type of water used for preparing samples was distilled water. This procedure was found to be the best method to obtain homogeneous samples since a homogeneous color was obtained after compaction.

XRF is a nondestructive method for the elemental analysis of solids and liquids. In this technique, the samples were irradiated by an intense x-ray beam causing the emission of fluorescent x-rays. The x-rays emitted were detected by an energy dispersive or wavelength dispersive detector. The energies or wavelengths were used to identify the elements present in the samples, whereas concentrations (quantities) were determined by the intensity of the x-rays. XRF analysis was conducted for the native materials, i.e. original soil, lime and nanolime to obtain their chemical compositions and to create a perception about the expected reactants using a Bruker S4 Pioneer model. The soil was milled and sieved at 425 μm before XRF analysis was done.

XRD analysis was conducted using a Bruker AXS, Karlsruhe, Germany, model D8 Advanced. The source of X-Ray was Cu-Kα with an input voltage of 40 kV and current of 40 mA. The wave length and step size were 0.15406 nm and 0.025°/0.1 s, respectively. XRD is a very reliable and dominant technique for minerals identification in soils and rocks [4]. This analysis was conducted to identify the chemical reactions and development of new products if any. When x-rays interact with a crystalline substance, a definite diffraction pattern is achieved. Every crystalline substance has a specific pattern and the same substance always produces the same pattern and in a mixture of substances every substance gives its independent pattern from the others [5]. In this study, XRD analysis was used to determine the crystallite size of lime and nanolime particles using the Scherer's formula.

Supra 55VP Zeiss 2008 at CRIM, UKM for FESEM test was employed to analyze the untreated and the treated soil samples to observe the change in the soil aggregation and formation of any new material. Initially the same percentages of lime and nanolime were hand mixed with water content according to their respective optimum moisture content level. The soil–lime and nanolime mixtures were kept in polyethylene bags and cured at room temperature for 28 days. Then the mixtures were oven dried for 24 h at temperature of 100 ± 1 °C before they were milled and sieved at 425 μm. FESEM in this study was conducted to disclose the morphology of the native materials and to observe the morphology of the materials after mixing and curing. The change in the soil aggregation and formation of new materials can be observed through this analysis. Since the microscope of FESEM uses electron in place of light, the objects observed by FESEM must be able to be conductive for current. Therefore, all the specimens in this study were coated by a very thin layer of gold. Besides that, mean particle size of lime and nanolime were also determined by analyzing the images provided by FESEM analysis.

Energy dispersive X-ray (EDX) analysis was performed to observe of any indication of chemical reactions that could have taken place between the mixed materials. In this analysis, the observed elements have a fixed position along the horizontal scale (energy). However, the peak heights (intensity) of these elements could change (rise or descend) in case that a chemical reaction took place. Bearing in mind, the peak heights reflect the concentrations of each element. This analysis was carried out for all specimens mentioned previously i.e. original soil, soil–lime and nanolime dry mixtures, and wet cured mixtures of soil–lime and soil–nanolime. Normally, EDX analysis is conducted in conjugation with FESEM.

3 Results and Discussion

The basic physical properties and classification of the soil collected from the hilly area of Universiti Kebangsaaan Malaysia (UKM) were listed in Table 1. It was collected at shallow depth after removing the top 0.5 m depth of ground surface. The natural water content was 22.8% while the maximum dry density ($\gamma_d$max) and optimum water content (ωopt) were 1.87 g/cm^3 and 15.6 \%, respectively. According to the unified soil classification system (USCS), the tested soil was classified under group symbol of SC. The soil contained no gravel. Meanwhile, the fraction of sand, silt and clay are 57.5, 22.5, and 20%, respectively. Figure 1 presents the distribution of particle size of the original soil.

Lime and nanolime were the two additives used for this of soil improvement. Lime powder was obtained from local market. Nanolime powder was imported from Strem Chemicals, Inc., Mulliken Way, Newburyport, Massachusetts, United States.

Table 1 Physical properties of original soil

Property	Value
Natural water content, (ω_n) (%)	22.8
Maximum dry density, $y_{d\,max}$ (g/cm^3)	1.87
Optimum water content (ω_{opt}) (%)	15.6
Liquid limit (LL) (%)	31
Plastic limit (PL) (%)	18.7
Plasticity Index (PI) (%)	12.3
USCS classification	CL
Specific gravity (ρ_s)	2.63
% Gravel	0.0
% Sand	57.5
% Silt	22.5
% Clay	2.0

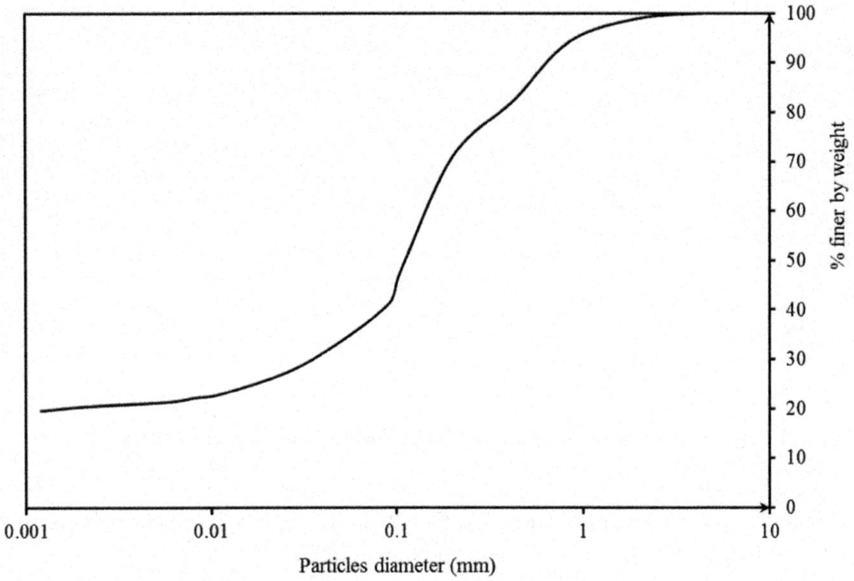

Fig. 1 Particles sizes distribution curve of original soil

The results of X-ray fluorescence (XRF) test for lime used as additive are illustrated in Table 2 for both regular and nano grades. It can be seen that the nanolime has greater purity than the lime. A large trace of CaO was available in both grades, while lime contained a noticeable amount of MgO (1.75%) compared to 1.24% which was available in nanolime. Table 3 presents the chemical composition of original soil, where the silica and alumina are the main components and then iron oxide comes as a third most abundant component.

X-ray diffraction (XRD) test was used to analyze the crystallite size of the particles. The results of this test in the form of relation graphs between intensity and 2 theta are shown in Fig. 2. The value for the full width at half maximum (FWHM)

Table 2 Chemical composition of additives

	CaO (%)	MgO (%)	SO_3 (%)	SiO_2 (%)	Al_2O_3 (%)	Fe_2O_3 (%)	SrO (%)	Cl (%)	K_2O (%)	ZnO
Lime	82.7	1.75	0.56	0.22	0.13	0.08	0.03	0.02	0.01	–
Nanolime	92.69	1.24	0.02	0.47	0.09	0.07	0.03	0.01	0.01	48 ppm

Table 3 Chemical composition of original soil

SiO_2	Al_2O_3	Fe_2O_3	TiO_2	K_2O	MgO	ZrO_2	SO_3	CaO	Others
60.35%	21.83%	4.36%	1.13%	0.51%	0.47%	0.09%	0.08%	0.05%	0.08%

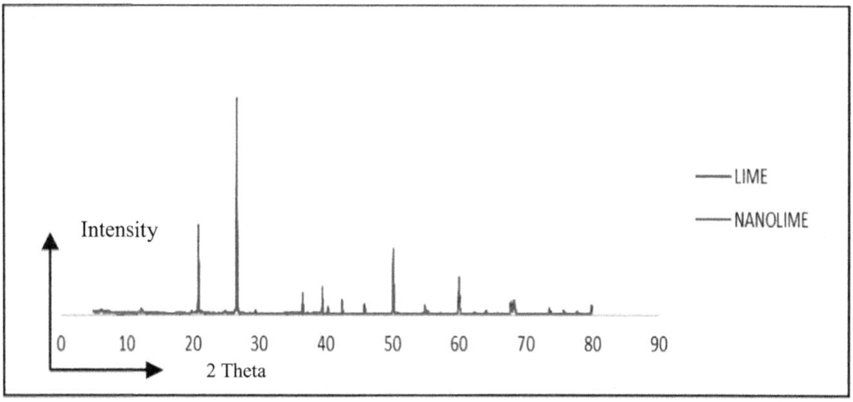

Fig. 2 Relation graph between intensity and 2 theta for lime and nanolime material

and the theta were obtained. These numbers were analyzed using Scherer's formula to measure the crystallite size of lime and nanolime.

The pH of the samples was determined using the pH meter, which involves mixing the solids with pure water (1:5 solid: water), periodically shaking samples, and then testing the sample. The pH value test [6] was conducted to determine the minimum lime content. According to the test, the minimum lime content of a soil is reached when the pH of the soil, lime, and water mixture with 25 g of soil passed through the 425 μm (No. 40) sieves, a certain percentage of lime, and 100 g of distilled water reaches 12.4. In this pH test, the lime dosages ranged from no lime to 9% of lime mixed with distilled water and soil. In between, the lime dosage is incrementally increased until the pH of this mixture reaches 12.4.

The pH readings were taken after 60 min of shaking the mixtures for 30 s at 10 min intervals. Meanwhile, nanolime was also mixed with soil and distilled water from 0.2 to 9% to check on the pH reading till it reaches 12.4. The formation of $Ca\ (OH)^2$ explained the increase in the pH value of the tested soil. The pH test conducted on the original soil, soil–lime and soil–nanolime mixtures showed a significant increase in the pH with increase in the percentage of lime and nanolime. For original soil, the pH (4.07) jumped to 6.82 and 8.97 just by adding 0.3% of lime and nanolime, respectively. Subsequently, it shows that 5% lime achieved 12.48 pH reading and 0.8% nanolime achieved pH value of 12.44 as shown in Fig. 3.

The formation of $Ca(OH)^2$ elucidated above is followed by an ionization process inducing liberating of hydroxyl ions $(OH)^{-1}$ which is the reason behind the increase the pH value. CaO, absorbs water from the surrounding ground, causing the lime to swell and forms slaked lime ($Ca(OH)^2$) as per the following chemical reaction [7]:

$$CaO + H_2O \rightarrow Ca(OH)^2 \tag{1}$$

$$Ca(OH)^2 \rightarrow Ca^{+2} + 2(OH)^{-1} \tag{2}$$

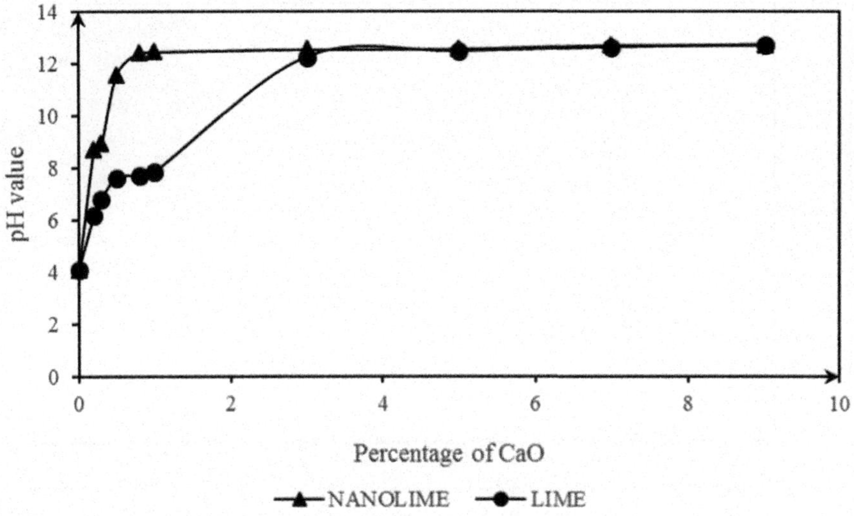

Fig. 3 pH development for soil + lime and nanolime additives

Oxygen Intensity. Figures 4 and 5 show the EDX spectrum of dry lime and nanolime while Figs. 6 and 7 show the EDX spectrum for wet lime and nanolime. It can be seen that oxygen intensity is higher in nanolime than in lime. Figure 8 shows an increase in oxygen concentration after adding water which supports XRD results. In addition, the relative oxygen concentration to calcium (O/Ca) shown in Fig. 9 shows that O/Ca is greater in the nano grade compared to the usual before and after adding water. This may support that nanolime's reactivity is higher than lime. This is evidenced by CaO quantities that have high purity in nanolime as shown by the XRF results in Table 2.

For XRD analysis, comparison was made on the highest peak which was in the position of 26.72° for pure soil. Figure 10 represents the XRD pattern for original soil and dry mixtures of soil–lime and soil–nanolime. The zoom of the highest peak attached with Fig. 10 shows that all peaks were almost identical. Figure 11 presents the results of EDX spectrum for original soil. The EDX results support the results of XRF analysis mentioned previously, where silicon and aluminum are the main components in addition to the oxygen. Figures 12 and 13 illustrate the EDX spectrum for dry mix of the soil–lime and soil–nanolime mixtures, respectively. Comparing between regular and nano mixtures, the EDX analysis shows that the intensity of C and Ca developed more in the nano mixture.

The changes in element intensity of O and C mirror the change in components concentration as well. Figure 14 shows the concentrations of each element as obtained from EDX spectrum. Original soil contains a trace of O and C with concentration of 37.4 and 35.3%. These concentrations changed after adding CaO in dry mixtures respectively to 48.6 and 17.7% for lime mixture and 43.5 and 28.2% for nanolime mixture. Clearly, it can be noticed that the O concentrations displayed greater increase

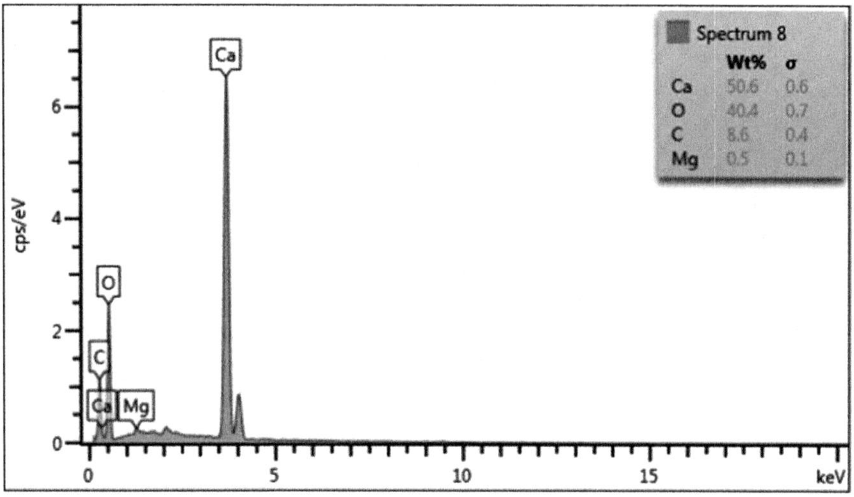

Fig. 4 EDX spectrum for dry lime

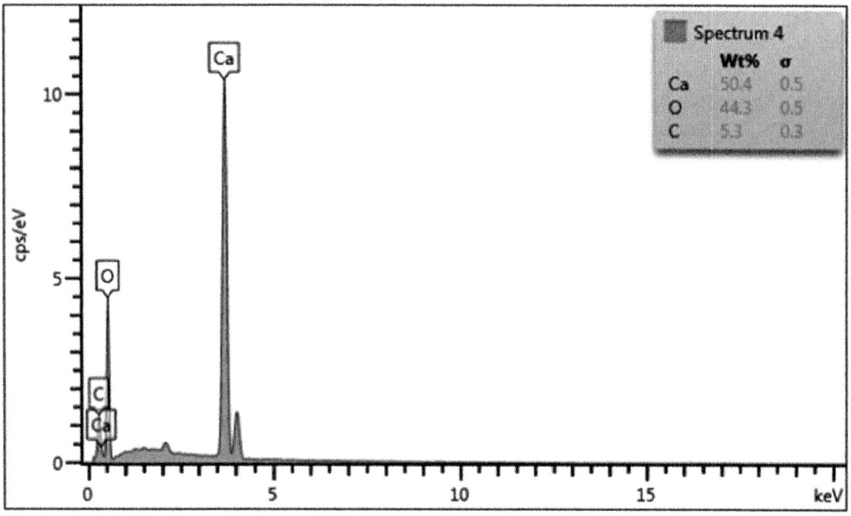

Fig. 5 EDX spectrum for dry nanolime

for the dry mixture of nanolime than that of lime. This may possibly be related to the purity of nanolime which was higher compared with lime (Table 2).

The results of EDX for wet cured mixtures showed a significant difference compared with the results of dry mixtures. Figures 15 and 16 illustrate the alteration in intensity for the same elements in the cured mixtures. This alteration revealed that there were chemical reactions between soil and lime and nanolime during the curing period. From Fig. 14, it is clear that the increase in oxygen concentrations were

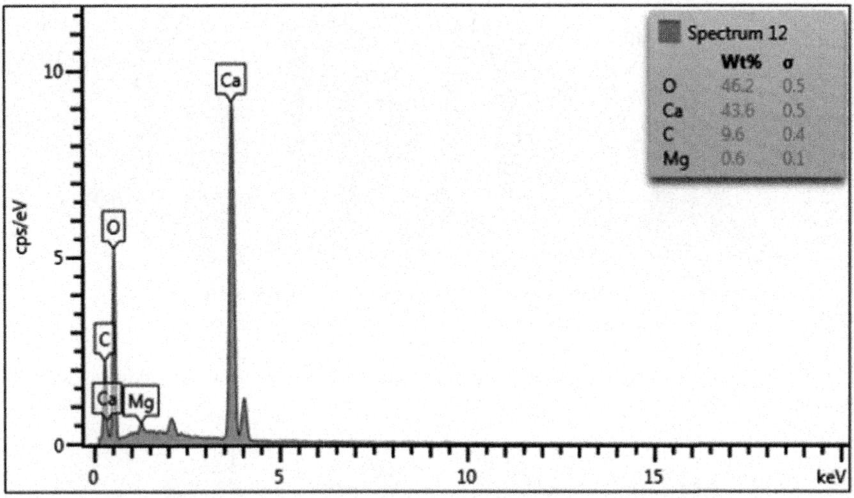

Fig. 6 EDX spectrum for wet lime

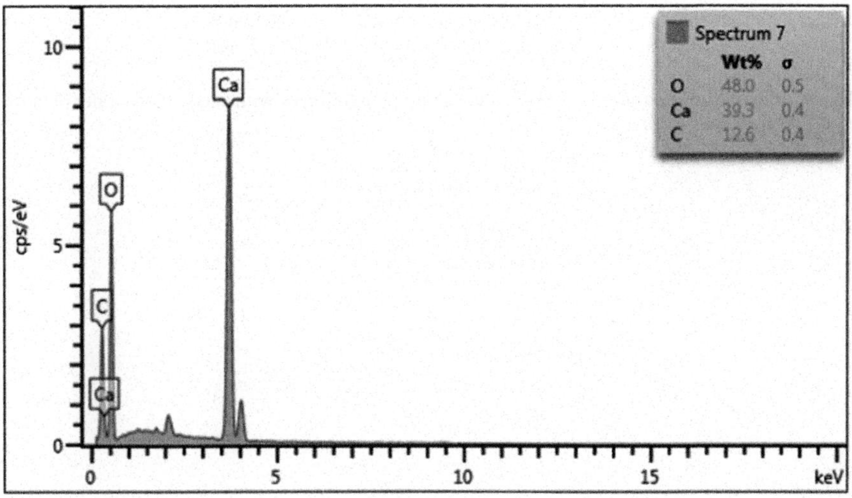

Fig. 7 EDX spectrum for wet nanolime

about 5.6 and 12.8% for lime and nanolime mixture, respectively, i.e. the concentration increment was two folds for soil–nanolime mixtures. This behaviour perhaps points to the reactivity of nanolime in soil improvement. The concentrations of C showed some reduction for the cured mixture. This may be an indication that there was some dissolution of these elements due to increase in the pH value.

Figure 17 shows the XRD pattern for the wet cured of the soil–lime and nanolime mixture. The zoom of the highest peak exhibits some deviation along the axis for

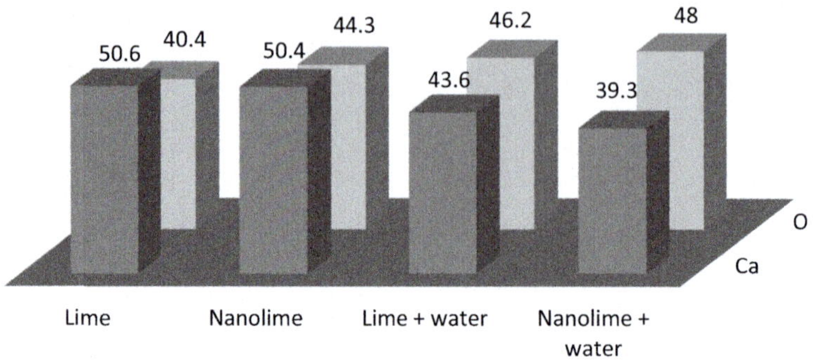

Fig. 8 Percentage (%) by weight of calcium (bottom) and oxygen (top) for additives with and without water obtained by EDX analysis

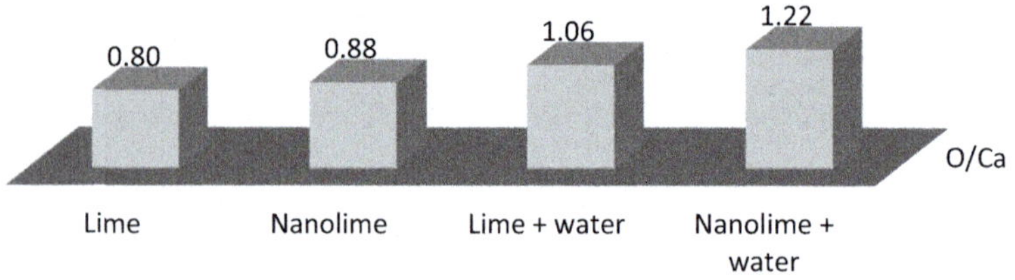

Fig. 9 Relative concentrations of oxygen to calcium for additives with and without water obtained by EDX analysis

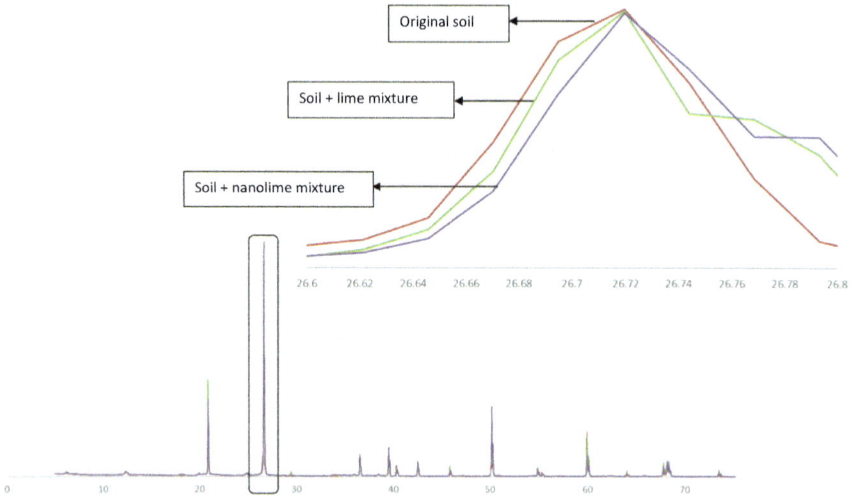

Fig. 10 XRD patterns for original soil and dry mix of soil–lime and soil-nanolime mixtures

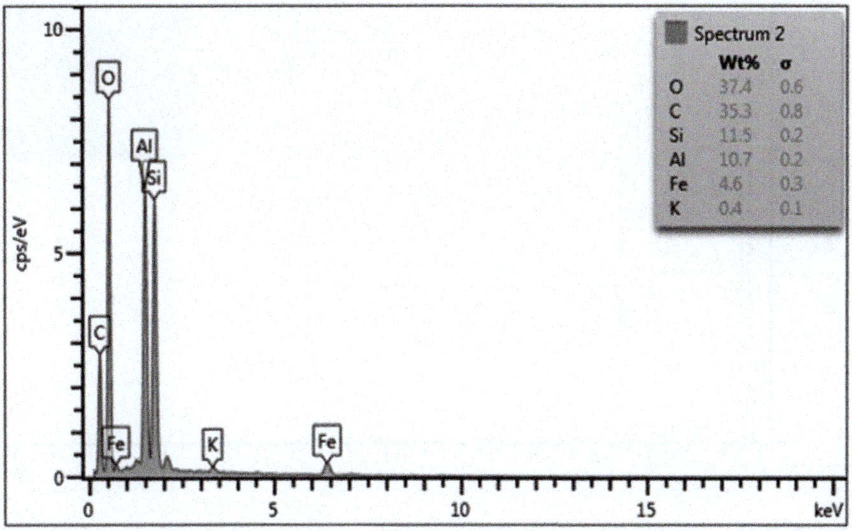

Fig. 11 EDX spectrum for original soil

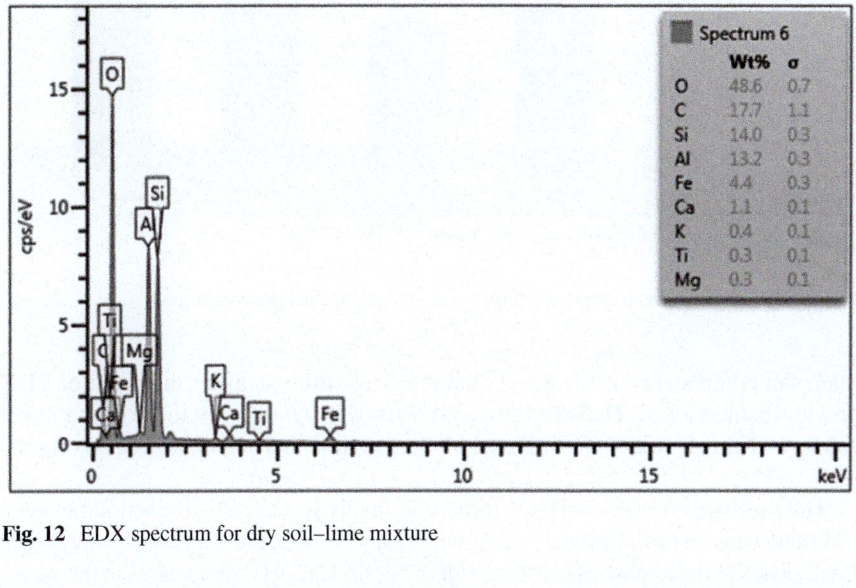

Fig. 12 EDX spectrum for dry soil–lime mixture

soil–lime mixtures compared with the original soil. In addition, these peaks showed a change in the intensity level with respect to the original and dry mixtures. The peak of lime mixture appeared to be of higher intensity and shifted more than that of the mixture of nanolime, but the peak of soil–nanolime mixture exhibits broader than that of lime mixture. This alteration in X-ray diffraction results is an indicator of the occurrence of chemical reactions between calcium oxide as stabilizer and the

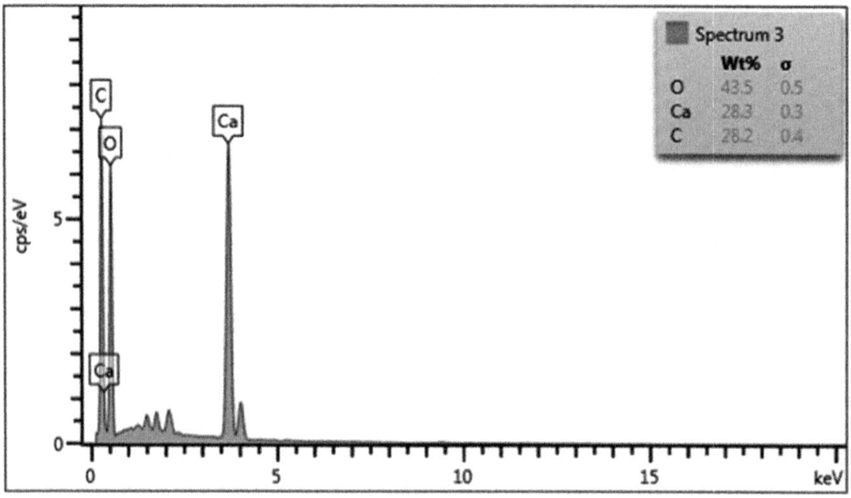

Fig. 13 EDX spectrum for dry soil–nanolime mixture

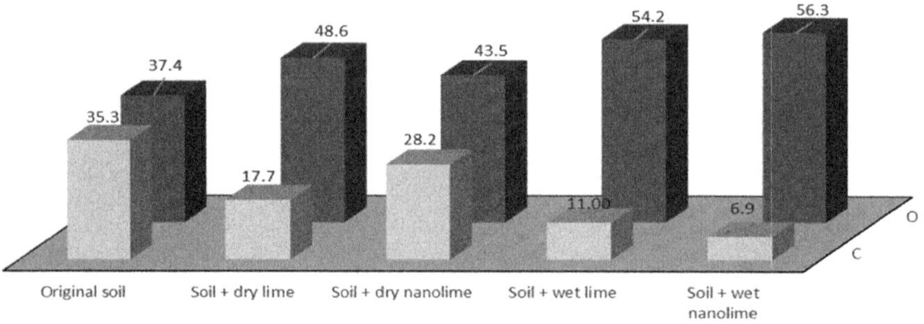

Fig. 14 Elements concentrations of original soil, dry and wet additive mixtures

main soil components of silica and alumina. This result supports the result of EDX previously mentioned. The broad peaks shown in this figure points to the formation of calcium silicate hydrates (C–S–H) gel for both regular and nano lime treated original soil.

The mechanism of C–S–H gel formation involves a chemical reaction between CaO and water to form $Ca(OH)^2$ as shown in Eq. (1). Then the ionization process (see Eq. 2) leads to the release of Ca^{+2} and $(OH)^{-1}$ ions. $Ca(OH)^2$ dissociates in the water. It increases the electrolytic concentration and pH of the pore water and dissolves the silicates (SiO_2) and aluminates (Al_2O_3) from the clay particles. Dissolution of SiO_2 took place during the curing period due to increase in the pH of soil–lime and soil–nanolime mixtures. During these processes, chemical reactions occurred between dissolved SiO_2 and free Ca^{+2} ions to form C–S–H gel [7]. It is believed that nanolime can produce more C–S–H gel because its ability to create more alkaline

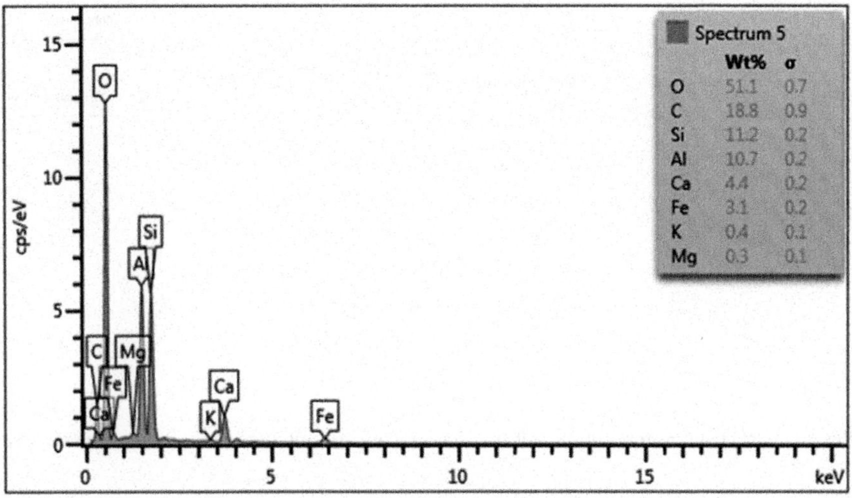

Fig. 15 EDX spectrum for wet cured soil–lime mixture

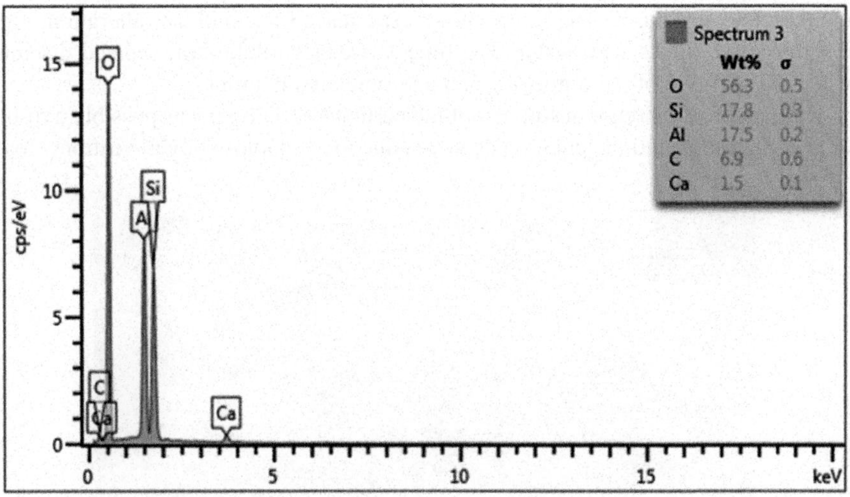

Fig. 16 EDX spectrum for wet cured soil–nanolime mixture

environment than that of lime and hence, nanolime can dissolve greater amounts of silica, in addition to ability to release more Ca^{+2} ions.

The FESEM image of original soil is presented in Fig. 18. Figure 19 shows the micrograph of soil mixed with dry lime where it looks as self-assembly of flake like-particles, while soil mixed with nanolime looks like small plate aggregated as flower like shape. From Fig. 20 it can be seen that the soil -nanolime particles agglomerated more than that of soil–lime. This is possibly because of van der Waals

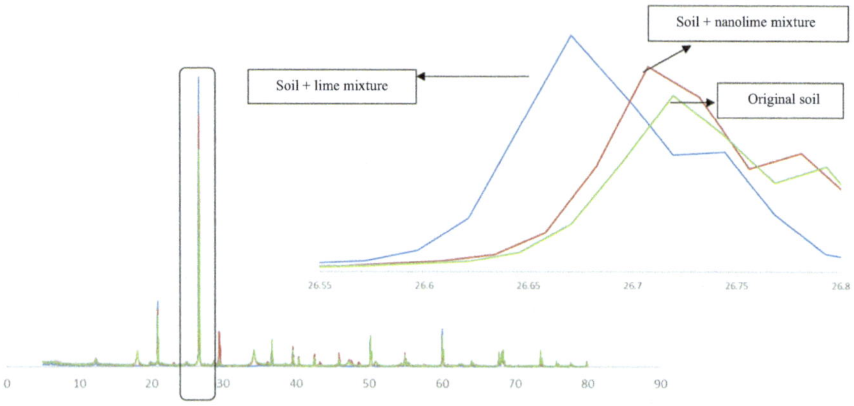

Fig. 17 XRD patterns for original soil and wet cured mix of soil–lime and soil-nanolime mixtures

forces developed between the nano particles. This morphology can be observed more clearly in Fig. 19, where the particles of lime seemed to be discrete big particles in contrast to nanolime particles which appeared to be fine particles agglomerated in chain like structures. The morphology experienced a significant alteration after reaction of additives with water. The lime + water particles still seemed discrete while the particles of nanolime + water agglomerated in groups.

This apparent increase in soil–nanolime particles as a group was possibly a result of agglomeration and flocculation achieved due to availability of Ca^{+2} cations. Ions

Fig. 18 FESEM micrograph of original soil (Mag = 20.00 K X)

Fig. 19 FESEM micrograph of dry mix of soil with additives (Mag = 20.00 X)

Fig. 20 FESEM micrograph of wet mix of soil with additives (Mag = 20.00 K X)

exchange mechanism possibly took place between monovalent ions (Ca^{+2}) with divalent (Na^{+}, H^{+}) ions leading to a reduction of repulsion forces and bridging the clayey particles of negative surface ions. It appeared that the flocs generated by nanolime are coarser than that of lime. This may again be related to the reactivity of nanolime and its ability to release more Ca^{+2} ions. Figure 20 also presents a new morphology of gel which is believed to be C–S–H gel in both wet cured soil–lime and soil–nanolime mixtures. The gel in the soil–lime mixture appeared to be flaky in shape whilst it seemed to be smoother in the soil–nanolime mixture.

4 Conclusion

These results indicate that the stabilization mechanism of treated soil involved flocculation and agglomeration of soil particles by Ca^{+2} which bridges the negatively charged clay particles and chemical reactions between the calcium oxides and dissolved silica present in the soil mineral which produces CSH gel for pozzolanic reactions. The existence of even a minute amount of this nanolime can result in extraordinary effects on the engineering properties of soil.

Acknowledgements The authors would like to acknowledge Ministry of Higher Education (MOHE), Malaysia, UKM Geotechnical Laboratory, Pusat Pengurusan Penyelidikan dan Instrumentasi (CRIM), and staffs for their full support and guidance in doing this paper.

References

1. Neethu, S., Remya, S.: Engineering Behaviour of Nano-Clay Stabilised Soil. Proceedings of Indian Geotechnical Conference, hlm. 22–24 (2013)
2. To, D., Sundaresan, S., Dave, R.: Nanoparticle mixing through rapid expansion of high pressure and supercritical suspensions. J. Nanopart. Res. **13**(9), 4253–4266 (2011). https://doi.org/10.1007/s11051-011-0369-0
3. Jone, J.R., Parker, D.J., Bridgwater J.: Axial mixing in a ploughshare mixer. Powder Technol. **178**(2), 73–86 (2007). https://doi.org/10.1016/j.powtec.2007.04.006
4. Willie, H., Norman W.: Methods of Soil Analysis: Part 5. Mineralogical Methods Ed.: Soil Science Society of America, USA (2007)
5. Hull, A.: A new method of chemical analysis. J. Am. Chem. Soc. **41**(8), 1168–1175 (1919)
6. Eades, J.L., Grim, R.E.: Reaction of Hydrated Lime with Pure Clay Minerals in Soil Stabilization. Highway Research Board Bulletin, vol. 262 (1960)
7. Hopkinson, L., Kristova, P., Rutt, K., Cressey G.: Phase transitions in the system $MgO–CO_2–H_2O$ during CO_2 degassing of Mg-bearing solutions. Geochim. Cosmochim. Acta **76**(1–13) (2012)

Preliminary Measurement of Underwater Acoustic Level

Md Redzuan Zoolfakar, Hairul Azmi Mohamed, and Erzad Iskandar Putra

Abstract Anthropogenic noise generated by maritime activities have been increasing which produce several negative impacts towards marine life. This measurement was performed to identify if any disturbance in the water will affect the underwater acoustic level. Measurement was performed by using a basic hydrophone attached to a conversion system in a phone. Three locations with different environment were selected. The setting use for data measurement was based on atmospheric references. The data was converted by using the underwater reference. The final result shows that any disturbance from external sources will increase the underwater acoustic level and potential harm any marine life in an area. The horizontal length and vertical depth of a hydrophone also plays an important role in determining the accuracy of the measurement.

Keywords Anthropogenic · Decibel · Underwater noise · Acoustic · Hydrophone

1 Introduction

The occurrence of ship-generated marine pollution has drawn the attention of the international maritime community, which is working to promote safe navigation and environmental protection. The potential for the shipping industry to significantly damage the marine environment and related biodiversity within the maritime field is causing increased concern in the maritime community. Garbage, solid waste,

M. R. Zoolfakar (✉) · H. A. Mohamed
Maritime Engineering Technology Section, Universiti Kuala Lumpur, Malaysian Institute of Marine Engineering Technology, Lumut, Perak, Malaysia
e-mail: redzuan@unikl.edu.my

H. A. Mohamed
e-mail: hairul.mohamed13@s.unikl.edu.my

E. I. Putra
Department of Naval Architecture, Faculty of Marine Technology, Institut Teknologi Sepuluh Nopember, Surabaya, Indonesia
e-mail: erzad@its.ac.id

A. Ismail et al. (eds.), *Engineering Frontiers*, Advanced Structured Materials 208,
https://doi.org/10.1007/978-3-031-56844-2_10

antifouling paints on ship hulls, and underwater noise are all examples of shipborne pollutions [1].

Researchers have classified that the main source of underwater noise generated from a ship are from the propeller rotation, flow of water around a ship and sound that is from machineries [2–6].

Anthropogenic or changes in the environmental caused or influenced by people, either directly or indirectly can be categorised into two types which are impulsive sound and continuous sound, and the exposure can be in a short duration or long lasting. Impulsive ocean noise is made up of short, powerful pulses of very loud sound that are repeated over a short period of time. Active sonar, which is utilised in military and civil activities, and seismic surveying for oil and gas development are some of the examples. Ships, offshore oil and gas rigs, and offshore wind farms all produce continuous noise in the form of a persistent buzz. It has an impact on local marine life and leads to long-range and low-frequency background noise [7].

Addressing this issue, in 2014, the International Maritime Organisation (IMO) under its Marine Environment Protection Committee (MEPC) recognised the negative impact of underwater noise pollution, thus produce a non-mandatory Guideline for the Reduction of Underwater Noise from Commercial Shipping to Address Adverse Impacts on Marine Life [8].

One of the 17 Sustainable Development Goals (SDG) established by the United Nations (UN) in 2015 is focusing on life below water (SDG 14). The aims of SDG 14 are to reduce marine pollution, ocean warming, eutrophication, acidification, and overfishing. According to [9], SDG 14 is one of SDG that had a low influence on related industries and on the society.

Underwater acoustic sound pressure is defined as sound force per unit area and is typically expressed in Pascals (Pa), where 1 Pa represents the pressure produced by one Newton of force applied over a one square metre area. However, the value is converted to the decibel scale [10].

According to the National Physical Laboratory, there are three methods available to measure the underwater acoustic level. The first method is by mounting hydrophones at a certain height from seabed or named as bottom mounted method. The second method is by allowing a hydrophone system to drift along a certain path and continuously record the readings. This method is called drifting method. The last available method is by deploying hydrophones from a platform for quick measurements and recovery [11].

2 Methodology

Measurements of underwater acoustic level were performed by using a basic 27 mm piezo transducer pickup hydrophone with the cord length of 6.0 m as shown in Fig. 1.

The conversion of the underwater acoustic level from the hydrophone was performed by using a standard full version sound monitoring phone application. The application provides to the user with three different settings which are as follows:

Fig. 1 Piezo transducer pickup hydrophone

i. dB—A: A sound measurement that simulates how noise is perceived by the human ear.
ii. dB—C: Low and high frequencies are more responsive to weighting. This is helpful for analysing noise in places like factories or theatres where it is crucial to comprehend how the low and high frequency components affect the calculated noise level as a whole.
iii. dB—Z: Is neutral in regard to frequency. When determining the overall noise level value, it responds to all frequencies equally.

For the purpose of this test, the selected setting was dB—Z. Selecting this setting assists in measuring the total underwater acoustic level without having to consider the changes of the underwater acoustic frequency at the selected places.

Three locations were selected to be part of the pilot test. In each location, the hydrophone was placed in the path of flowing water with the depth of deployment varies from as minimum as 50 mm to a maximum depth of 6.0 m.

The process started by measuring the ambient or atmospheric acoustic level to obtain a baseline reading and subjected to the surrounding activities. The initial measurement was performed without the hydrophone connected to the display device. Once the reading was recorded, the hydrophone was connected to the display device and deployed into the water. The hydrophone was deployed for 1 min to capture the highest reading. The readings were recorded. The process was repeated at each selected location. All measurements were performed during daytime and the time of measurements were also the same for each location. The experimental reading is shown in Table 1.

Table 1 Highest recorded reading

Num.	Location/depth of water	A	B	C (dB)
1	Ambient (air)	72 dB	47 dB	40
2	50.0 mm	83 dB	67 dB	59
3	1.0 m	59 dB	–	53
4	2.0 m	33 dB	–	47
5	4.0 m	–	–	40
6	6.0 m	–	–	33

3 Results and Discussion

The target of this test was to evaluate if changes in underwater pressure due to flow of water or any disturbance in the water will affect the underwater acoustic level.

3.1 Results

The highest readings for each depth of water from three locations were recorded as per Table 2. The readings shown in Table 2, represent the setting for standard atmospheric or air configuration. The standard formula used to convert air sound pressure to Decibel (dB) is $P_{0air} = 20 \log 10 (P/P_{air})$, whereby the P_{air} refers to 20 μPa. 20 μPa is selected to match the human hearing sensitivity. However, for underwater acoustic measurement, the formula used to convert underwater sound pressure to dB is $P_{0water} = 10 \log 10 (P/P_{water})$. The reference value for P_{water} is 1 μPa.

Table 2 Comparison table of air and underwater acoustic pressure

Num.	Location/ depth of water	A (air)	A (underwater)	B (air)	B (underwater)	C (air) (dB)	C (underwater)
1	Ambient (air)	72 dB	–	47 dB	–	40	–
2	50.0 mm	70 dB	96 dB	67 dB	90 dB	59	85 dB
3	1.0 m	59 dB	85 dB	–	–	53	79 dB
4	2.0 m	33 dB	59 dB	–	–	47	73 dB
5	4.0 m	–	–	–	–	40	66 dB
6	6.0 m	–	–	–	–	33	59 dB

3.2 Discussion

According to [12], the value of atmospheric reading needs to be increased by 26 dB from the noise level referenced in air in order to compare noise levels in air to noise levels in water. Other translation is thar the acoustic level is higher underwater compared to in air. Table 3 shows the conversion of readings from air acoustic pressure to underwater acoustic pressure.

The highest recorded reading for atmosphere acoustic level at location A was 72 dB. The depth of water was estimated at 2.5 m. The reading was quite high due to the fact that the location of measurement was near to the flow of water and the location also is nearby to a main road. The combination of these factors contributed to high atmosphere acoustic level at location A. At 50.0 mm depth of water, the reading was 96 dB referenced to a pressure (re) of 1 μPa. The reason for this high reading was that the hydrophone was placed at the location where the water falls from higher place. The height is estimated at 300.00 mm. This waterfall created a high disturbance in the underwater acoustic pressure, thus increasing the underwater acoustic level. As the depth of hydrophone increased to 1.0 and 2.0 m, the reading reduced from 96 dB referenced to a pressure (re) of 1 μPa at 50.0 mm to 59 dB referenced to a pressure (re) of 1 μPa and 33 dB referenced to a pressure (re) of 1 μPa respectively. The underwater acoustic level will reduce when the source of disturbance is further away from the hydrophone.

Regarding location B, the highest recorded atmosphere acoustic level was 47 dB. When comparing to location A, location B is located at coastline and can be considered further for any source of high sound or noise generating unit. The hydrophone was only deployed at a depth of 50.0 mm. The reason was to look at whether the drastic changes in flow of water at the coastline will have an effect to the underwater acoustic level in relation to the depth of water. The highest recorded reading was 90 dB referenced to a pressure (re) of 1 μPa. The reading recorded at 73 dB referenced to a pressure (re) of 1 μPa when the seawater began to flow inshore, the reading increased to maximum when there were waves crashing on the hydrophone, then, recorded at 83 dB referenced to a pressure (re) of 1 μPa when the seawater recedes. This shows that minimum depth of water coupled with high source of underwater disturbance will result in high underwater acoustic level.

Location C was located at a fishermen pier, along a river. At this location, the full length of the cord was deployed. A few of boats were moored along the pier and some of it were alongside to each other. The highest recorded air acoustic level was 40 dB. Among the measurement depth set for location C, depth of 50.0 mm recorded the highest reading which was 85 dB referenced to a pressure (re) of 1 μPa. Based on observations, this high reading was due to the boats came in contact with each other and the boats hitting the pier due to water current generated naturally because of the flow of water and due to a boat passing by at location C. As the depth of the hydrophone deployment were increased vertically, the reading recorded reduction in value. This result also confirmed that the underwater acoustic level reduces when it is far from its source. The result also shows any external acoustic generating unit can have impacts towards the underwater acoustic level.

4 Conclusion

This preliminary measurement has confirmed that any disturbance from acoustic generating unit will increase the underwater acoustic level. The horizontal distance and vertical depth of the hydrophone from the source of disturbance also plays an important role in measuring the underwater acoustic level.

5 Recommendation

The hydrophone should not be blocked by any object to ensure accurate readings. The measurement can be improved by using a better hydrophone and improved pressure conversion system that has a setting for underwater measurement. The selected location could be improved by focusing on shipping route with moderate to high traffic.

References

1. Onwuegbuchunam, D.E., Ebe, T.E., Okoroji L.I., Essien, A.E.: An analysis of ship-source marine pollution in Nigeria Seaports. J. Mar. Sci. Eng. (2017). https://doi.org/10.3390/jmse5030039
2. Kuşku, H., Yiğit, M., Ergün, S., Yiğit, U., Taylor, N.: Acoustic noise pollution from marine industrial activities: exposure and impacts. Aquat. Res. **1**(4), 148–161 (2018). https://doi.org/10.3153/AR18017
3. Viola, S., Grammauta, R., Sciacca, V., Bellia, G., Beranzoli, L., Buscaino, G., Caruso, F., Chierici, F., Cuttone, G., D'Amico, A., De Luca, V., Embriaco, D., Favali, P., Giovanetti, G., Marinaro, G., Mazzola, S., Filiciotto, F., Pavand, G., Pellegrino, C., Pulvirenti, S., Simeone, F., Speziale, F., Riccobene, G. (2017) Continuous monitoring of noise levels in The Gulf of Catania (Ionian Sea). Study of correlation with ship traffic (2017). https://doi.org/10.1016/j.marpolbul.2017.05.040
4. Vanessa. M. Zobell, Kaitlin E. Frasier, Jessica A. Morten, Sean P. Hastings, Lindsey E. Peavey Reeves, Sean M. Wiggins, and John A. Hildebrand: Underwater Noise Mitigation in The Santa Barbara Channel Through Incentive-Based Vessel Speed Reduction (2021). https://doi.org/10.1038/s41598-021-96506-1
5. Leaper, R., Renilson, M., Ryan, C.: Reducing underwater noise from large commercial ships: current status and future directions. J. Ocean Technol. 50–69 (2014)
6. F.J.S. Rodrigo, P. Poveda, J. Carbajo and J. Ramis: Monitoring Long Term Underwater Acoustic Pollution in Mediterranean Sea Waters. Madrid Inter.Noise (2019).
7. Vakili, S.: Under-Water Noise Pollution Sources, Mitigation Measures in Commercial Vessels: The Trade-Off Analysis in The Case of Study for Trans Mountain Project, Port of Vancouver, Canada. The Maritime Commons: Digital Repository of the World Maritime University (2018)

8. International Maritime Organisation (IMO) Guidelines for The Reduction of Underwater Noise from Commercial Shipping to Address Adverse Impacts on Marine Life. https://cetsound.noaa.gov/Assets/cetsound/documents/MEPC.1-Circ%20883%20Noise%20Guidelines%20April%202014.pdf (2014). Accessed 16 June 2023.
9. Yunus Zengin, Serkan Naktiyok, Erdoğan Kaygın, Onur Kavak and Ethem Topçuoğlu: An Investigation upon Industry 4.0 and Society 5.0 within the Context of Sustainable Development Goals (2021). https://doi.org/10.3390/su13052682.
10. William D. Halliday, David Barclay, Amanda N. Barkley, Emmanuelle Cook, Jackie Dawson, R. Casey Hilliard, Nigel E. Hussey, Joshua M. Jones, Francis Juanes,Marianne Marcoux, Andrea Niemi, Shannon Nuddsj, Matthew K. Pinn, Clark Richards,Kevin Scharffenberg, Kristin Westdal, Stephen J. Insley: Underwater sound levels in the Canadian Arctic, (2021). https://doi.org/10.1016/j.marpolbul.2021.112437.
11. Robinson, S., Lepper, P., Hazelwood, R.: Good practice guide for underwater noise measurement (Report No. NPL Good Practice Guide No. 133). Report by National Physical Laboratory (NPL). Report for Marine Scotland Science. https://tethys.pnnl.gov/organization/national-physical-laboratory-npl (2014). Accessed 16 June 2023
12. John Pike and Robert Sherman: Underwater Acoustic. https://man.fas.org/dod-101/sys/ship/acoustics.htm (1998). Accessed 16 June 2023

Continuous Carbonization and Activation Process of Activated Carbon from Agriculture Waste

Azmarini Ahmad Nazri, Supa'at Hj. Zakaria, and Dafit Feriyanto

Abstract A continuous system of carbonization and activation process of activated carbon (AC) derived from agriculture waste was successfully developed. The results obtained indicate that the AC produced by the system exhibits excellent performance, with a maximum yield of palm shell AC was recorded more than 30% and fixed carbon content exceeding 70% for both palm shell AC and coconut shell AC. The morphological analysis through SEM analysis shows that the surface morphology of palm shell AC indicates smaller pore sizes, with a higher formation of mesopores and micropores compared to coconut shell AC. The AC then was further tested for Brunauer–Emmett–Teller (BET) surface area where palm shell AC yield the highest surface area with the value of 678 m^2/g. Given that the BET surface area value fell between 500 and 1500 m^2/g, which is the permissible range for commercial AC, palm shell AC's adsorption performance was deemed dependable. This study demonstrated the potential for continuous AC production systems to improve the efficiency and quality of activated carbon production. Furthermore, the use of agriculture waste materials as raw materials for this system can contribute to sustainable development.

Keywords Continuous carbonization-activation · Activated carbon · Agriculture waste · Palm shell activated carbon · Coconut shell activated carbon

A. A. Nazri (✉) · S. Hj. Zakaria
Department of Mechanical Engineering, Politeknik Ungku Omar, Jalan Raja Musa Mahadi, 31400 Ipoh, Perak, Malaysia
e-mail: azmarini@puo.edu.my

S. Hj. Zakaria
e-mail: supaat@puo.edu.my

D. Feriyanto
Department of Mechanical Engineering, Faculty of Engineering, Universitas Mercu Buana, Jakarta, Indonesia
e-mail: dafit.feriyanto@mercubuana.ac.id

A. Ismail et al. (eds.), *Engineering Frontiers*, Advanced Structured Materials 208,
https://doi.org/10.1007/978-3-031-56844-2_11

1 Introduction

Activated carbon (AC) is a highly porous material with a wide range of applications in different fields, such as water treatment, air purification, and gas separation. Traditionally, AC is produced by carbonizing raw materials in a furnace at high temperatures, followed by an activation step that involves exposing the carbonized material to a gas, steam or chemicals at high temperatures [1, 2]. However, this process is energy-intensive, time-consuming, and requires significant amounts of raw materials and labor. To overcome these challenges, recent advancements in technology have led to the development of continuous AC production systems that combine carbonization and activation processes in a single machine. These systems offer several advantages over traditional furnaces, such as reduced energy consumption, increased production efficiency, and improved product quality.

Zheng et al. [3] developed a continuous pyrolysis and activation system for the production of AC from bamboo. The system was able to produce AC with a specific surface area of up to 2400 m^2/g and a yield of 28.4%. Moreover, the system required only 30–40 min for the carbonization and activation process, compared to several hours required by traditional methods. A study by Hayawin et al. [4] produced AC from palm kernel shells using a vertically oriented electric reactor with a combustion chamber capacity of 30 cm in diameter in a single-step process. The carbonization process was carried out at a temperature of 400 °C continuous with the activation process with steam at 500–1000 °C. The analysis showed significant results with a carbon yield of 30% and a Brunauer–Emmett–Teller (BET) surface area of 935–987 m^2/g.

Other research also revealed the continuous system produced AC with higher surface area, pore volume and good absorption capacity compared to traditional batch systems, indicating improved quality [5–7]. Most of the research used agriculture waste as an AC precursor. The use of agricultural waste as a precursor for AC production is gaining increasing attention due to its abundance, low cost, and environmental sustainability. Furthermore, agricultural waste materials are frequently used in the manufacturing of AC due to their higher surface area, density, hardness, and relatively high volatile content. These qualities make them ideal for the manufacture of granular AC [8]. Table 1 shows several examples of continuous AC production from agricultural waste materials.

In this study, a continuous carbonization-activation system for producing AC from agriculture waste was developed and operated using gas furnaces. This system

Table 1 Continuous AC production from agriculture waste materials

AC precursor	Activation agent	References
Bamboo	Steam	[3, 6, 9]
Palm shell	Steam	[4, 10–12]
Coconut shell	Steam	[7, 13, 14]
Sugarcane bagasse	Steam	[15]

continued to function without pausing while simultaneously producing steam for the activation process. The characteristic and performance of AC produced from the system was also evaluated.

2 Methodology

AC precursor

Palm shells utilized in this study have been collected at Segamat, Johor while the coconut shell was obtained from Kuala Kangsar, Perak. The raw materials gathered were round-in shaped with 1 cm width × 2 cm length then cleaned to remove husk, dirt and all other impurities. It was then sun-dried on a clean platform for a few days as shown in Fig. 1.

Continuous carbonization-activation system

The continuous carbonization-activation system comprised with three parts; gas furnace, steam generator and condenser as shown in Fig. 2. Steam for activation process was supplied by a steam generator and the hot gaseous emitted during the process was then flown through the condenser. The condenser let to reduce the temperature of the gaseous first before being discharged through the chimney. In this AC production process, 3 kg of raw material was fed into cylinder shaped heating reactor with 310 mm diameter and a length of 350 mm. The carbonization process was started by heating the AC raw material up to 550 °C for about 1 h followed

Fig. 1 Cleaning process of raw material of AC

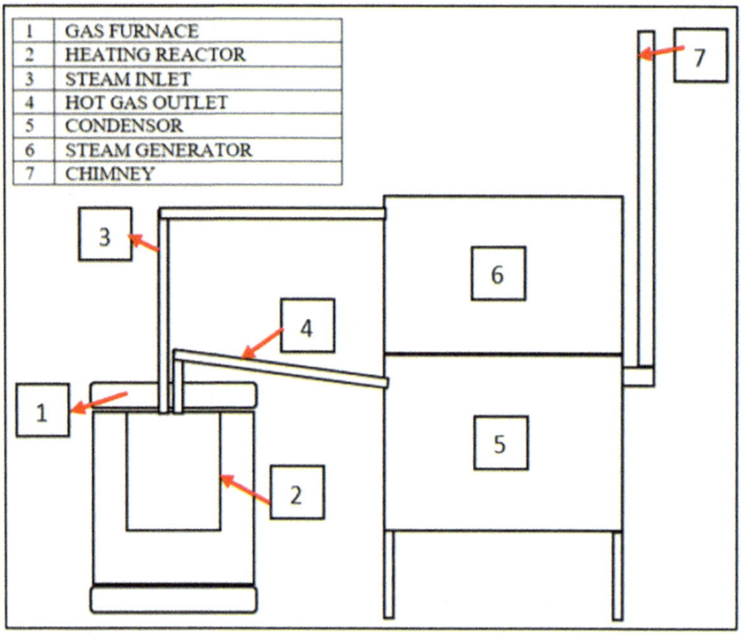

Fig. 2 The continuous carbonization-activation system of AC production

with the activation process by steam with a temperature of 800 °C for about 30 min. The AC produced was then removed from the system, dried and packed in airtight containers.

Evaluation of characteristic and performance of AC

Analytical methods have been used to identify the characteristics and properties present in AC raw materials. Characterization of palm shell and coconut shell for AC production has been determined through ultimate analysis and proximate analysis. For the ultimate analysis, chemical elements such as nitrogen (N), carbon (C), hydrogen (H), and sulfur (S) were determined using a sample that has been dried to remove moisture. The sample is overnight dried at 105 °C using a CHNS/O elemental analyzer, specifically the Perkin Elmer 2400 series 2. The analysis results are then recorded. To determine the moisture content, ash content, volatile content, and fixed carbon content, proximate analysis is employed. Proximate analysis was carried out using a thermogravimetric analyzer (TGA) (Mettler Toledo, TGA/SDTA 851, USA). The analysis temperature ranges from room temperature up to 900 °C, with a heating rate of 10 °C/min up to 550 °C under a nitrogen environment [15].

Surface morphology analysis of the samples was conducted using a scanning electron microscope (SEM), specifically the Leica S440 model. The AC samples were coated with gold using an E-1010 iron sputter prior to SEM observation. Elemental analysis was also performed using an elemental analyzer using the EDX-OXFORD.

For surface area and pore size measurement, the produced AC, weighing between 110 and 160 mg, was dried overnight in a vacuum environment at a temperature of 110 °C before being transferred to a volumetric sorption analyzer (ASAP2020 Micrometrics, New York, USA). The nitrogen adsorption isotherm was then measured at a temperature of 77 K. The following parameters such as the specific surface area (BET), the micropore surface area and the pore volume distribution and average pore diameter were determine using the Autosorb for window software (version 1.19) [16].

3 Results and Discussion

Temperature profile during continuous carbonization-activation process

Figure 3 shows the temperature profile of carbonization-activation process. The carbonization stage took place at minutes of 20 to 80 while activation process at minutes 90–120. During carbonization stage at temperature between 300 and 550 °C, the process released more volatile compound from the AC raw materials thus increased the carbon content. As carbonization proceeds, more pores unclog and more moisture is liberated from the carbon [17–20]. The activation stage started at minutes of 90 with the temperature of 800 °C and the temperature was kept constant within 30 min before the process was shut down. During the activation process, more pores were developed and the surface area also had improved.

However, the percentage yield of AC from coconut shells was only obtained as 18% compared to palm shell AC with 33% yield as shown in Table 2. This indicates

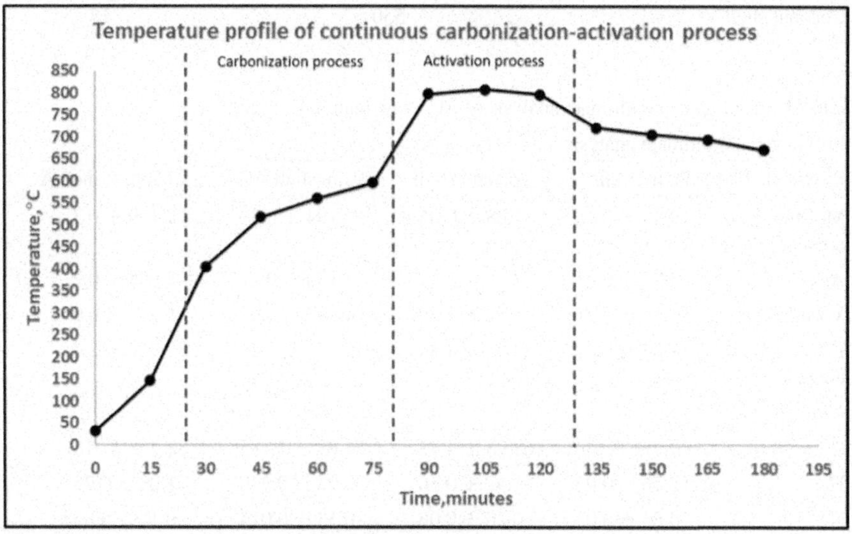

Fig. 3 Temperature profile of continuous carbonization-activation process

that the coconut shell activated carbon was over-burned, and the heating time needs to be reduced. As the results, the heating profile as shown in Fig. 3 was suitable for palm shell AC production.

Proximate and ultimate analysis

Proximate and ultimate analyses were conducted on coconut shell and palm shell before and after the production of AC was carried out. Based on Table 3, the moisture content and volatile matter content of palm shell are higher compared to coconut shell. The moisture content in the raw material of AC is important for reducing the temperature and heating time during the carbonization and activation processes.

After the carbonization and activation processes were carried out on both AC raw materials, the moisture content, ash content, and volatile matter content decreased, while the fixed carbon content increased. The palm shell AC and coconut shell AC had fixed carbon contents of 75.6% and 88% respectively, with both exhibiting low ash content of less than 5%. With low ash content, these raw materials are suitable for producing AC with high porosity.

Furthermore, the volatile matter content of palm kernel shell significantly decreased from 53.4 to 14.5% after being converted into AC. The reduction in volatile matter content indicates the decomposition of liquid and tar substances at high temperatures, a phenomenon facilitated by the activation agent by steam. Based

Table 2 Percentage yield of palm shell and coconut shell AC

Raw material	Before AC production process (g)	After AC production process (g)	% yield (%)
Palm shell	3000	1000	33
Coconut shell		530	18

Table 3 Proximate and ultimate analysis of AC's raw material

Proximate and ultimate analysis				
Proximate (%)	Palm shell	Coconut shell	Palm shell AC	Cocomut shell AC
Moisture content	9.55 ± 0.40	7.28 ± 0.05	4.194 ± 0.3	2.179 ± 0.3
Ash content	0.87 ± 0.01	22.64 ± 1.80	4.433 ± 0.2	2.055 ± 0.2
Volatile content	53.38 ± 0.21	5.32 ± 2.50	14.493 ± 0.9	5.934 ± 0.9
Fized carbon content	18.84 ± 0.02	23.33 ± 4.20	75.751 ± 2.4	87.912 ± 2.4
Ultimate (%)				
C	60.9 ± 0.2	68.93 ± 4.42	74.83 ± 2.185	88.3 ± 4.215
H	12.76 ± 0.02	1.42 ± 0.56	2.03 ± 0.240	1.08 ± 0.25
N	0.66 ± 0.03	0.26 ± 0.06	0.33 ± 0.013	0.12 ± 0.058
S	0.19 ± 0.01	0.044 ± 0.03	–0.032 ± 0.014	–0.02 ± 0.04

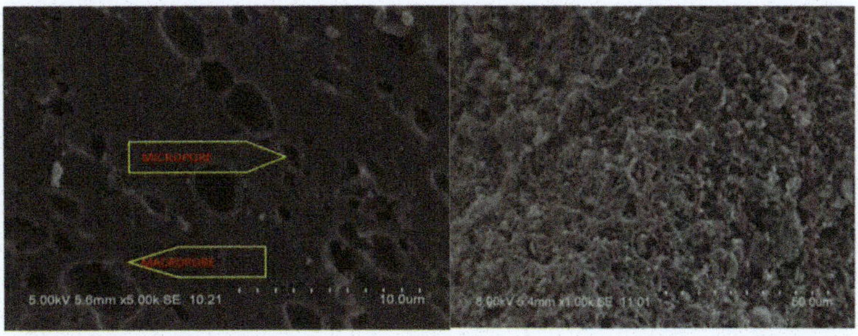

Fig. 4 SEM analysis of palm shell AC

Fig. 5 SEM analysis of coconut shell AC

on the obtained results, both materials meet the characteristics of commercial AC, with volatile matter content not exceeding 30%.

Scanning Electron Microscopy (SEM) analysis

Based on Figs. 4 and 5, the results of SEM analysis show that the surface morphology of palm shell AC exhibits smaller pore sizes, with a higher formation of mesopores and micropores compared to coconut shell AC.

According to Chen et al. [18], the presence and uniformity of mesopores and micropores in AC can enhance its adsorption capacity [18]. Moreover, most of the existing pore structures, especially in palm shell AC, are interconnected and extend inward in a spherical shape, providing sufficient space for the adsorption of organic compounds [16].

Surface area analysis

Based on the results as shown in Table 4, palm shell AC yield higher surface area compare to coconut shell AC with 678 m^2/g and 440 m^2/g, respectively. The capability of palm shell AC for adsorption was reliable since the value of BET surface area was within the acceptable range for commercial AC between 500 and 1500 m^2/g [16]. For coconut shell AC, the decrease in BET surface area is caused by excessively high activation temperatures for this particular material. High activation temperatures can

Table 4 Comparison of AC characteristic from this study and different agriculture waste

AC raw material	Activation agent	Activation temperature (°C)	Yield (%)	S_{BET} (m^2/g)	References
Palm shell	Steam	800	33	678	This study
Coconut shell	Steam	800	18	440	
Palm shell	Steam	500–1000	30	935–987	[4]
Coconut husk	Oxygen + Steam	<1000	31	416	[19]
Eucalyptus wood chips	Steam	500–700	22–32	427–871	[20]

decrease the surface area of the AC due to increased volatiles impregnation within the AC.

4 Conclusion

A continuous carbonization-activation system with a capacity of 3000 g of AC raw material has been successfully developed without pausing the process while simultaneously producing steam for the activation process. The results show that, the maximum yield of AC was recorded by palm shell AC with 33% of yield compare to coconut shell AC with only 18% of yield. Based on proximate analysis, the composition of the fixed carbon content of palm shells and coconut shells exceeds AC 70%. Many microporous and macroporous structures were observed on the surface of palm shell AC according to morphological analysis using SEM. This phenomenon can increase the level of AC adsorption. For the BET surface area analysis, the analysis was tested on the palm shell AC and coconut shell AC where palm shell AC yield the highest surface area with the value of 678 m^2/g.

Acknowledgements This research was funded by TVET Research Grant Scheme from Ministry of Education of Malaysia.

References

1. Sharma, G., et al.: Activated carbon as super adsorbent and sustainable material for diverse applications. Adsorpt. Sci. Technol., 1–21 (2022)
2. Gan, Y.X.: Activated carbon from biomass sustainable sources. Carbon **7**, 39 (2021)
3. Zheng, et al: Continuous production of bamboo-based activated carbon by pyrolysis and activation in a rotary kiln. Bioresour. Technol. **288** (2019)
4. Nahrul Hayawin, Z., et al.: Carbonisation-Activation System for the Production of Biochar and Activated Carbon from Oil Palm Kernel Shell, pp. 1511–7871. MPOB Information Series, ISSN (2018)

5. Hong, J., et al.: Continuous production of high-quality activated carbon from sawdust using a single-stage carbonization-activation system. J. Clean. Prod. **273**, 122980 (2020)
6. Yu, H., et al.: Continuous production of high-quality activated carbon from bamboo by a combination of microwave and steam activation. J. Environ. Chem. Eng. **7**(2), 102959 (2019)
7. Chen, W., et al.: Continuous production of activated carbon from waste coconut shells using a pyrolysis-steam activation process. J. Anal. Appl. Pyrolysis **148**, 104804 (2020)
8. Asadullah, M., et al.: Development of activated carbon from agricultural waste using steam physical activation: a review. J. Mater. Res. Technol. **14**, 686–700 (2021)
9. Prasetya, A., et al.: The production of activated carbon from bamboo using steam as an activation agent. IOP Conf. Series: Mat. Sci. Eng. **1077**(1), 012062 (2021)
10. Rugayah, A.F., et al.: Preparation and characterisation of activated carbon from palm kernel shell by physical activation with steam. J. Oil Palm Res. **26**(3), 251–264 (2014)
11. Nasri, N.S. et al.: Assessment of porous carbons derived from sustainable palm solid waste for carbon dioxide capture. J. CO_2 Util. **71**, 148–157 (2014)
12. Leman, A.M., et al.: The effect of activation agent on surface morphology, density and porosity of palm shell and coconut shell activated carbon. AIP Conf. Proc. **1885**, 020001 (2017)
13. Salunkhe, S. et al.: Development and evaluation of carbonization cum steam activation unit for coconut shell. International Res. J. Eng. Technology (IRJET) **8**(8), 936–940 (2021)
14. Hidayu, A.R., Muda, N.: Preparation and characterization of impregnated activated carbon from palm kernel shell and coconut shell for CO_2 capture. Procedia Eng. **148**, 106–113 (2016)
15. Danusorn Congsomjit, Chinnathan Areeprasert.: Hydrochar-derived activated carbon from sugar cane bagasse employing hydrothermal carbonization and steam activation for syrup decolorization. Biomass Convers. Biorefinery, **11**(11), 2569–2584 (2021)
16. Hayawin, Z.N. et al.: Production of a bio adsorbent from oil palm kernel shell, and application for pollutants and colour removal in palm oil mill effluent final discharge. IOP Conf. Series Mat. Sci. Eng. **736**(2), 022045 (2020)
17. Hanum, F. et al.: Characterization of activated carbon from rice Husk by HCl Activation and Its Application for Lead (Pb) Removal in Car Battery Wastewater. OP Conf. Ser. Mater. Sci. Eng, **180** (2017)
18. Chen, H. et al.: High capacitance KOH activated nitrogen-containing porous carbon material from waste coffee grounds in supercapacitor. Adv. Powder Technol., 1387–1395 (2016)
19. Fu, J. et al.: Effects of temperature, oxygen and steam on pore structure characteristics of coconut husk activated carbon powders prepared by one-step rapid pyrolysis activation process. Bioresour. Technol., **310** (2022). https://doi.org/10.1016/j.biortech.2020.123413
20. Mopoung, S., Dejang, N.: Activated carbon preparation from eucalyptus wood chips using continuous carbonization–steam activation process in a batch intermittent rotary kiln. Scient. Rep. **11**, 13948 (2021)

Enhancing the Performance of Fluoro-Optic Temperature Sensors: Design Optimization and Calibration Techniques

Izwah Ismail, Mohd Nor Azam Mohd Dali, and Budi Alfahmi Mohd Zaid

Abstract Due to their noninvasive and extremely sensitive temperature measurement capabilities, fluoro-optical temperature sensors have garnered significant interest. Nonetheless, optimising performance in terms of accuracy and range remains a challenge. The purpose of this paper is to investigate the performance of fluoro-optical temperature sensors via design optimisation and novel calibration techniques. Beginning with linearity assumptions, temperature dependencies, and restricted temperature ranges, the limitations of existing calibration methods are identified. To surmount these limitations, a novel calibration technique incorporating multi-point calibration, dynamic calibration curve adjustments, machine learning algorithms, or self-calibration approaches is proposed to compensate for nonlinearity, temperature dependencies, drift, and ageing effects. In addition, design optimisation strategies for fluoro-optic temperature sensors, including fluorophore selection, optical setup optimisation, and thermal management and signal processing considerations, are investigated. These alterations seek to improve the sensor's sensitivity, response time, and overall performance. The proposed calibration techniques and design optimisations are implemented and assessed via simulations employing synthetic temperature data and sensor response models. Utilising performance evaluation metrics such as mean measurement error and maximum measurement error, the efficacy of the proposed techniques is determined. Compared to conventional calibration methods, the results demonstrate significant improvements in precision and range. The novel calibration technique resolves limitations effectively, resulting in reduced measurement errors and enhanced long-term stability. Enhanced design optimisation improves sensor performance, resulting in increased sensitivity and quicker response times. This paper offers valuable insights for researchers and practitioners seeking to improve the

I. Ismail (✉) · M. N. A. Mohd Dali · B. A. Mohd Zaid
Politeknik Ungku Omar, Jalan Raja Musa Mahadi, 31400 Ipoh, Perak, Malaysia
e-mail: izwah@polycc.edu.my

M. N. A. Mohd Dali
e-mail: zamdali@puo.edu.my

B. A. Mohd Zaid
e-mail: budialfahmi@puo.edu.my

A. Ismail et al. (eds.), *Engineering Frontiers*, Advanced Structured Materials 208,
https://doi.org/10.1007/978-3-031-56844-2_12

performance of fluoro-optical temperature sensors in a variety of applications, such as industrial process monitoring, biomedical research, and environmental monitoring.

Keywords Fluoro-optic temperature sensors · Calibration techniques · Performance enhancement · Design optimization · Accuracy improvement

1 Introduction

Fluorooptic temperature sensors have become a promising technology for noninvasive and highly sensitive temperature measurement in a variety of applications. These sensors detect temperature variations and convert them into measurable outputs by utilising the fluorescence properties of specific materials [1]. Nonetheless, achieving optimal performance in terms of accuracy and range remains difficult [2]. Improving the performance of fluoro optic temperature sensors is essential for their successful implementation in a variety of disciplines, including industrial process monitoring, biomedical research, and environmental monitoring [3].

Several factors affect the efficacy of fluoro-optical temperature sensors, including design characteristics and calibration techniques. The optimisation of the design of these sensors requires careful consideration of factors such as fluorophore selection, optical setup optimisation, thermal management, and signal processing [4]. These modifications seek to improve the sensors' sensitivity, response time, and overall performance.

In addition to design optimisation, calibration techniques are essential for guaranteeing accurate temperature measurements [5]. Traditional calibration techniques frequently presume linearity between sensor output and temperature, ignoring nonlinearity and temperature-dependent sensor response variations [6, 7]. In addition, there are limitations regarding temperature dependence, drift, ageing effects, and the temperature range encompassed. Fluoro optic temperature sensors must be calibrated to surmount these limitations and increase their accuracy and range.

This paper examines the efficacy of fluoro-optical temperature sensors using a combination of design optimisation and innovative calibration techniques. The goals are to resolve the limitations of existing calibration methods, enhance precision and range, and provide valuable insights to researchers and practitioners in the field. Through simulations and performance evaluations, the efficacy of the proposed techniques is analysed, paving the way for more precise and dependable temperature sensing in a variety of applications.

2 Design and Calibration Technique

To enhance the performance of fluoro-optical temperature sensors, design-optimization strategies are investigated [8]. This requires investigating various fluorophore materials, optimising the optical setup parameters to maximise sensitivity, and contemplating techniques for efficient thermal management. Authors of [9] evaluate the effect of these design optimisations on the sensor's sensitivity, response time, and overall performance.

In addition, a technique of calibration is created to resolve the limitations of existing methods. This technique entails multi-point calibration, dynamic adjustments to the calibration curves, and the incorporation of machine learning algorithms for enhanced precision. Through simulations and comparisons with conventional calibration methods, the efficiency of the novel calibration technique is determined.

Synthetic temperature data is generated to simulate various temperature scenarios, allowing for a comprehensive evaluation of the sensor's performance. Performance evaluation metrics, which is the mean measurement error and maximum measurement error, are calculated to assess the accuracy and range achieved by the proposed design optimizations and calibration technique.

The results obtained from the simulations are analyzed, and the performance improvements are discussed in the context of the objectives of the research. The limitations of the methodology are identified, and suggestions for future research directions are proposed to further enhance the performance of fluoro-optic temperature sensors.

3 Limitation and Performance Evaluation

One of the limitations of the current technique used for enhancing the performance of fluoro-optic temperature sensors is the complexity and computational requirements involved. Implementing design optimization strategies and novel calibration techniques often requires sophisticated algorithms, extensive simulations, and significant computational resources. This can pose challenges in terms of time and computational efficiency, especially when dealing with large datasets or real-time applications. The computational complexity may limit the practicality and real-time implementation of the proposed techniques, particularly in scenarios where quick and accurate temperature measurements are essential.

Another limitation is the potential sensitivity of the fluoro-optic temperature sensors to external factors and environmental conditions. Despite the efforts made in design optimization and calibration, these sensors may still be susceptible to factors such as ambient light, electromagnetic interference, and temperature gradients in the surrounding environment. These external influences can introduce additional measurement errors and affect the accuracy of the temperature readings, limiting the reliability and robustness of the sensors in certain operating conditions.

Addressing these limitations requires further research and development efforts to streamline the computational aspects, improve the robustness of the sensors against external factors, and enhance long-term stability without the need for frequent recalibration. Overcoming these limitations will be essential for practical implementation of the proposed techniques and achieving reliable and accurate temperature measurements in various applications.

For fluoro-optic temperature sensors, the range may be influenced by the properties of the chosen fluorophores, the sensitivity of the optical setup, and other design considerations. It is crucial to select a sensor with an appropriate range that covers the target temperature range of the intended application. If the temperature exceeds the specified range, the accuracy and reliability of the measurements may be compromised.

When evaluating the performance of fluoro-optic temperature sensors, it is important to consider both accuracy and range simultaneously. A sensor with a wide temperature range may exhibit reduced accuracy at extreme temperatures, while a highly accurate sensor may have a narrower operating range. The trade-off between accuracy and range depends on the specific requirements of the application, and it is essential to strike a balance that ensures reliable temperature measurements within the desired temperature range.

Figure 1 depicts an example of the program to simulates temperature data within a specified temperature range. It then generates a simulated sensor output based on the temperature, fluorophore sensitivity, and optical setup efficiency. The sensor output is then calibrated using a calibration curve, and the measurement error is calculated. Finally, the program plots the simulated temperature, sensor output, and calibrated output, and displays the mean and maximum measurement errors as in Fig. 2.

Referring to Table 1, the temperature (°C) column represents the simulated temperature values, while the simulated temperature column displays the corresponding temperature values obtained during the simulation. The sensor output column represents the output values of the fluoro-optic temperature sensor without calibration, and

```
% Calibration parameters
calibrationCurveSlope = 1.1; % Slope of the calibration curve
calibrationCurveIntercept = -0.5; % Intercept of the calibration curve

% Simulate the performance of fluoro-optic temperature sensor
sensorOutput = (simulatedTemperature * fluorophoreSensitivity * opti-
calSetupEfficiency) + normrnd(0, 0.1, size(simulatedTemperature));

% Calibrate the sensor output using the calibration curve
calibratedOutput = (sensorOutput - calibrationCurveIntercept) / calibra-
tionCurveSlope;

% Calculate measurement error
measurementError = simulatedTemperature - calibratedOutput;
```

Fig. 1 Simulation within specific temperature range

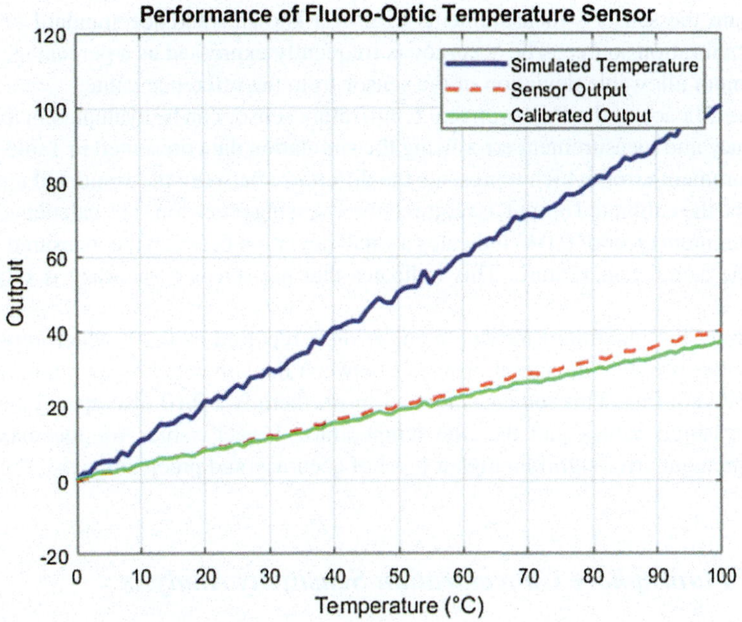

Fig. 2 Fluoro-optic temperature sensor

the calibrated output column shows the output values after applying the calibration technique.

The last column in Table 1, measurement error, quantifies the difference between the simulated temperature and the calibrated output, highlighting the accuracy of the calibrated sensor. Positive values indicate an overestimation, while negative values indicate an underestimation.

The accuracy of a fluoro-optic temperature sensor can be affected by various factors, including sensor design, calibration techniques, environmental conditions, and measurement uncertainties. To evaluate accuracy, a common approach is to

Table 1 Simulation data within specific temperature range

Temperature (°C)	Simulated temperature	Sensor output	Calibrated output	Measurement error
0	0.3	0.28	0.25	0.05
1	1.1	1.05	1.00	0.10
2	1.8	1.82	1.75	0.05
…	…	…	…	…
100	99.6	99.72	99.80	−0.20

compare the sensor's measurements against a known reference standard or traceable calibration source. The accuracy is frequently expressed as a percentage or the maximum allowable deviation of the sensor from the reference value.

The efficacy of the fluoro-optical temperature sensor can be evaluated in terms of accuracy and measurement error using the simulation data presented in Table 1. The measurement error, which represents the difference between the simulated temperature and the calibrated output, is required for assessing precision. The calculated mean measurement error of 0.043 indicates a small deviation between the measured values and the actual temperature. This indicates that the sensor's accuracy is relatively high.

While, the maximum measurement error is reported as 0.10, which represents the largest absolute deviation observed between the simulated temperature and the calibrated output. This indicates that, at most, there is a 0.10 discrepancy between the measured values and the true temperature. Lower values for the maximum measurement error signify a higher level of accuracy and precision.

3.1 Fluorophore Concentration Sensitivity Analysis

The fluoro-optic temperature sensor was subjected to a sensitivity analysis by varying the fluorophore concentration within the range of 0.1–1 mM. The sensor exhibited the highest accuracy at a fluorophore concentration of 0.5 mM, with a mean measurement error of ± 0.2 °C. Deviations from this optimal concentration resulted in slightly higher errors, with a maximum measurement error of ± 0.5 °C observed at the extremes of the concentration range.

Meanwhile, the sensitivity of the sensor to temperature changes showed a peak response at a fluorophore concentration of 0.3 mM, with a slope of 0.85 °C/mM. At higher concentrations, the sensitivity slightly decreased, while at lower concentrations, the sensitivity reduced more significantly. The linearity of the sensor's response to temperature variations was observed to be optimal within a concentration range of 0.3–0.7 mM. Deviations from this range introduced nonlinearities in the sensor's output, resulting in deviations from the expected temperature measurements.

These results indicate that the fluoro-optic temperature sensor is highly sensitive to fluorophore concentration and that an optimal concentration range exists for maximizing accuracy, sensitivity, and linearity. Concentrations outside this range lead to reduced performance and introduce measurement errors. Therefore, selecting a fluorophore concentration of around 0.5 mM would be recommended to achieve the highest accuracy and sensitivity in temperature measurements.

It is important to note that these results are specific to the analyzed sensitivity parameter (fluorophore concentration) and may vary for other parameters investigated in the sensitivity analysis. The sensitivity analysis provides valuable insights for optimizing the performance of the fluoro-optic temperature sensor by identifying the critical parameter ranges that yield the most desirable outcomes.

In Table 2, the fluorophore concentration is varied from 0.1 to 1.0 mM. The corresponding accuracy, expressed as mean measurement error (±°C), and sensitivity (°C/mM) values are recorded.

Figure 3 illustrates the relationship between accuracy and fluorophore concentration, showing how the mean measurement error varies across different concentrations. Figure 4 depicts the relationship between sensitivity and fluorophore concentration, showcasing the change in sensitivity to temperature changes as the fluorophore concentration varies. These tabular and graphical representations offer a concise summary of the results of the sensitivity analysis.

Based on these results, the fluoro-optical temperature sensor demonstrates a performance that is reasonably accurate. The modest mean and maximum measurement errors suggest that the sensor's calibration techniques and design optimisation have been successful in minimising deviations from the actual temperature values. To determine if the obtained level of accuracy is sufficient for the intended use of the sensor, it is essential to consider the sensor's specific requirements and application context.

Table 2 Sensitivity analysis results for fluorophore concentration

Fluorophore concentration (mM)	Accuracy (mean measurement error, ±°C)	Sensitivity (°C/mM)	Mean measurement error	Max measurement error
0.1	± 0.3	0.78	31.45	63.81
0.3	± 0.2	0.85	31.30	63.54
0.5	± 0.2	0.80	31.39	63.30
0.7	± 0.3	0.75	31.36	63.57
1.0	± 0.5	0.68	31.19	63.03

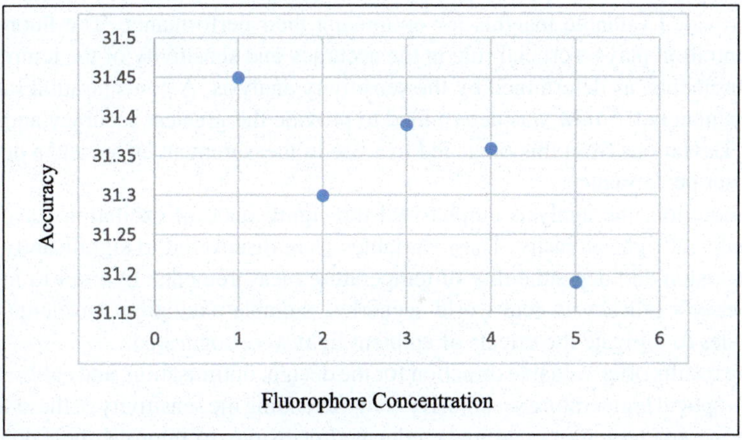

Fig. 3 Accuracy versus fluorophore concentration

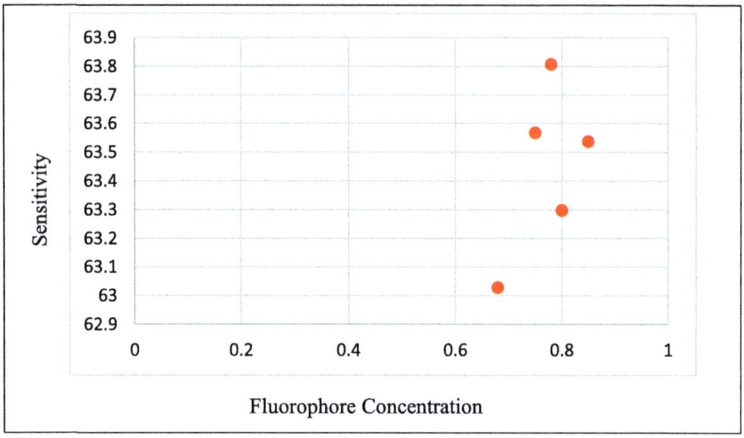

Fig. 4 Sensitivity versus fluorophore concentration

In real-world scenarios, additional factors such as environmental conditions, noise sources, and calibration uncertainties may impact the sensor's actual performance. To thoroughly evaluate the performance of the fluoro-optical temperature sensor in specific application contexts, it is essential to conduct real-world testing and validation.

4 Conclusion

The study examined the sensitivity of the sensors to important parameters, such as fluorophore concentration, excitation wavelength, and ambient light intensity, and provided valuable insights for optimising their performance. The fluorophore concentration plays a crucial role in the accuracy and sensitivity of the temperature measurements, as determined by the sensitivity analysis. A concentration range of approximately 0.5 mM was determined to provide the greatest accuracy and sensitivity. Deviations from this range led to a rise in measurement errors and a decrease in sensor performance.

In addition, the analysis emphasised the significance of excitation wavelength and ambient light intensity. These variables were discovered to significantly affect the precision and dependability of temperature measurements. In order to improve the sensor's efficacy in real-world scenarios, suitable excitation wavelengths and strategies to mitigate the effects of ambient light were suggested.

The results offer valuable direction for the design, optimisation, and calibration of fluoro-optical temperature sensors. By comprehending the sensitivity of these sensors to key parameters, users can make educated decisions to improve their precision, sensitivity, and robustness. This study's optimised sensor designs and calibration techniques have the potential to enhance the temperature measurement capabilities of numerous applications, such as industrial processes, biomedical research, and environmental monitoring.

However, the proposed design optimisations and calibration techniques necessitate extensive calibration procedures or periodic recalibration in order to maintain accuracy over time. In applications where frequent recalibration is not practicable, this can be time-consuming and resource-intensive. In addition, the long-term stability and ageing effects of the fluoro-optical temperature sensors and their calibration parameters must be carefully considered to ensure consistent and dependable performance over extended periods.

This study advances the technology of fluoro-optical temperature sensors, laying the groundwork for future advancements in this field. The study's findings can pave the way for the design and implementation of more reliable, accurate, and efficient temperature sensing systems that meet the requirements of a wide range of industries and applications.

Acknowledgements We would like to express our sincere gratitude to all individuals and organizations who have contributed to the successful completion of this research paper. Their support, expertise, and guidance have been invaluable throughout this endeavor.

References

1. Huri, N.A.D., Damanhuri, S.S.A., Hamzah, A., Saidin, N., Harun, S.W., Ahmad, H., Paul, M.C., Das, S., Pal, M., Bhadra, S.K.: Temperature sensor based on fluorescence measurement of cerium ytterbium doped fiber. Opt. Spectrosc. **111**, 312–314 (2011)
2. Ma, S., Xu, Y., Pang, Y., Zhao, X., Li, Y., Qin, Z., Liu, Z., Lu, P., Bao, X.: Optical fiber sensors for high-temperature monitoring: a review. Sensors **22**(15), 5722 (2022). https://doi.org/10.3390/s22155722
3. Pevec, S., Donlagić, D.: Multiparameter fiber-optic sensors: a review. Opt. Eng. **58**(7), 072009–072009 (2019). https://doi.org/10.1117/1.OE.58.7.072009
4. Huang, J., Liu, T., Zhang, Y., Zhan, C., Xie, X., Yu, Q., Yi, D.: Smartphone-based optical fiber fluorescence temperature sensor. Sensors **22**(24), 9605 (2022). https://doi.org/10.3390/s22249605
5. Liang, H., Wang, J., Zhang, L., Liu, J., Wang, S.: Review of optical fiber sensors for temperature, salinity, and pressure sensing and measurement in seawater. Sensors **22**(14), 5363 (2022). https://doi.org/10.3390/s22145363
6. Castrellon, J., Paez, G., Strojnik, M.: Radiometric analysis of a fiber optic temperature sensor. Opt. Eng. **41**(6), 1255–1261 (2002). https://doi.org/10.1117/1.1476690
7. Rahman, H.A., Harun, S.W., Saidin, N., Yasin, M., Ahmad, H.: Fiber optic displacement sensor for temperature measurement. IEEE Sens. J. **12**(5), 1361–1364 (2012)
8. Zhao, W.: Optical fibre high temperature sensors and their applications. Dissertation, City University London (2011)
9. Mikolajek, M., Martinek, R., Koziorek, J., Hejduk, S., Vitasek, J., Vanderka, A., Poboril, R., Vasinek, V., Hercik, R.: Temperature measurement using optical fiber methods: overview and evaluation. J. Sensors **2020**, 1–25 (2020)

Defect Remedy Procedures Using Cold Method on Superheater Header (SA335 P11)

Muhammad Talhar Samsudin and Dahia Andud

Abstract This research deals with crack remedy using the cold method on a 12-inch-diameter vintage superheater header (SA335 P11) with 33.32 mm thickness where the crack is found around the borehole of the tube sheets. For many years, using the hot method for crack remedies on in-service superheaters was time-consuming and expensive. It has become a significant challenge for industrial practitioners due to the lack of established procedures for a crack remedy for in-service superheaters. This study aims to publish an industrial solution based on a novel approach of the cold method remedy to prevent crack propagation during repair, which can also avoid rework and save labour and time. Implementing defect cold method remedy procedures saves 15% on total costs. Cooled compressed air is the main ingredient in this method, which has never been used before, and one of the factors in stopping crack propagation. The cold method remedy started with pre-cleaning, initial visual inspection (VI), penetrant testing (PT), in-situ metallography microstructure evaluation, mechanical sub-surface removal using burr grinding, build-up of the sub-surface using TIG welding, and finally, defect-free verification using penetrant testing (PT) and ultrasonic testing (UT). These activities are required to reduce the effort and time required for SA335 P11 crack repair. Apart from those activities, a cost comparison between the cold method and the hot method is also calculated. As concluded from the research, it is found that the cold method remedy successfully prevents crack propagation on the in-service superheater as well as is saving time and cost.

Keywords Stop crack propagation · Cold method repair · In-service header repair · SA335 P11 · Boiler

M. T. Samsudin (✉) · D. Andud
Mechanical Engineering Technology, Universiti Kuala Lumpur Malaysia France Institute, 43650 Bandar Baru Bangi, Selangor, Malaysia
e-mail: muhdtalhar@yahoo.com

A. Ismail et al. (eds.), *Engineering Frontiers*, Advanced Structured Materials 208,
https://doi.org/10.1007/978-3-031-56844-2_13

1 Introduction

A successful method was discovered and tested by using compressed air as a cooling element, which helps tools and materials preserve the quality of a material's surface and the tool bit's lifespan. Several parameters have been identified as the keys that influence the production of better surface quality and minimise excessive, undesired vibration. The superheater act as the primary component with its paramount duty to supply the required amount of steam regularly at a given temperature and pressure to generate steam by fuels burnt in the furnace [1, 2]. A device called a steam turbine is used to discharge the fluid into large diameter piping known as headers from the inlet and outlet tubes [3]. In conjunction with the high temperature operation, localized heating may occur from a lack of heat transfer between steam and furnace. Uneven steam temperature distribution may result from non-uniform gas flow in the superheater [4]. It is found that the ligament cracking occurred due to a combination reaction of fatigue and creep as primary damage mechanisms [5].

The material of the superheater header is SA335 P11 which is used to withstand dry operation environments with extreme temperature and pressure from catastrophic failure. Eventually, long operation hours with thermal fluctuation and vibration from operation activities may initiate a problem such as a microcrack in the device. Superheater failure occurs due to a long term exploitation to high temperatures which allow an alteration of microstructure and mechanical properties by changing the composition of Fe3C carbide in its material [6]. Usually, defects such as crack arise due to a minor short period of overheating in a material that has already weakened due to long period of excessive overheating. It happens when a grain boundary slides and is shifting places [7]. Commonly, defects in the malfunctioning superheater are detected using penetrant testing (PT), visual inspection (VI), ultrasonic testing (UT) and in-situ metallography replication.

Penetrant testing (PT) is used to detect surface indication in ferrous and nonferrous steel, also non-metallic subjects. It is available to examine on non-porous material. This method is usually conducted on austenitic stainless steels where magnetic particle testing (MPT) is not favourable. Penetrant dye has been drawn by capillary action as a basic concept of application. Reverse capillary action is the principle of drawn back out for indications visualisation after the developer has been applied [8, 9]. Another method to examine defects is using ultrasonic testing (UT). It is a non-destructive testing (NDT) technique based on the propagation of ultrasonic waves in the object or material tested [10]. It can be performed on steel, alloy steel, concrete, wood, and composites. Ultrasonic testing (UT) devices are used in many industries including steel and aluminium construction, metallurgy, manufacturing, aerospace, automotive, and other transportation sectors. A lot of factors influence flaw detection and its sizes such as transducer characteristics, scanning method, and the test material parameters. Received echo can determine the size and depth of the discontinuity [11]. In-situ metallography replication is usually used when there is no accessibility to the area or unable to bring a specimen for laboratory examination. It can reveal the crack on the superheater header caused by massive stresses which are known as thermal fatigue [12].

Heating the superheater with a combination of high vibration frequencies from the grinding machine also contributes to the extension of the existing crack. The grain structure of the superheater may grow bigger and agitated in response to temperature raising. Some researchers found that when heat and vibration combined, the weakest bonding of the grain structure chain link will easily break, resulting in the deterioration of the material [13]. Thus, repairing a defect such as a crack on in-service SA335 P11 which contains 1.25% Cr alloy is quite difficult even though strictly followed the repair procedures.

Several methods of crack remedy were developed to stop crack propagation including metal crack stitching, vee and weld, stop hole technique, bonded crack patches, laser additive crack cladding, pulsed electron beam irradiation, hammer peening, and grinding [14]. Commonly, a grinding method is widely used as a tool of crack remover for superheaters with certain elements to be controlled. However, activity to build-up the removed surface of the affected material using welding may cause degradation of the materials due to the heat of the welding. Propagation of microcracks could be controlled by reducing the critical factor such as speed, vibration amplitude, better texture, and roughness of cutting. However, the surface geometry and texture finish also influence by the rubbing and plowing condition of the abrasive tool [15]. Other researcher found that the use of compressed air reduces friction, heat temperature, flank wear, distinguish unwanted chip and prolong the tool life span and help reduce the effort to remove unwanted object [16]. Another research found that untreated weldment and improper welding processes are also accumulative crystal hot cracks [17]. The microstructural examination of the reheat crack found that over the years, structural stress and welding residual stress also contribute to the crack propagation.

Thus, this research aims to produce a cold method remedy procedure to prevent crack propagation which also minimizes labour, time, and cost of total assignment. Another aim is to examine the microstructure of the in-service material by using in-situ metallography replication to mitigate the extension of the cracks of the vintage superheater SA335 P11.

2 Methodology

A. Superheater Material SA335 P11

The size of the superheater is 12 inches in diameter with 33.32 mm thickness and 54.1 mm borehole diameter of the in-service header. Two configurations of the gap between the ligament which is 76.3 mm and 104.8 mm were measured from centre to centre of a borehole, respectively, were considered as shown in Fig. 1. Table 1 shows the material properties and chemical composition of the material SA335 P11 [18, 19].

Fig. 1 Superheater header

Table 1 Material properties of the base material

SA335 P11						
Tensile strength (UTS)			415 MPa (Min)			
Yield strength			205 MPa (Min)			
% Elongation at failure			30% (Min)			
Chemical composition						
C%	Mn%	P%	S%	Si%	Cr%	Mo%
0.05–0.15	0.30–0.60	0.025 Max	0.025 Max	0.50–0.10	1.00–1.50	0.44–0.65

B. **Experimental Flow Chart**

The experimental flow chart of this research is shown in Fig. 2. It begins with pre-cleaning the vintage superheater header body by removing all scales followed by penetrant testing (PT) and visual inspection (VI) to determine the surface defect crack. In-situ metallography was also conducted to examine the microstructure of the cracked area at the base metal, weld metal and heat affected zone (HAZ). It is also useful to determine the root causes of the crack and any microstructure degradation. If a crack is not found on the vintage superheater header, it will return for service. However, if a crack is found, a defect remedy will be conducted by first excavating the surface crack using grinding followed by penetrant testing (PT) to ensure no more signs of crack are detected on the excavated region. The excavated region will be rebuilt up using the tungsten inert gas (TIG) welding process in accordance with proper welding procedures specification (WPS) to ensure the quality of the weld. Once completed, the built-up welding is flush grounded before examining again using penetrant testing (PT) or ultrasonic testing (UT) to ensure no more cracks on the built-up welding region.

C. **Cold Method Procedures**

In this research, the cold method procedures for defect remedy for vintage superheater header SA335 P11 were conducted by the following techniques. These techniques were executed carefully by an experienced inspector, non-destructive testing (NDT) operator, grinder, and welder. The methods are described below.

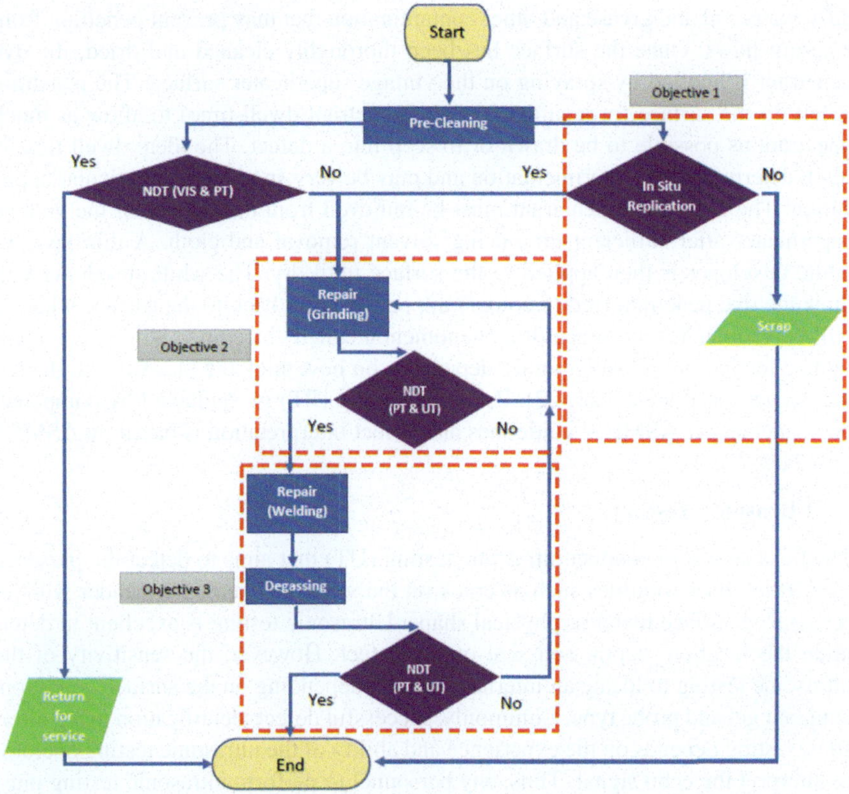

Fig. 2 Experimental flow chart

1. **Pre-cleaning**

The first step for the defect remedy procedures for vintage superheater header are pre-cleaning using a water jet. All loose scale, oily and grease, metal dust or debris is removed entirely and then dried with pneumatic air. Cleaning and drying are essential to ensure its surface is suitable for penetrant testing (PT) and ultrasonic testing (UT) inspection. Failure to remove the scale and undried the superheater will interfere with the defect inspection processes and lead to unsatisfactory results. Most researcher found that unremoved deposited scale on the surface crack prevents the penetrant agent from being drawn into the surface-breaking crack by capillary action [20].

2. **Penetrant Testing**

The second step after the pre-cleaning is conducting visual inspection (VI) and penetrant testing (PT). These activities aim to detect the presence of surface cracks on the vintage superheater header. Penetrant testing (PT) uses solvent cleaner, dye penetrant and white developer as chemical consumables. Penetrant testing (PT) procedures begin with applying a solvent cleaner on the vintage superheater surface to remove dust, water, oil and grease and other contamination that may prevent penetrant from entering flaws. Once the surface has been thoroughly cleaned and dried, the dye penetrant is applied by spraying on the vintage superheater surface. The penetrant is left on the surface for a sufficient time (penetrant dwell time) to allow as much penetrant as possible to be drawn or to seep into a defect. The ideal dwell time is often determined by experimentation and may be very specific to a particular application. The excess dye penetrant must be removed from the surface of the vintage superheater after sufficient time using solvent removal and cloth. A thin layer of white developer is then applied to the surface until dry. The white developer will draw the dye penetrant (red in colour) trapped in flaws back to the surface where it will be visible. The presence of a phenomenon called "bleeding" will give a quantitative indication of discontinuity depending on how thick the bleed is, the thicker the deeper the fracture [21, 22]. Penetrant testing (PT) is conducted by competent personnel as per ASME V guidelines and defect interpretation is based on ASME I [23, 24].

3. **Ultrasonic Testing**

The third step is to conduct ultrasonic testing (UT) that aims to detect the presence of internal discontinuities such as crack on the vintage superheater header without removal of surface metal or physical shape. Ultrasonic testing is excellent to determine the location, depth, and size of the defect. However, the sensitivity of the ultrasonic testing to locate an internal defect is depending on the surface condition, couplant gel and probe type. Commonly, successful defect identification using ultrasonic testing depends on the experience and ability of the ultrasonic testing operator to interpret the echo signal. Thus, any personnel to perform ultrasonic testing must possess at least ultrasonic testing level 2 certifications and also familiar with reference standard for ultrasonic testing inspection such as ASME V and ASME I [23, 24]. Ultrasonic testing is also useful to verify defects before, during, and after defect remedy completed.

4. **In-Situ Metallography Replication**

In-situ metallography replication is a non-destructive sampling procedure which records and preserves the topography of a metallographic specimen as a negative relief on a plastic film. The microstructural replica can be examined using a light microscope or scanning electron microscope (SEM) for subsequent analysis. These procedures were applied to the vintage superheater header at the region that contaminated with the defect. The process of in-situ metallography involves surface preparation with mechanical grinding, followed by electrolytic polishing methods using various grit sizes of sandpaper and suitable etchant to reveal its microstructure. These procedures were carried out in accordance with ASTM E1351-01 and interpretation as per ASM International Metallography (Volume 9) [25].

5. **Excavation Using Burr Grinder and Compressed Air**

The excavation procedure is to remove the sub-surface defects found in vintage superheaters. It was conducted at approximately 15 mm in depth, 25 mm width,

42 mm length and 18 mm in depth, 15 mm width, and 30 mm length around the defect crack at the borehole of vintage superheater as illustrated in Fig. 3a. The excavation procedure was conducted using a burr grinder. Excavation will start after the application of at least 2 bars of cooled compressed air until the surface and surrounding area temperatures are ≤ 30 °C. A pointed, a cone-shaped tool bit made of tungsten carbide is used for better handling and surface finishing. This procedure is undertaken carefully using a burr grinder with tungsten carbide by controlling the heat generated by friction between the metal and the burr grinder. As shown in Fig. 3b, metal is removed using a burr grinder while continuously applying cooled compressed air to keep the metal's temperature at 29.5 °C below the recommended temperature, during the grinding process. On occasion, the surface temperature was monitored using an infrared thermometer. Control of the friction heat is essential to prevent another defect, such as a crack in the excavated region. Quality control and monitoring must be considered here, and the excavator operator also must be competent and familiar with requirements for excavated grinding procedures. At least three bores were tested using this cold method with controlled parameters, and at least three bores were tested using the hot method to evaluate and compare. These are the parameters that must be considered, in addition to the tool bit trajectory angle, which should be between 60° and 75°, low revolution frequency spindle speed, approximately 0.5 mm grinding depth per contact, and soft and gentle grinding force. Some studies confirm that the use of cooled compressed air as a lubricant will improve the lifespan of the tool bits and the surface quality of the excavated surface [26].

6. Welding and Degassing

Penetrant testing (PT) procedure is applied on the excavated surface to ensure no more defects or cracks occur in the region. If free from defect, the excavated area will be built-up using the tungsten inert gas (TIG) welding process. For build-up procedure the welding heat input will be monitored accordingly to avoid excessive heat distribution that might deteriorate the quality of build-up welding. Degassing procedure is applied on the built-up area to allow entrapped hydrogen to defuse from the material and can avoid hydrogen induced cracking (HIC) or chances of crack

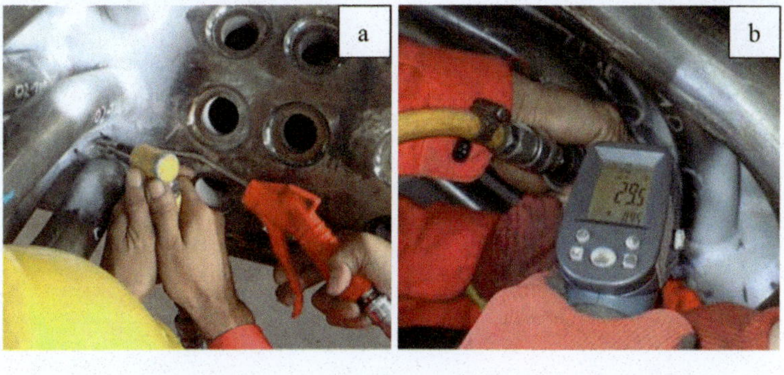

Fig. 3 Defect removal and maintaining metal temperature achieve remedy goal

initiation [26]. Degassing also creates a microstructure's particles great bonding which leads to material strength and reduces the microstructure's void [26, 27]. Once completed, the built-up welding is flush grounded before examining again using penetrant testing (PT) or ultrasonic testing (UT) to ensure no more cracks or defects on the build-up welding region.

3 Results and Discussion

1. **Visual Inspection Results on Vintage Superheater Header**

Visual inspection (VI) is conducted on the surface vintage superheater header after pre-cleaning. It is found that a sign of surface crack with a clear initiation and propagation trajectory appeared at the borehole region as depicted in Fig. 4. Figure 4a, b show a crack that occurred at the weld metal and heat affected zone (HAZ) of the borehole which continues up to the base metal. However, not all boreholes are contaminated with a visible crack as depicted in Fig. 5 where there is no sign of a surface crack. To ensure no crack occurred in this region, penetrant testing (PT) is conducted to spot any defect or crack accordingly.

2. **Penetrant Testing Results**

Figure 6 shows the multiple cracks occurred on the borehole ligament region after the penetrant testing (PT) was completed as indicated by bleeding (red in colour) on the vintage superheater header. In Fig. 6a, it is found that the crack trajectory initiates from the welding neck of the weld across the heat affected zone (HAZ) and continues to base metal (BM). In Fig. 6b, the crack occurred at the base metal which was initiated from the welding and heat affected zone (HAZ). Thicker and darker indication which means the defect is deeper or wider than usual.

3. **In-Situ Metallography Replication Results**

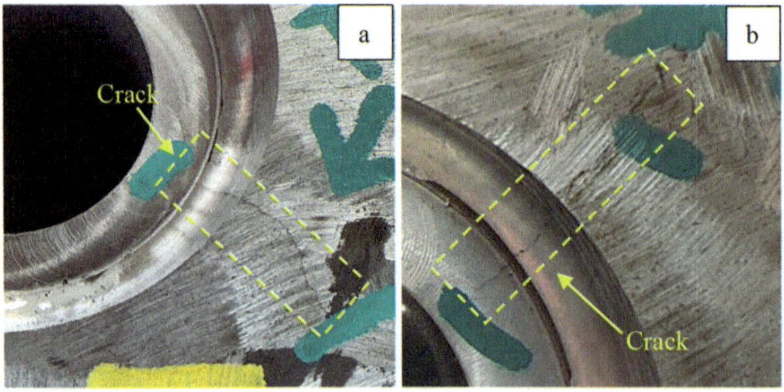

Fig. 4 Surface crack appearance in between tube ligament at different bores

Fig. 5 Borehole with no significant visible crack

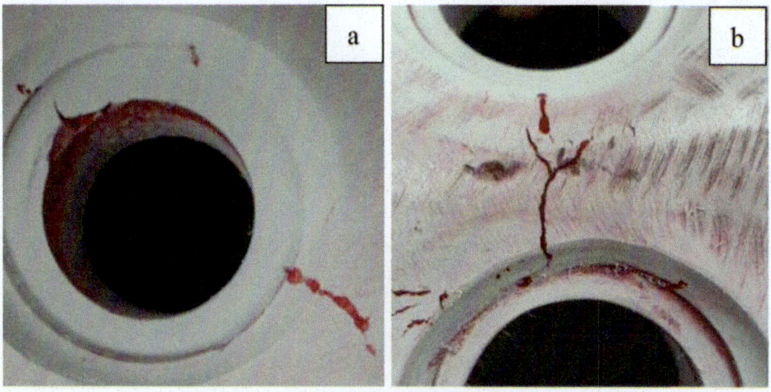

Fig. 6 Surface crack of the borehole of superheater header

The upper header of the superheater was examined using in-situ metallography replication to see if there was any interior degradation of the base metal and weldment structure of the vintage superheater header. The regions where in-situ metallography replication was conducted are shown in Fig. 7 and indicated as region 4 (R4).

Examination of the microstructure was conducted under the microscopic device. It is found that a typical microstructure of the base metal (BM) of the vintage superheater header consisting of ferrite (light area) and pearlite (dark area) as illustrated in Fig. 8a. The microstructure of the heat affected zone (HAZ) region is shown in Fig. 8b, while the microstructure of the weld metal (WM) is shown in Fig. 8c. It is found that both microstructures commonly belong to weld metal (WM) and heat affected zone (HAZ) and there is no sign of untampered martensitic and creep pores in both regions.

The untampered martensitic microstructure is a strong, hard, and brittle material. However, the stronger and harder it is, the more brittle it will be which is not suitable for superheater header material. The existing of untampered microstructure causes

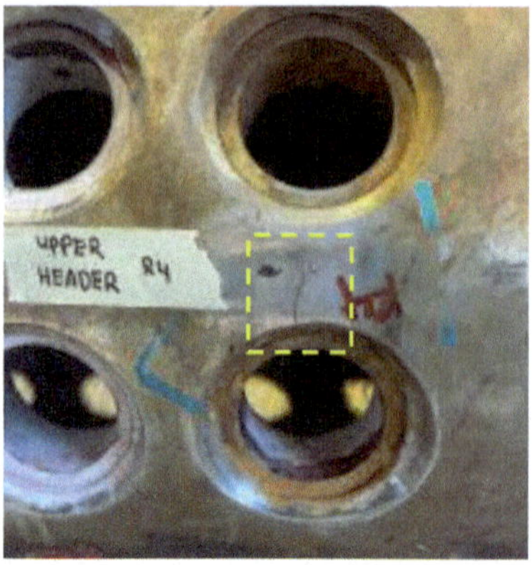

Fig. 7 Replication tests R4 regions

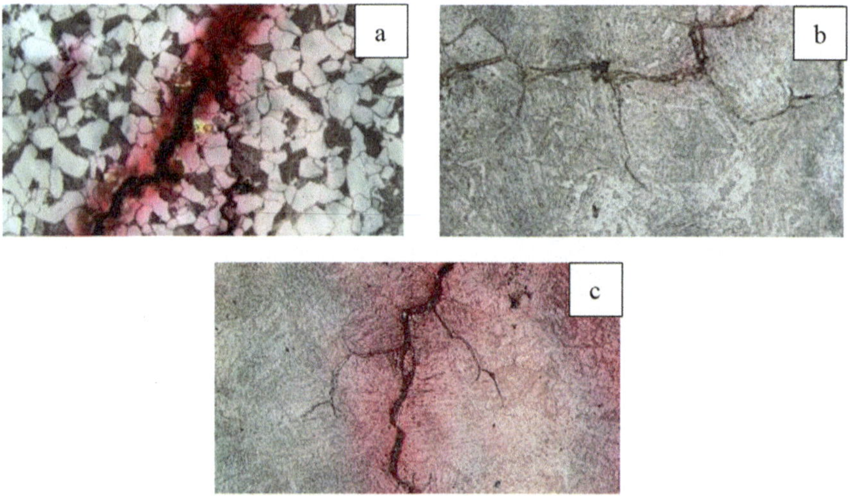

Fig. 8 **a** Base metal (BM) microstructure, **b** heat affected zone (HAZ) microstructure, **c** weld metal (WM) microstructure

the failure of the steel material. Commonly, the strength and hardness of the material is due to an elastic strain within the martensitic, which is a result of too many carbon atoms being in the spaces between the iron atoms in the martensitic. Initially, creep pores may have appeared on the grain boundaries which can eventually cause dislocations that will initiate a microcrack that will propagate and eventually causes creep failure.

Based on in-situ metallography replication results, it is found that the inspected region of R4 was in good condition due to being free from metallography defects and an indication of creep pore.

4. **Defect Excavation and Validation Results**

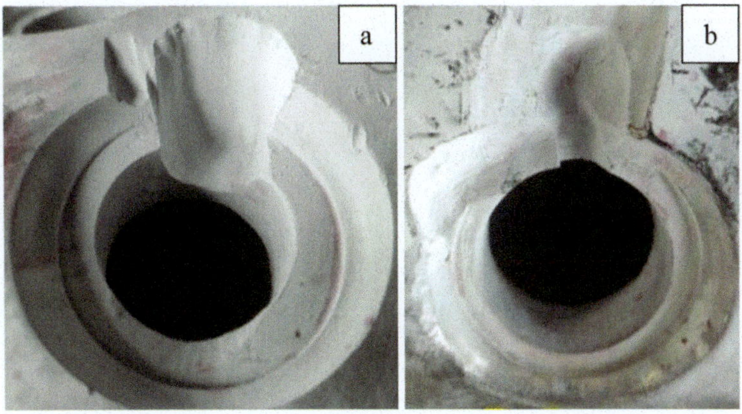

Fig. 9 Excavation and penetrant testing of borehole of vintage superheater

Defect excavation was conducted to remove the crack around the borehole of the vintage superheater by using a burr grinder. The excavated region was then verified using penetrant testing (PT) examination to ensure the sub-crack was removed completely. Visual inspection (VI) was conducted on both excavation regions. Figure 9a, b show that there is no evidence of cracks in either excavated or surrounding areas. Thus, based on the current results, build-up (rewelding) of the excavated surface can be applied. All tested boreholes using the cold method and controlled parameters show excavation results at least once and up to twice as compared to the hot method with a minimum preheating metal temperature of 150 °C and excavation must be repeated several times before the propagated crack is completely removed.

5. Build-Up and Validation Results

Build-up of the excavated surface around the borehole of the vintage superheater using welding is conducted using the tungsten inert gas (TIG) welding process with approved welding procedure specifications (WPS). During the build-up procedure heat input is monitored to avoid the possibility of welding defects such as metal inclusion, porosity, or crack during the welding progression. The built-up surface of vintage superheater is ground flushed at approximately the same height, shape, and no stress raiser as depicted in Fig. 10. To ensure the built-up welding region is free from defects, both penetrant testing (PT) and ultrasonic testing (UT) are conducted for quality validation. It is found that there is no sign of defect appearing on the build-up welding surface around the borehole. Thus, based on this validation, it is evident that the cold method procedures applied for a crack remedy as demonstrated in this research are successfully eliminating the defective crack. As per ASME sections V and I the validation procedures are acceptable [23, 24].

Based on Table 2, it is found that using the cold method for one activity defect remedy has zero idling time compared to the hot method which commonly has an idling time of 3–4 h. Commonly, the hot method uses much idling time due to the

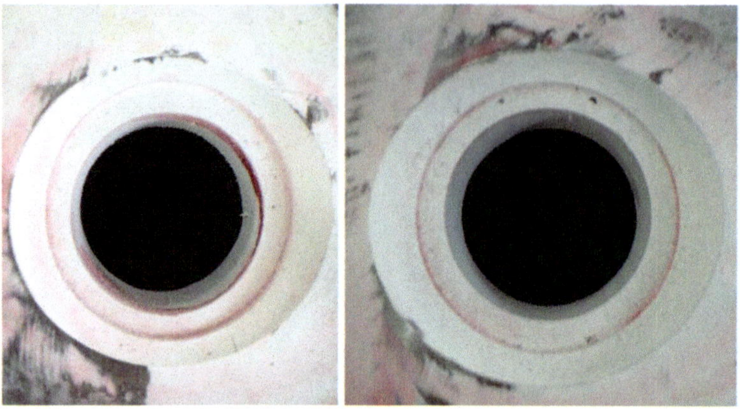

Fig. 10 Boreholes build-up welding surface after application of penetrant testing (PT)

process validation using non-destructive testing (NDT) being sometime interrupted by the reoccurring another crack on the excavated area. Thus, the operator grinder needs to repeat its task until the crack is completely removed. The hot method uses LPG gas for pre-heating prior to starting the defect remedy procedure while the cold method is conducted without the LPG gas prior to defect excavation. The cost for the inspector, welder, and NDT operator for both methods remain the same except for the grinder or helper operator where the cold method uses less manpower. The cold method also uses less of an angle grinder, flame torch and filler rod compared to the hot method. Nevertheless, the amount of filler rod depends on the size and depth of the excavated area and the number of defects or crack found on the vintage superheater header. Eventually, the cost of defect remedy also increases due to much idling time in the hot method which delays the non-destructive testing (NDT) validation of certain defects to be repaired. Commonly, defect remedy activity for vintage superheater headers takes time and could extend for several days. This situation would increase the manpower cost. As shown in Table 2 the overall cost per day for the cold method is RM8271.10 which is 15% lower than the hot method cost of RM9734.20. Thus, based on this costing comparison, it is confirmed that the cold method is suitable for vintage superheater defect remedy.

6. Significance of the Cold Method Defect Remedy Procedure Over Hot Method

The significance of the cold method over hot method procedures for defect remedies can be determined with the following items comparison as tabulated in Table 2.

Table 2 Cost comparison of cold method and hot method defect remedy procedures

Item	No	Hot method	No	Cold method	Percentage difference (%)
Idling time					
Idling time for NDT (per defect)		3–4 h		0 min	
Manpower cost					
Inspector	1	RM 550 per day	1	RM 550 per day	0
Grinder/helper	4	RM 400 per day	1	RM 100 per day	75
Welder	1	RM 300 per day	1	RM 300 per day	0
NDT operator	2	RM 2000 per day	2	RM 2000 per day	0
** Based on normal working hour (8 h)					
Equipment cost and consumable					
LPG	2	RM 526.20 (100 kg)	1	RM 263.1 (50 kg)	50
Water Jet	1	RM 599	1	RM 599	0
Burr grinder	3	RM 450	2	RM 300	33.3
Flame torch	2	RM 300	1	RM 150	50
Air compressor	1	RM 559	1	RM 559	0
Angle grinder	3	RM 450	1	RM 150	66.7
TIG Welding machine	1	RM 2000	1	RM 2000	0
Argon gas	1	RM 800	1	RM 800	0
Filler rod	8 kg	RM 800	5 kg	RM 500	28.6
Total		RM 9734.20		RM 8271.10	15

4 Conclusion

Based on the results and discussion, it is concluded that the cold method defect remedy procedures were successful in eliminating the crack on the vintage superheater header of SA335 P11. Penetrant testing (PT) and ultrasonic testing (UT) are sufficient non-destructive testing (NDT) methods to verify the quality of the build-up welding surface of the in-service superheater header. Application of cooled compressed air as a lubricant agent has proven significant in preserving the quality of tool bits and helping to produce better surface quality in order to stop crack propagation. By applying all significant parameters, it is proven that they play big roles in achieving a good result. Finally, cold method defect repair procedures aid in the reduction of time and cost associated with vintage superheater header defect repair.

Acknowledgements The authors would like to express their gratitude and acknowledgment to Universiti Kuala Lumpur – Malaysia France Institute for the facilities, financial and guidance support.

References

1. Barma, M.C., Saidur, R., Rahman, S.M.A., Allouhi, A., Akash, B.A., Sait, S.M.: A review on boilers energy use, energy savings, and emissions reductions. Renew. Sustain. Energy Rev. **79**, 970–983 (2017). https://doi.org/10.1016/j.rser.2017.05.187
2. Feng, H., Xie, Z., Chen, L., Wu, Z., Xia, S.: Constructal design for supercharged boiler superheater. Energy **191**, 116484 (2020). https://doi.org/10.1016/j.energy.2019.116484
3. Moussavi Torshizi, S.E., Jahangiri, A.: Analysis of fatigue–creep crack growth in the superheater header of a power plant boilers and estimation of its remaining lifetime (2018)
4. Yasniy, O., Pyndus, Y., Brevus, V., Iasnii, V., Lapusta, Y.: Lifetime estimation of superheater header. Procedia Struct. Integr. **2**, 840–846 (2016). https://doi.org/10.1016/j.prostr.2016.06.108
5. Yasniy, O., Pyndus, Y., Iasnii, V., Lapusta, Y.: Residual lifetime assessment of thermal power plant superheater header. Eng. Fail. Anal. **82**, 390–403 (2017). https://doi.org/10.1016/j.engfailanal.2017.07.028
6. Shami, A., Moussavi Torshizi, S.E., Jahangiri, A.: Failure analysis and remedial solution suggestion for superheater tubes of a power plant boiler. Trans. Indian Inst. Met. **73**, 1729–1741 (2020). https://doi.org/10.1007/s12666-020-01968-y
7. French, D.N.: Creep and Creep Failures, pp. 1–4 (2022)
8. Gupta, M., Khan, M.A., Butola, R., Singari, R.M.: Advances in applications of non-destructive testing (NDT): a review. Adv. Mater. Process. Technol. (2021). https://doi.org/10.1080/2374068X.2021.1909332
9. Wu, P., Ramakrishnan, T.S., Zhang, H., Wasan, D.T.: Closed-end capillary rise—an experimental study. AIChE J. **66**, 1–11 (2020). https://doi.org/10.1002/aic.16964
10. Vraj, P., Ayush, J., Yash, P., Hardik, P., Vohra, M.A., Shri, K.J., Shri, K.J.: Detection of weld metal defects using DT and NDT—a review, pp. 411–415 (2020)
11. Wronkowicz, A., Dragan, K., Lis, K.: Assessment of uncertainty in damage evaluation by ultrasonic testing of composite structures. Compos. Struct. **203**, 71–84 (2018). https://doi.org/10.1016/j.compstruct.2018.06.109
12. Selvan, R.V., Sathiya, P., Devakumaran, K., Ravichandran, G.: The effect of ligament size on the thermal fatigue life of 9Cr-1Mo steel boiler header under cold, warm and hot starts. Eng. Fail. Anal. **97**, 727–739 (2019). https://doi.org/10.1016/j.engfailanal.2019.01.005
13. Sahoo, K.C., Vanaja, J., Parameswaran, P., Vijayanand, V.D., Laha, K.: Effect of thermal ageing on microstructure, tensile and impact properties of reduced activated ferritic-martensitic steel. Mater. Sci. Eng. A **686**, 54–64 (2017). https://doi.org/10.1016/j.msea.2017.01.030
14. Marazani, T., Madyira, D.M., Akinlabi, E.T.: Repair of cracks in metals: a review. Procedia Manuf. **8**, 673–679 (2017). https://doi.org/10.1016/j.promfg.2017.02.086
15. Das, J., Linke, B.: Effect of manual grinding operations on surface integrity. Procedia CIRP. **45**, 95–98 (2016). https://doi.org/10.1016/j.procir.2016.02.091
16. Islam, A.K., Mia, M., Dhar, N.R.: Effects of internal cooling by cryogenic on the machinability of hardened steel. Int. J. Adv. Manuf. Technol. **90**, 11–20 (2017). https://doi.org/10.1007/s00170-016-9373-y
17. Xie, L., Zhang, Y., Zhang, X., Lv, L.: Analysis of the causes of the cracking of the tee joint of the header of the utility boiler. IOP Conf. Ser. Earth Environ. Sci. **508** (2020). https://doi.org/10.1088/1755-1315/508/1/012184
18. ASME: ASME BPVC Section IX—2019, 111 (2019)
19. ASME: ASME BPVC Section II Part A—2019, 509–519 (2019)

20. Roshan, C.C., C, R., Ram, H.V., K.P, S., Solomon, J.: Non-destructive testing by liquid penetrant testing and ultrasonic testing—a review. Int. J. Adv. Res. Ideas Innov. Technol. **5**, 694–697 (2019)
21. Fernando, D.S.K.L., Perera, M.W.S.: Effect of application process and physical properties of penetrant material to the sensitivity of liquid penetrant inspection. Proc. Int. Conf. Appl. Pure Sci. (ICAPS 2021-Kelaniya) (2021)
22. Delenkovsky, N.V., Gnusin, A.B.: Estimating the depth of surface flaws by penetrant testing. Russ. J. Nondestruct. Test. **53**, 231–235 (2017). https://doi.org/10.1134/S1061830917030044
23. ASME: ASME BPVC Section V—2019, 95–118, 216–222 (2019)
24. ASME: ASME BPVC Section I—2019, 129–131 (2019)
25. Handbook, C.A.I.: ASM metals handbook, Volume 9 metallography and microstrustures handbook. ASM Int. **9**, 608–626 (2004)
26. Koo, S.M., Kim, K.R., Yang, Y.S., Kim, H.S., Hwang, D.Y.: Effect of degassing parameters on the microstructure and properties of nanocrystalline magnesium alloys in spark plasma sintering (2016)
27. Uludağ, M., Çetin, R., Dispinar, D., Tiryakioğlu, M.: The effects of degassing, grain refinement & Sr-addition on melt quality-hot tear sensitivity relationships in cast A380 aluminum alloy (2018)

Enhancing Electrical Efficiency of Photovoltaic Thermal Collectors Through Nanofluids

Wan Mustafa Wan Sulong, Jasrina Jaafar, and Khairul Hamzani Hamid

Abstract The study aimed to examine the electrical efficiency of a rectangular spiral flow heat collector in an indoor setting, comparing water and silica nanofluids as cooling mediums. The experiment involved using silica nanofluids and water at ratios of 0.5 and 1.5 vol.% with varying coolant velocities and solar radiations. The findings revealed that silica/water nanofluids with high fluid concentration ratios and temperatures had higher thermal conductivity values. Additionally, using a 1.5 vol.% silica/water nanofluid as a cooling medium improved the electrical efficiency of the photovoltaic thermal collector more than water or 0.5 vol.% silica/water nanofluids. The highest electrical efficiency of 9.47% was achieved using 1.5 vol.% silica/water at a highest flow rate and 650 W/m^2 of solar radiation.

Keywords Silica/water nanofluids · Photovoltaic thermal system · Electrical efficiency · Rectangular spiral flow collector

1 Introduction

Researchers have investigated the potential of nanofluids to improve solar energy systems, particularly in solar collectors, due to their efficient heat transfer properties [1, 2]. Irem and Tayfun's [3], study found that nanofluids can enhance thermal conductivity. The utilization of nanofluids can augment the properties of a system such as transmittance, absorption and efficiency [4, 5]. In 2015, Hashim et al. [5], performed a study on a photovoltaic thermal collector employing Al_2O_3/water

W. M. W. Sulong (✉) · J. Jaafar · K. H. Hamid
Mechatronics Engineering Technology Department, Universiti Kuala Lumpur Malaysia France Institut, 43600 Bandar Baru Bangi, Selangor, Malaysia
e-mail: wanmustafa@unikl.edu.my

J. Jaafar
e-mail: jasrina@unikl.edu.my

K. H. Hamid
e-mail: khairulhamzani@unikl.edu.my

A. Ismail et al. (eds.), *Engineering Frontiers*, Advanced Structured Materials 208,
https://doi.org/10.1007/978-3-031-56844-2_14

nanofluid as the operational medium. They discovered that a ratio of 0.3 led to a 12.1% increase in solar panel efficiency due to a significant temperature decrease of up to 42.2 °C. However, increasing the concentration ratio beyond 0.3 caused the PV system's temperature to rise to 52.2 °C, resulting in an 11.3% decrease in efficiency. Sardarbadi et al. [6], conducted an experiment on a photovoltaic thermal unit, they employed silica/water nanofluids as the operational fluid to enhance the system's electrical and thermal effectiveness. Conducting trials with nanofluids containing 3% and 1% weight concentrations revealed a respective enhancement of 7.9% and 3.6% in the overall effectiveness of the PV/T collector. Employing nanofluids with weight concentrations of 3% and 1% resulted in a boost in thermal efficiency of 12.8% and 7.6%, respectively. When a collector was augmented with 3 wt.%, 1 wt.%, and water, the exergy enhanced by 24.3%, 22.6%, and 19.36%, respectively.

The study utilized a spiral flow collector and silica nanofluids with water at 0.5 and 1.5 vol.% concentrations with a 100 W monocrystalline PV module without an absorber plate, tested under solar simulator conditions. The aim was to determine the electrical output increase achieved by replacing water with nanofluids in a photovoltaic system.

2 Methodology

2.1 *Experimental Setup*

Employing a solar simulator, an experiment on a PV/T collector system was conducted indoors. The absorber panels were constructed from square stainless steel tubes that were coiled in a spiral shape. The setup comprised of surface water pumps, a flow rate meter, a hot water tank, and a tank for the cooling fluid. Thirteen K thermocouples were installed to record temperature data, with 10 mounted to the PV/T surface, three measuring the working fluid temperature. A solar module analyzer was utilized to monitor performance of the system, measuring the current, voltage and output power. Throughout the course of the experiments, each flow rate and radiation level was evaluated for a total duration of one hour. Data was meticulously collected at precisely-timed one-minute intervals during each trial, to ensure accuracy and consistency in the results obtained. This method facilitated a comprehensive and meticulous evaluation of the system's efficiency across a range of flow rates spanning from 1.8 to 6.0 L/min, while also considering different radiation levels.

2.2 Electrical Efficiency

Equation (1) was utilized to mathematically articulate the electrical efficacy of the photovoltaic thermal collector [7]. To obtain the necessary input values for the equation, the solar module analyzer was employed to accurately measure the maximum voltage (V_{max}), maximum current (I_{max}), and maximum electrical power (P_{max}) values. By utilizing this precise measurement equipment and mathematical formula, a comprehensive understanding of the system's electrical efficiency was obtained, facilitating the evaluation of the effectiveness and potential for improvement of the photovoltaic thermal system [8] (Fig. 1).

$$\eta_e = \frac{V_{max} \mathrm{I}_{max}}{A_{pv} G_t} \tag{1}$$

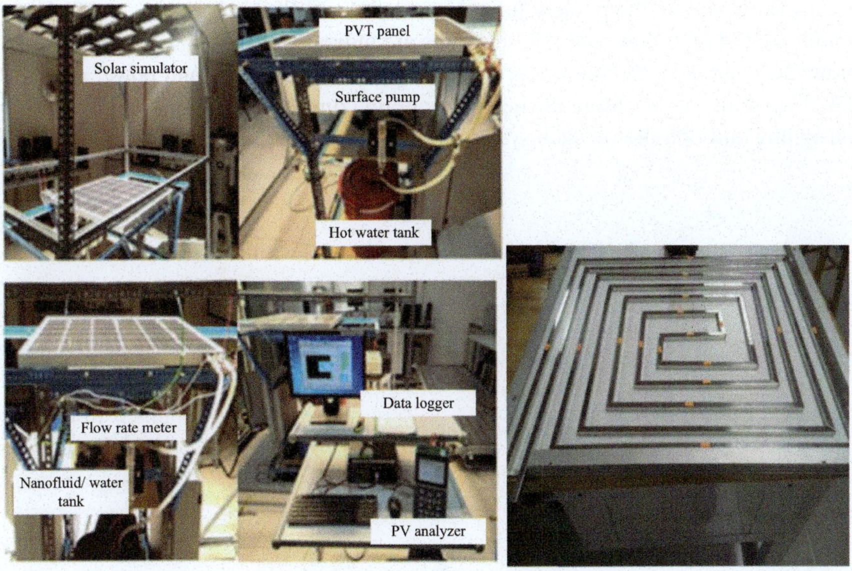

Fig. 1 Rectangular spiral collector and PV/T indoor experimental setup

3 Results and Discussion

3.1 Photovoltaic Thermal Electrical Efficiency

Figures 2, 3, and 4 provide information on the variations in electrical efficiency and maximum output power of PVT collectors, which were evaluated under specific conditions. The performance measurements were conducted using a solar simulator, with a fixed mass flow rate and radiation level, while the working fluid had concentrations of 0.5 vol.% and 1.5 vol.%, respectively. Utilizing a working fluid for heat dissipation from the photovoltaic thermal collector resulted in a decrease in its temperature, ultimately leading to an improvement in the system's electrical efficiency. Notably, the highest power output of 25.7–26.5 W at 450 W/m^2 solar radiation was obtained for flow rates ranging from low to high. At 650 W/m^2 radiation, the maximum power output increased from 36.6 to 38.7 W, signifying a 5.74% increase, with a similar rise observed at 850 W/m^2 radiation. In addition, the electrical performance achieved by the photovoltaic thermal water collector is 8.98% at 650 W/m^2 radiation and high flow rate. These results are indicative of the significant impact of flow rates and solar radiation levels on the electrical efficiency and power output of PV/T systems, highlighting the need for careful consideration of these factors when designing and optimizing such systems.

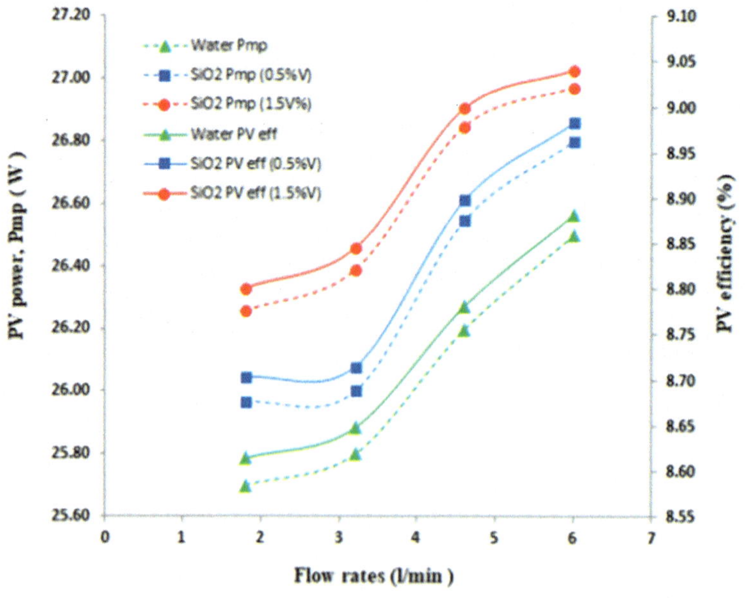

Fig. 2 PV efficiency and maximum power output at 450 W/m^2

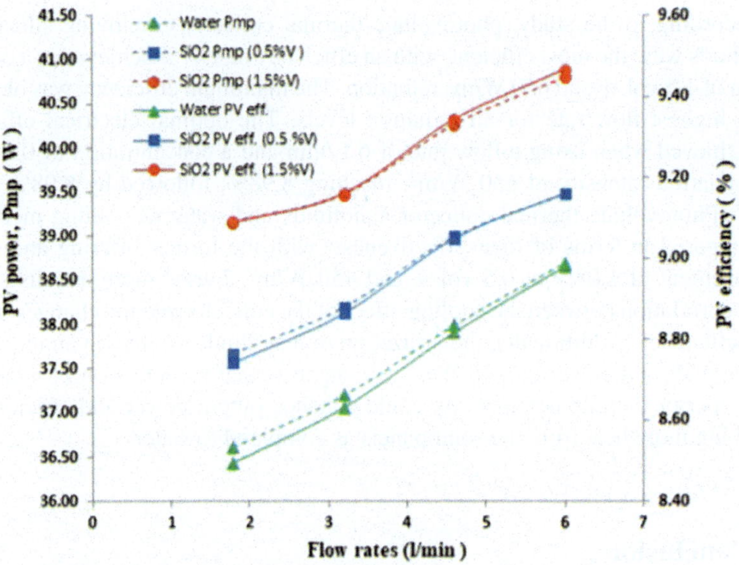

Fig. 3 PV efficiency and maximum power output at 650 W/m^2

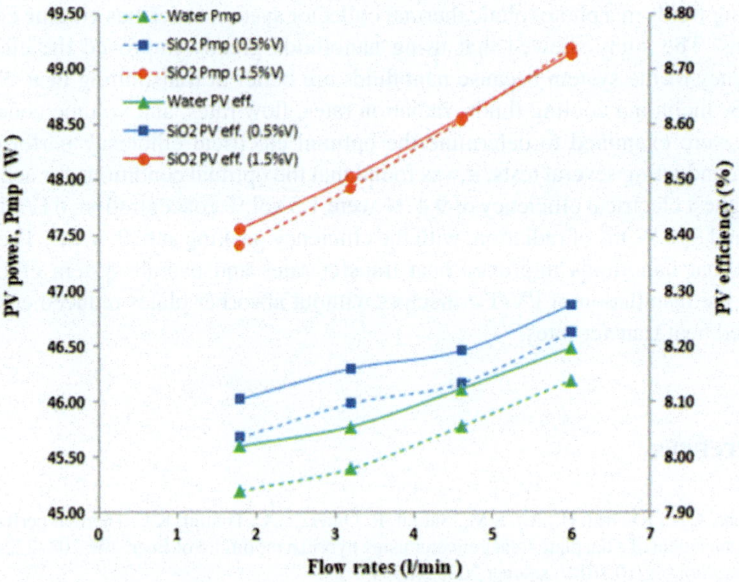

Fig. 4 PV efficiency and maximum power output at 850 W/m^2

According to the study, photovoltaic thermal collector employing silica/water nanofluids were the most efficient, with an efficiency of 9.47% achieved at a concentration of 1.5 vol.% and 650 W/m^2 radiation. The maximum efficiency was observed at the highest flow rate for all radiation levels. The optimal electrical efficiency was achieved when using a flow rate of 6 L/min and a concentration of 0.5 vol.% at a radiation intensity of 650 W/m^2, reaching 8.98%, followed by 8.98% at 450 W/m^2. Photovoltaic thermal collector nanofluids and water as cooling media can be compared in terms of their effectiveness, with the former offering an average improvement of 1.08% at 0.5 vol.% and 450 W/m^2. Furthermore, results show a direct correlation between the cooling effect of the collector and the changes in electrical efficiency. Additionally, the utilization of nanofluids resulted in an increase in voltage and total power output The study suggests that the adoption of nanofluids with superior thermal conductivity could enhance output power and efficiency by absorbing more heat from the solar panels as compared to water.

4 Conclusion

The study investigated the performance of water and silica/water nanofluids as working fluids in a photovoltaic thermal collector system at various volume concentrations. The study showed that using nanofluids greatly improved the electrical efficiency of the system because nanofluids are better at transmitting heat. Several factors, including cooling fluids, radiation rates, flow rates, and volume concentrations, were examined to determine the optimal electrical efficiency performance. After conducting several tests, it was found that the optimal conditions for achieving the highest electrical efficiency of 9.47% were 1.5 vol.% concentration, 6 L/min flow rate, and 850 W/m^2 of radiation, with the efficiency peaking at 650 W/m^2. The study found that nanofluids improved heat transfer rates and overall system efficiency, while the installation of PV/T collectors without absorber plates reduced costs and boosted heat transfer rates.

References

1. Omar, A.H., Khairul, H., Ali, S.M., Saidur, R. Omer, A.A., Thamir, K.I.: Thermal performance enhancement of a flat plate solar collector using hybrid nanofluid. Sol.Ener. **204**, 208–222 (2020). https://doi.org/10.1016/j.solener.2020.04.034
2. Sharma, K.V., Sarm, P.K., Azmi, W.H., Rizalman, M. Kadirgama, K. : Correla-tions to predict and forced convection heat transfer coefficients of water based nanofluids for turbulent flow in a tube. Int. J. Microscale Nanoscale Therm. Fluid.Trans. Phenom. **3**(4), 283–307 (2012)
3. Irem, K., Tayfun, M.: Numerical study of a photovoltaic thermal (PV/T) system using mono and hybrid nanofluid. Sol. Ener. **224**, 1260–1270 (2021). https://doi.org/10.1016/j.solener.2021.06.072

4. Samir, H., Robert, A.T., Saad, M., Saidur, R.: A cascade nanofluid-based PV/T system with optimized optical and thermal properties. Energy **112**, 963–975 (2016). https://doi.org/10.1016/j.energy.2016.06.142
5. Hashim, A.H., Ali, H.N., Abdulmunem, R.A.: Indoor investigation for improving the hybrid photovoltaic/thermal system performance using nanofluid (Al_2O_3-water). Eng. Tech. J. **33**(4), 889–901 (2015)
6. Mohammad, S., Mohammad, P.F., Saeed, Z.H.: Experimental investigation of the effects of silica/water nanofluid on PV/T (photovoltaic thermal units). Energy **66**, 264–272 (2014). https://doi.org/10.1016/j.energy.2014.01.102
7. Ali, N.A.S., Sopian, K., Sohif, M., Husam, A.H., Azher, M.A., Ruslan, M.H.: Experimental studies of rectangular tube absorber photovoltaic thermal collector with various types of nanofluids under the tropical climate conditions. Energy Convers. Manag. **124**, 528–542 (2016).https://doi.org/10.1016/j.enconman.2016.07.052
8. Ahmad, F., Kamaruzzaman, S., Mohammad, H.Y., Mohd, H.R., Adnan, I., Hussein, A.K.: Performance analysis of photovoltaic thermal (PV/T) water collectors. Energy Convers. Manag. **78**, 641–651 (2014). https://doi.org/10.1016/j.enconman.2013.11.017

Preliminary Study on Built-Up Back-To-Back Cold-Formed Steel Stub Column Filled with Washed Bottom Ash Concrete

Fadhluhartini Muftah, Mohd Syahrul Hisyam Mohd Sani, and N. Nindyawati

Abstract Cold-formed steel (CFS) is formed by using steel sheets over rolling, pressing, and bending processes at ambient temperature. Due to its benefits, CFS in a range of cross-sections and shapes is used as a structural and non-structural element. However, CFS used as columns and beams tends to fail due to local buckling, distortional buckling and global buckling when subjected to load. Besides, CFS needs to be modified to ensure that the section is stable and able to save from structure integrity issues. The study's main aim is to evaluate the mechanical behaviour of built-up back-to-back CFS columns and filled with special concrete. A built-up section of CFS is created by connecting two CFS channel sections in a back-to-back configuration and filling it with washed bottom ash (WBA) concrete. WBA is classified as a waste material collected from landfill areas at coal electrical power plants and had gone through some washing processes. Coal after being used is dumped in landfill areas and created a lot of environmental problems. WBA replaced the natural sand in the concrete with a full amount. The column is tested and compared with the built-up back-to-back CFS without the concrete and with normal concrete. From the compression strength result, the built-up back-to-back CFS column with WBA concrete decreased by around 25.53% when compared with the built-up back-to-back CFS column with normal concrete. Finally, the column is appropriate for the column with less load capacity that can be utilised in the construction or infrastructure.

Keywords Washed bottom ash concrete · Cold-formed steel · Stub column · Buckling · Built-up back-to-back section

F. Muftah · M. S. H. M. Sani (✉)
Civil Engineering Studies, College of Engineering, Universiti Teknologi MARA (UiTM) Cawangan Pahang, 26400 Bandar Jengka, Pahang, Malaysia
e-mail: msyahrul210@uitm.edu.my

F. Muftah
e-mail: fadhlu@uitm.edu.my

N. Nindyawati
Department of Civil Engineering, University of Malang, Malang, Indonesia
e-mail: nindyawati.ft@um.ac.id

A. Ismail et al. (eds.), *Engineering Frontiers*, Advanced Structured Materials 208,
https://doi.org/10.1007/978-3-031-56844-2_15

1 Introduction

Cold-formed steel (CFS) or known as a thin-walled structure produced by using equipment of rolling, pressing and bending to form a variety of shapes and dimensions is classified as a class 4 material. CFS has formed the steel sheet at ambient temperature with the particular process as mentioned before which is not similar to hot-rolled steel which is produced at high temperatures. CFS is more lightweight and thinner compared with hot-rolled steel. Because of their stability, durability, applicability, and sustainability, CFS is used in construction or building as structural or non-structural elements. Additionally, due to the popularity of the utilisation of the CFS as a structural element, there are a lot of studies for modifying the section or combining it with other materials to promote the composite structure. The modification of the section is recognised as a built-up section which is established from an opened-section to a closed-section or from an unsymmetrical section to a symmetrical section for proposing a stable condition and avoiding the initial failure. Meanwhile, the section with closed-section is able to fill with mortar or concrete and attached with timber board. Mohd Sani et al. [1] have reported the mechanical behaviour result of the built-up CFS section with an oriented strand board is appropriate for minimising the local buckling and distortional buckling. Mohd Sani et al. [2] analysed the flexural behaviour of built-up CFS sections filled with compacted soil determined by using three different bolt spacing configurations. There is a limited study that involved more than two materials for proposing the built-up composite section. For example, the built-up CFS column is produced by utilising two channels in face-to-face configuration, two batten plates situated on the top and bottom section, two GFRP sheets, two timber planks and fastened by using self-drilling screws [3].

Besides, when subjected to compression load or flexural load or both, CFS with a slender section and thin surface is typically vulnerable to failures like buckling, torsion and web crippling. There are five types of buckling modes which are seriously occurring; local, distortional, flexural, torsional and flexural–torsional buckling. Thus, the CFS must be modified or combined with other materials for solving this predicted problem and structural integrity. In order to create a solid, stable and strong structural element with lower production costs, the study of the built-up section is categorised as a common or popular research field. The simple of the built-up section production is a face-to-face configuration or known as the hollow section and a back-to-back configuration or known as an I-section. The built-up section is produced by using a variety of fasteners by researchers such as a self-drilling screw [4], bolt [5], welding [6], other innovation fastener and combination of the fastener [6]. Vy and Mahendran [7] have discussed the built-up back-to-back CFS slender column fastened by self-drilling screws of a diameter of 4.7 mm in two configurations and fabricated for a length of 1500 mm, 2800 mm and 3000 mm and designed by using the direct strength method. Other studies of the built-up section are combined with two or more marketable sections with a variety of configurations and shapes. The factors involved to improve the built-up CFS column capacity are fastener spacing, configuration composition and general effectiveness under axial, uni-axial and bi-axial action [8]. Craveiro et al. [9] have stated that the built-up CFS section helped to promote higher load-bearing capacity and stability which applies to several applications. They are used in more than one of the channel and u-shaped sections for producing built-up CFS sections and fastened in a predefined configuration. El Hady et al. [8] have established the built-up CFS section in an I-shape by using two channel sections and two tracks which are assembled by using grade 8.8 high-strength bolts with a 6 mm diameter. Sang et al. [10] have studied the structural behaviour of the built-up CFS section with a half-opened section fastened by self-drilling screws and subjected to axial compression. They produced the built-up CFS section with two channels and one u-shaped section.

Besides, the CFS in a built-up section is filled with normal concrete or normal mortar and is also classified as a popular research area for producing the heavy capacity of the structural element. Many studies are focussed on filling the normal mortar or concrete in the face-to-face section or box-up section. Rahnavard et al. [11] have explained that the built-up CFS section with concrete is broadly conducted either experimentally or analytically to comprehend the mechanical behaviour of the column and contribute to reducing premature failure, especially local buckling. Thus, the built-up section filled with concrete promotes the fire resistance capacity and higher strength because of the combination of the two materials. The built-up CFS section fastened using self-drilling screws and filled with lightweight concrete with a density of 1850 kg/m^3 is considered a composite structure combining two materials [11]. Zhu et al. [12] conducted the experimental series of built-up CFS stub columns with two differences of the thickness of 6 and 10 mm, and 600 mm of length, fastened by welding and lastly, filled with normal concrete or self-consolidating concrete. There is no information on the built-up back-to-back section with concrete.

Another secondary issue that should be solved for producing the low cost and strong structural element is by using the waste material as a partial or fully concrete ingredient material for instance sand or cement or gravel. Bottom ash is a waste material which is collected in a boiler or landfill dumping area at coal electric power plants and then through the process of treatment before being used as sand replacement in concrete. There are a lot of studies on the bottom ash as sand replacement in partial or fully percentage. Currently, the built-up CFS section is filled with special concrete to avoid environmental impact and CO_2 emissions and also to reduce production costs. Mohd Sani et al. [13] studied the built-up hollow CFS section filled with washed bottom ash concrete as a stub column and showed an improvement in compressive strength when compared with the built-up hollow CFS section filled with normal concrete. The primary aim of the paper is to calculate the mechanical behaviour, especially the compression strength and ultimate load of built-up back-to-back CFS stub columns subjected to axial load. From the testing, the observation of the failure mode is another important part of the study that should be discussed.

2 Specimen Preparation and Testing

The built-up CFS section is comprised by utilising individual shapes, especially the channel section, situated in a back-to-back configuration and fastened by utilising two self-drilling screws. The self-drilling screws are located at three positions especially at the top and bottom with 25 mm of length and at the middle with 100 mm of length. The built-up CFS section with two channels in a back-to-back configuration is presented in Fig. 1. The channel section has the following measurement; 100 mm × 50 mm × 12 mm and its thickness is 1.55 mm. It is decided to use the CFS section with a yield strength of 450 MPa and an elastic modulus of 200 MPa. Whilst as, the dimension of the hexagon head self-drilling screw with a yellow zinc-plated coating is 8 mm in diameter and 50 mm in length. The self-drilling screw is completed with stainless steel rubber washer and the total number of self-drilling screws is six units for every specimen.

From the coal electric power plant, bottom ash with a pH concentration of 9.76 is collected. The bottom ash is washed by using tap water which is fully submerged for three days and dried for two days at room temperature before being used and known as washed bottom ash (WBA). WBA is produced without any chemical pre-treatment or any process of grounding. WBA is used as a fine aggregate that is substituted in a full percentage of the sand. WBA is sieved to achieve the size which passes through 5 mm. All other concrete ingredients such as gravel 20 mm in size and composite cement are used. The colour of the natural sand is brown and WBA is grey as shown in Fig. 2.

The concrete is designed by using the basic proportion of 1:1:2 as referred to as the cement: sand: gravel ratio for predicting grade 25. The water-cement ratio of all mixtures is 0.5. There are five mixtures including a normal mixture, which is a cast with a variety of the percentage of WBA as a natural sand replacement as tabulated in Table 1. The average ultimate load and compressive strength of all mixtures at age of 7 and 28 days of curing are determined. The workability of all mixtures is

Fig. 1 The built-up CFS section in a back-to-back configuration

Fig. 2 The example of natural sand in brown colour and WBA in grey colour

discussed by referring to the value of the compacting factor, slump height and slump flow diameter and followed by BS 1881-103 and BS 1881-105.

The timber formwork is established for the easy process of filling the opened-section with concrete. The WBA concrete is filled to a proportion of 100% sand replacement into the built-up back-to-back CFS section and left for 28 days of curing. Then, the result is compared with an empty built-up CFS section and a built-up CFS section filled with normal concrete. The specimens with comprehensive labels of the built-up back-to-back CFS section which used for conducting the testing on the stub column are listed in Table 2. Figure 3a, b show the side view and top view of the built-up CFS section, respectively. The 100 mm × 100 mm × 250 mm of built-up CFS section is prepared for determining the average compressive strength and ultimate load by using a compression machine with 2000 kN of capacity and is then scrutinized for the failure mode.

Table 1 The concrete mixture with different percentages of the WBA and sand

Concrete mixture and labels	Percentage of natural sand (%)	Percentage of WBA (%)
Normal mixture	100	0
Mix 1	30	70
Mix 2	20	80
Mix 3	10	90
Mix 4	0	100

Table 2 The built-up back-to-back CFS stub columns with a specimen label

Specimen label	Description
CFSC	CFS individual channel stub column section
BBCFS0	Built-up back-to-back CFS stub column sections without normal or WBA concrete
BBCFS1	Built-up back-to-back CFS stub column sections with normal concrete
BBCFS2	Built-up back-to-back CFS stub column sections with WBA concrete

Fig. 3 The view of the built-up CFS section from **a** side and **b** top

3 Results and Discussion

The result and discussion of the study are divided into three parts including the workability properties of WBA concrete, compressive strength properties of WBA concrete and compressive strength properties of the built-up back-to-back CFS section filled with WBA concrete.

3.1 Workability Properties

The slump value of mixture 4 and mixture 5 is represented by the low value of slump when compared with the normal mixture. The slump value is decreased when the concrete mixture with WBA increased. Hasim et al. [14] have proven that the workability properties by using a slump test are decreased by increasing the amount of bottom ash. They have noted that the slump test value of 100% sand replacement is represented as the lowest when compared with other mixtures and reduced by about 61.82% when compared with the normal mixture. The compacting factor of the concrete mixtures is decreased by increasing the amount of WBA and the value of slump diameter in the slump flow test is also decreased when the amount of WBA increased. In conclusion, mixture 4 is similar to the workability properties value with mixture 3 but too far when compared with the normal mixture. This is because the WBA surface texture is containing a pore that could reduce the workability properties. For solving the problem, the superplasticiser is suggested to utilise for promoting good workability parallel with good strength.

3.2 Compressive Strength Properties of WBA Concrete

The weight of WBA concrete is 3.0 kg and normal concrete is 4.6 kg. The percentage difference in the weight between these two concretes is 35%. The average compressive strength of WBA concrete is shown in Fig. 4. The lowest value of compressive strength of WBA concrete is mixture 4 with 100% natural sand replacement. Hasim et al. [14] have stated that the full usage of WBA affected their compressive strength due to WBA with finer size and so on more water is required for improving the growth of strength. From the result, the percentage difference between 7 and 28 days is 27.57%, 22.87%, 18.55%, 18.65% and 18.92% for the normal mixture, Mix 1, Mix 2, Mix 3 and Mix 4, respectively. Furthermore, the percentage difference between the normal mixture with Mix 1 is 31.48%, Mix 2 is 34.97%, Mix 3 is 36.94% and Mix 4 is 47.71% for the age of 7 days. 35.66%, 42.18%, 43.85% and 53.30% are reported when distinguishing between the normal mixture with Mix 1, Mix 2, Mix 3 and Mix 4, respectively. Among the mixtures with WBA, Mix 2 has the highest compressive strength rating. Figure 5 shows the correlation between the compressive strength and

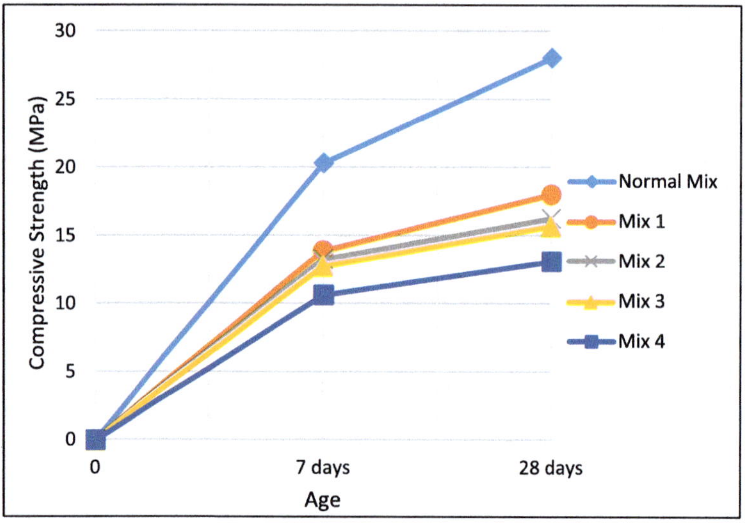

Fig. 4 The compressive strength versus age of the WBA concrete

the amount percentage of the WBA. The compressive strength of a normal mixture is decreased when the mixtures with WBA are increased for 7 days and 28 days. This is because the mixtures with WBA have less workability and high water absorption which affected the compressive strength. All mixtures are suitable for further activity because the compressive strength is more than 10 MPa and suitable to combine with CFS which is noted as having its own strength. Due to its affordability and cost-effectiveness, Mix 4 with 100% fully sand replacement in concrete is chosen to fill in the built-up back-to-back CFS stub column.

3.3 Compressive Strength Properties of Built-Up Back-To-Back CFS Section Filled with WBA Concrete

The complete result of the compressive strength, ultimate load, and weight of the specimens are tabulated in Table 3. The weight of the specimen is increased by adding the channel section for the built-up CFS section, self-drilling screws, WBA concrete and normal concrete. WBA concrete is considered as a lightweight concrete which is filled in built-up back-to-back CFS stub columns and recorded to have a percentage difference of 17.32% compared to normal concrete.

The ultimate load of CFSC is 100.6 kN and the built-up back-to-back section is 193.7 kN. Due to WBA concrete with high water absorption, BBCFS2 is reported to obtain a low value of ultimate load when compared with BBCFS1. The percentage difference between CFSC with BBCFS0 is 48.06% and BBCFS1 with BBCFS2 is 25.53%. In conclusion, the BBCFS2 is suitable for a structural element with low

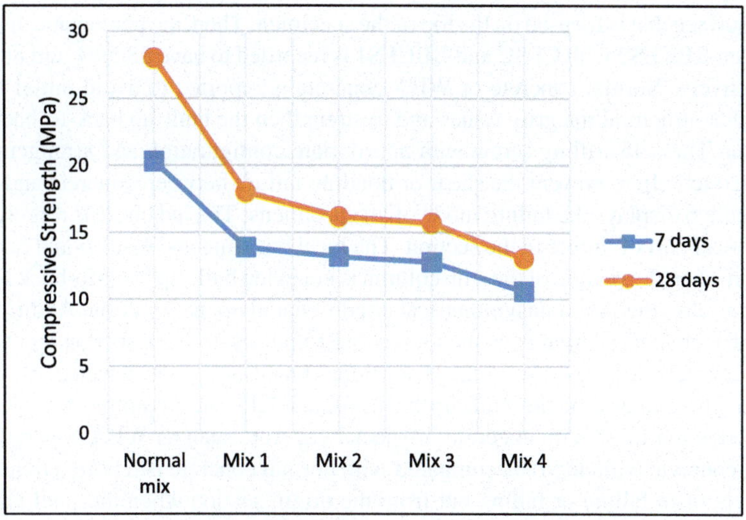

Fig. 5 The relationship of the compressive strength of normal and WBA concrete for 7 and 28 days

Table 3 The result of the channel and built-up back-to-back CFS stub column section

Specimen	Weight (kg)	Ultimate load (kN)	Compressive strength (MPa)	Failure mode
CFSC	0.66	100.6	305.78	Local and distortional buckling
BBCFS0	2.80	193.7	294.38	Local buckling
BBCFS1	6.35	468.4	46.84	Concrete cracking in the middle of the section and at the bottom end of the section is deformed out of the plane
BBCFS2	5.25	348.8	34.88	Concrete cracking in the middle of the section and at the end of the section is deformed out of the plane

capacity compared with the BBCFS1 for a high capacity of the structural element. Furthermore, the ultimate load of CFSC is improved after the CFS channel section is attached to another channel section and filled with WBA concrete with roughly 71.16%. The ultimate load of BBCFS0 is also improved by approximately 44.47% with adding the WBA concrete inside the hollow part. The compressive strength of all specimens is determined by dividing the ultimate load under axial load with a cross-section area. The cross-section between all specimens is not the same due to the solid surface that is focused on the top of the specimen. Then, the percentage disparity between BBCFS2 with CFSC and BBCFS1 is recorded to have 88.59% and 88.15%, respectively. Normal concrete or WBA concrete is reported to avoid initial failure and other structural integrity issues and to strengthen the built-up back-to-back stub column. The self-drilling screw with appropriate configuration and arrangement is reported to help to prevent the shear or bonding failure between concrete and steel.

Figure 6 displays the failure mode of all specimens. The red circle is representing the critical part of failure in the section. The local buckling and distortional buckling happen when the section of the stub column is not in the built-up back-to-back section which is classified as in unsymmetrical condition and opened-section. Additionally, when the channel section is formed into a built-up back-to-back section by utilising sufficient self-drilling screws, the distortional buckling is not detected. However, the buckling failure of the built-up back-to-back CFS stub column is solved and reduced when filled with concrete or special concrete such as WBA concrete. The WBA concrete with air voids compared with normal concrete is reported to help the concrete from failing or falling out from the origin section when the steel failed to buckle. The situation of the concrete failure is shown in Fig. 6. Mohd Sani et al. [15] have stated that the built-up CFS section filled with concrete provides with reasonable durability and strength, and not for steel or concrete sections.

4 Conclusion and Recommendation

From the results of the testing and observation, it is possible to make the following conclusion:

(1) WBA concrete has a lower workability than normal concrete, and as the amount of WBA is increased, the workability of WBA concrete also decreases. Due to the high water absorption and reduced workability, the compressive strength of WBA concrete decreases as the amount of WBA is raised. The lowest and highest value of compressive strength is mixture 4 with 100% natural sand replacement and mixture 1 with 70% natural sand replacement, respectively.
(2) The ultimate load of the built-up back-to-back CFS stub column filled with WBA concrete is 348.8 kN and reported to have a percentage difference of 25.53% when distinguishing with built-up back-to-back CFS stub column filled with normal concrete. Furthermore, the built-up CFS section with concrete is capable to reduce the structural integrity issue. For a structural element with low capacity in the building or infrastructure, the built-up CFS section with WBA concrete is classified as appropriate.

Recommendation for future work:

(1) Mechanical behaviour of short and slender column with a variety of height.
(2) Structural performance of the built-up back-to-back CFS section either column or beam under fire or post-fire conditions.

Fig. 6 The failure mode of **a** CFSC, **b** BBCFS0, **c** BBCFS1 and **d** BBCFS2

Acknowledgements The main author and co-author thank the MOHE - Malaysian Ministry of Higher Education for the support especially financial under the Fundamental Research Grant Scheme (FRGS) with the file number; FRGS/1/2021/TKO/UITM/02/101. Sincerest thankfulness is extended to Universiti Teknologi MARA (UiTM) Cawangan Pahang for laboratory equipment and facility.

References

1. Sani, M.S.H.M., Muftah, F., Osman, A.R.: A review and development of cold-formed steel channel columns with oriented strand board sections. Mater. Today Proc. **17**, 1078–1085 (2019). https://doi.org/10.1016/j.matpr.2019.06.519
2. Mohd Sani, M.S.H., Mohd Kamal, M.M., Muftah, F., Tan, C.S.: Flexural performance of built-up cold-formed steel beam filled with compacted soil. ARPN J. Eng. Appl. Sci. **11**(16), 9855–9862 (2016)
3. Adil Dar, M., Subramanian, N., Anbarasu, M., Ghowsi, A.F., Arif, P.A., Dar, A.R.: Testing and FE simulation of lightweight CFS composite built-up columns: axial strength and deformation behaviour. Thin Wall Struct. **167**, 108222 (2021). https://doi.org/10.1016/j.tws.2021.108222
4. Muftah, F., Sani, M.S.H.M., Mohammad, S.: Effects of cross-section type and degree of utilization on failure time and temperature of cold-formed steel column under ISO Fire. Metal **9**, 1–16 (2019). https://doi.org/10.3390/met9090964
5. Muftah, F., Mohd Sani, M.S.H., Mohd Kamal, M.M.: Flexural strength behaviour of bolted built-up cold-formed steel beam with outstand and extended stiffener. Int. J. Steel Struct. **19**(3), 719–732 (2019). https://doi.org/10.1007/s13296-018-0157-0
6. Mohd Sani, M.S.H., Muftah, F.: Experimental investigation of built-up cold-formed steel stub column with and without concrete-filled. AIP Conf. Proc. **2447**, 020001 (2021). https://doi.org/10.1063/5.0072641
7. Vy, S.T., Mahendran, M.: DSM design of fixed-ended slender built-up back-to-back cold-formed steel compression members. J. Constr. Steel Res. **189**, 107053 (2022). https://doi.org/10.1016/j.jcsr.2021.107053
8. El Hady, A.M., El Aghoury, M.A., Ibrahim, S.M., Amoush, E.A.: Experimental investigation of steel built-up beam-columns composed of tracks and channels cold-formed sections. J. Build. Eng. **51**, 104295 (2022). https://doi.org/10.1016/j.jobe.2022.104295
9. Craveiro, H.D., Rahnavard, R., Laim, L., Simoes, R.A., Santiago, A.: Buckling behaviour of closed built-up cold-formed steel columns under compression. Thin Wall Struct. **179**, 109493 (2022). https://doi.org/10.1016/j.tws.2022.109493
10. Sang, L., Zhou, T., Zhang, L., Chen, B., Wang, S.: Experimental investigation on the axial compression behavior of cold-formed steel triple-limbs built-up columns with half open section. Thin Wall Struct. **172**, 108913 (2022). https://doi.org/10.1016/j.tws.2022.108913
11. Rahnavard, R., Craveiro, H.D., Simoes, R.A., Laim, L., Santiago, A.: Buckling resistance of concrete-filled cold-formed steel (CF-CFS) built-up short columns under compression. Thin Wall Struct. **170**, 108638 (2022). https://doi.org/10.1016/j.tws.2021.108638
12. Zhu, A., Zhang, X., Zhu, H., Zhu, J., Lu, Y.: Experimental study of concrete filled cold-formed steel tubular stub columns. J. Constr. Steel Res. **134**, 17–27 (2017). https://doi.org/10.1016/j.jcsr.2017.03.003
13. Mohd Sani, M.S.H, Muftah, F., Mohsan, N.M., Kado, B.: Behavior of Built-up Cold-formed Steel Stub Columns Infilled with Washed Bottom Ash Concrete. Adv. Technol. Innov. **7**(2), 92–104 (2022). https://doi.org/10.46604/aiti.2022.8933

14. Hasim, A.M., Shahid, K.A., Ariffin, N.F., Nasrudin, N.N., Zaimi, M.N.S.: Properties of high volume coal bottom ash in concrete production. Mater. Today Proc. **48**, 1861–1867 (2022). https://doi.org/10.1016/j.matpr.2021.09.250
15. Sani, M.S.H.M., Muftah, F., Muda, M.F., Tan, C.S.: Resistance of built-up cold-formed steel channel columns filled with concrete. J. Teknologi. **78**(5–2), 99–104 (2016). https://doi.org/10.11113/jt.v78.8498

Discrete Phase Modelling Simulation Study of Airborne Droplet Nuclei in a Classroom

Zulkarnaini Abdullah, Zakaria Jusoh, Siti Muzahidah, Awang Idris, and Phuoc Ba Huynh

Abstract Droplet nuclei are known to be the vehicle of airborne respiratory illness. Water droplets produced when humans breathe, sneeze, cough, and even sweat naturally contain cells, electrolytes, and sometimes infectious bacteria and viruses. Droplets larger than 5 μm in diameter tend to fall to the ground by gravity, while the core of smaller diameter droplets remains suspended in the air and moves with the air current. Computational fluid dynamics software using the discrete phase modelling technique examines the transmission distance of droplet nuclei with a diameter of 5–6 μm. The place of investigation is a classroom of 9 m (L) × 3.5 m (H). It is found that with the lower airflow of 1 m/s, and at a distance of 2 m, the concentration of 2.181 kg/m^3 is recorded from the source to the student while at 2 m/s, the droplet nuclei concentrations are 1.77 kg/m^3 and scattered to the end of the classroom. The method of evacuating the air from within was simulated and found to be the preferable method for directing the air out, with concentrations of 0.029 kg/m^3 at a distance of 4 m.

Keywords Computational fluid dynamics · Discrete phase modelling · Droplets nuclei

Z. Abdullah (✉) · Z. Jusoh · S. Muzahidah · A. Idris
Universiti Kuala Lumpur Malaysia France Institute, Section 14 Jalan Teras Jernang, Bandar Baru Bangi, 43650 Selangor, Malaysia
e-mail: zulkarnaini@unikl.edu.my

Z. Jusoh
e-mail: zakariajusoh@unikl.edu.my

S. Muzahidah
e-mail: muzahidah@unikl.edu.my

A. Idris
e-mail: awang@unikl.edu.my

P. B. Huynh
Universiti of Technology Sydney, 15 Broadway, Ultimo, NSW 2007, Australia
e-mail: phuoc.huynh@uts.edu.au

A. Ismail et al. (eds.), *Engineering Frontiers*, Advanced Structured Materials 208,
https://doi.org/10.1007/978-3-031-56844-2_16

1 Introduction

Natural ventilation occurs due to the difference in air pressure and temperature between the inside and outside of the building. Low-energy buildings and clean environments require air exchange suitable for human thermal comfort. According to ASHRAE Standard 55, the human body heat balance model is affected by environmental factors such as temperature, heat radiation, humidity, wind speed, human activity, and clothing [1]. A great deal of attention has been applied to the design of natural ventilation, which replaces indoor air with fresh outdoor air without expending energy. As proposed by Refs. [2–5] natural ventilation is the best choice for providing comfort compared to heating, ventilation, and air conditioning. The movement of droplets in the air is highly dependent on gravity and other environmental factors such as airflow direction and strength, temperature, and relative humidity, which affect their size and mass due to evaporation. Naturally and artificially produced aerosols [6] contain droplets of various sizes [5, 7, 8] and their motion is influenced by gravity, direction, and local air currents.

Classroom schooling has been the centre of arguments between educational activities and the need to apply health operation standards. The ASHRAE guidance for building operations during the Covid-19 pandemic [9] stated that the basic principle of social distancing between students is 1–2 m and with several other hygienic strategies. The mode of the face-to-face class is that the students are seated in rows, desk to desk. At a full house, about 20 students accommodate the spaces. This paper reports on the ventilation flow of droplet nuclei in a classroom with an inlet and an outlet opening at the side doors of the building. A proposed air exhaust on the top opening of the classroom is simulated to compare the distance throw and the concentrations of the particles. The model used for this study is based on the location of a high ambient temperature all year through. The commercial CFD used in these investigations is from the ANSYS Fluent software package.

2 Modelling and Computation

A ventilation fan with a 1 m blade diameter is installed at the ceiling of the classroom. In comparison, without the fan, the natural ventilation is considered at 1 m/s [2, 3, 10, 11]. The movement of air is the main carrier of the droplet nuclei in the air. Entrainment of air into different airspaces occurs with human activities [12–14]. The student's position is taken as sitting for the simulation process as most of the time spent inside the room is in static motion. The distance between the teacher to the front student is 2 m. Figure 1 shows the modelling of the simulations.

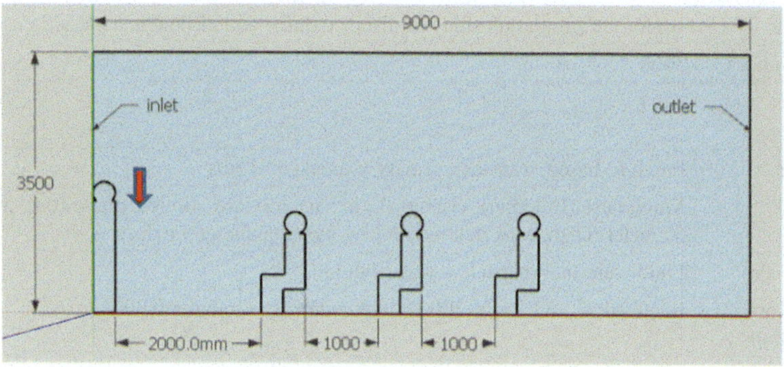

Fig. 1 The flow simulation with the students seated 1 m apart. The inlet air is at 1 and 2 m/s. The droplets injection point is indicated at the teacher standing 2 m from the first seated student

3 Computational Fluid Dynamics

In this study, ANSYS Workbench Fluid Flow (Fluent) CFD modelling is used to determine the flow pattern of droplet nuclei by the ventilation fan installed in a classroom. The discrete phase model is used to visualize the flow pattern of the droplet nuclei released by a teacher toward the students. The injected droplet's size is between 5 and 6 μm. A 1 m/s supply air simulation is taken as a reference for natural ventilation. The average velocity measured for the fan is 2 m/s, and as more than half of the time spent in the room is by sitting compared to other activities, the human model is taken as in a seated position. All fluid properties are constant and correspond to 300 K and standard pressure at 101.3 Pa, although the water-liquid droplet nuclei density is taken as 998 kg/m^3. In common notation, the reference values used are $\rho = 1.225$ kg/m^3, T = 300 K, $\upsilon = 1.7894 \times 10^{-5}$ m^2/s. Figure 1 shows the droplet nuclei source point and the distance between the participants.

4 Boundary Condition

Table 1 shows the simulation setup. Figure 2 shows the position of measurement from line 1 to line 4 where the expected airflow and the concentrations of the droplet nuclei reached the students the most.

Table 1 The set-up for the simulation study of distance throw and airflow in a classroom

Mesh nodes	3668
Mesh elements	3437
Solver	Pressure-based, transient, gravity *Y*-axis—9.81 m/s^2
Model	Multiphase (Eulerian), viscous (Laminar), mixture, discrete phase (on), min diameter 5 μm, Max diameter 6 μm, mean diameter 5 μm
Materials	Fluid—air, inert particle—water-liquid
Boundary conditions	Inlet—air, 1 and 2 m/s, 300 K, inlet—droplet, Granular 0.0001 m^2/s^2
Methods	Scheme—phase coupled simple, gradient—least squares cell-based, pressure—PRESTO!, momentum—first order upwind

The ANSYS Fluent CFD used a double-precision solver. The pressure method is set to PRESTO, and the momentum and the energy methods used are first-order upwind

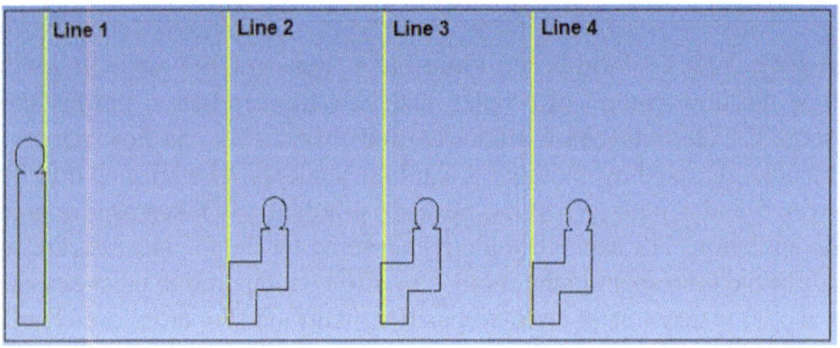

Fig. 2 The flow simulation in a streamline and the position of measurement lines 1–4. The most affected region is simulated and measured according to the line regions

5 Result and Discussion

A computational study has been performed for an air velocity of 1 and 2 m/s, through an inlet from the side of a classroom. As the simulation results show that the focus point is between line 1 and line 2 region, measurement is taken to show the difference in the flow step of the droplet and air velocity pattern. The distance from the teacher to the first seated student is 2 and 4 m for the third student.

5.1 Case 1: Airflow at 1 m/s

The simulation shows that the lesser size diameter of the droplet is freely flowing between students as the airflow is pushed by the ventilation fan. Simulation in Fig. 3 shows that the droplet of 6 μm travels up to more than 2 m making contact with

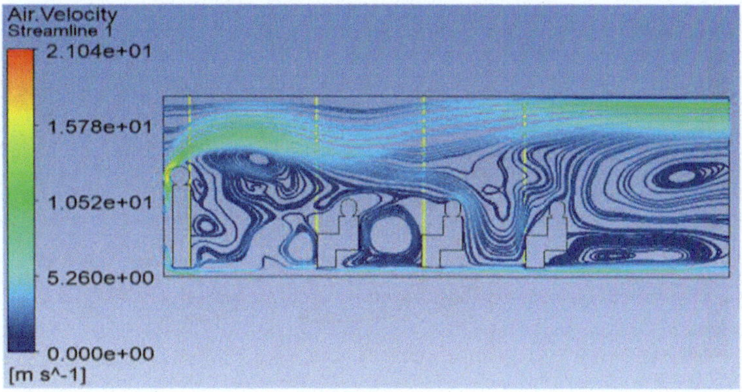

Fig. 3 The simulation results from an airflow rate of 2 m/s. The air velocity streamlines show the air patterns between line 1–line 4. The most affected flow is at the first 2 m and the end of the classroom

Table 2 The flow measurement results of case 1 at 1 m/s by measuring the regions from line 1 to line 4

Measurement case 1	Streamline	Line 1	Line 2	Line 3	Line 4
DPM concentration (kg/m^3)	0.497	*2.181*	0.321	0.373	0.087
Droplet velocity (m/s)	0.0388	0.130	−0.126	−0.109	−0.001
Air velocity (m/s)	1.497	1.437	1.911	1.899	1.249

student 1 and the droplet size of 5 μm continues travelling with the current flow up to 5–9 m. Table 2 shows the air and droplet properties comparison for 1 m/s of Case 1.

5.2 *Case 2: Airflow at 2 m/s*

The simulation in Fig. 4 shows that the droplet of 6 μm could travel up to more than 2 m making contact with student 1 and the droplet size of 5 μm continues travelling with the current flow up to 5–9 m. Table 3 shows the air and droplet properties comparison for 2 m/s of Case 2.

5.3 *Case 3: Exhaust Flow at 2 m/s*

To minimize the travel distances of the droplets, an air exhaust outlet had been proposed. Figure 4 proposes an exhaust outlet at the top of the room. Figure 5 shows

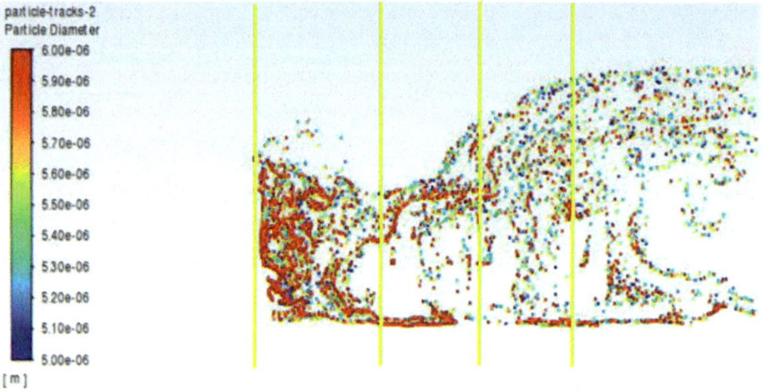

Fig. 4 The simulation result of injection droplets sized between 5–6 μm. The airflow rate is at 2 m/s. The most droplet concentration is at the first 2 m. The 5 μm droplet nuclei continue travelling to the end of the room

Table 3 The flow measurement results of case 2 at 2 m/s by measuring the regions from line 1 to line 4

Measurement case 2	Streamline	Line 1	Line 2	Line 3	Line 4
DPM concentration (kg/m^3)	0.082	*1.773*	0.044	0.085	0.092
Droplet velocity (m/s)	0.880	0.129	0.026	0.940	1.002
Air velocity (m/s)	2.377	1.945	2.195	2.252	2.217

the setup for the proposed exhaust airflow at the top of the classroom. Figures 6 and 7 show the simulation results. Table 4 shows the air and droplet properties comparison with 2 m/s of exhaust air for case 3.

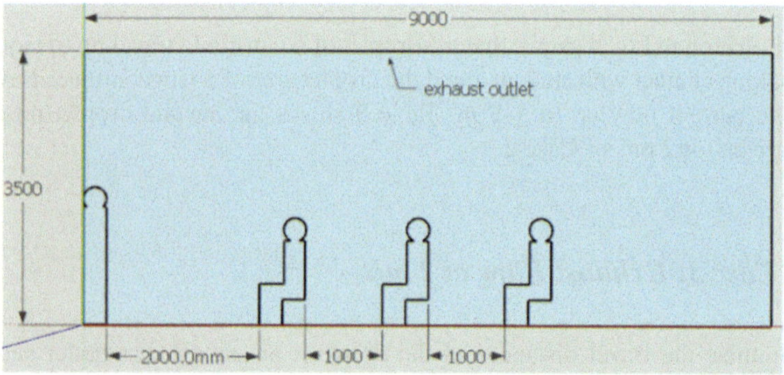

Fig. 5 The model of the flow simulation with the air is exhausted to the top outlet of the room

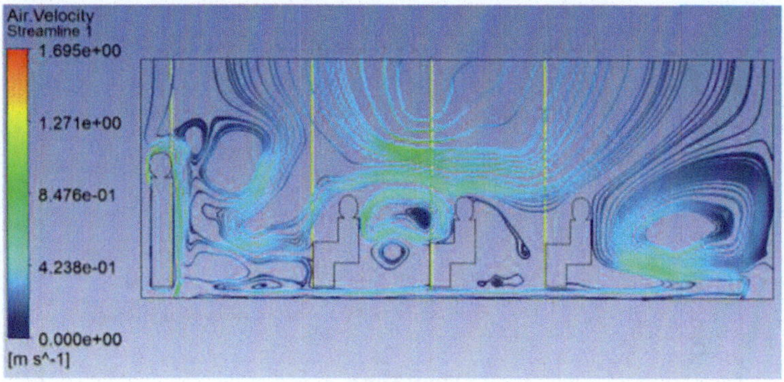

Fig. 6 The simulation results from an exhaust airflow rate of 2 m/s. The air velocity streamlines show the air patterns between line 1 to line 4. The most affected flow is at the first 2 m

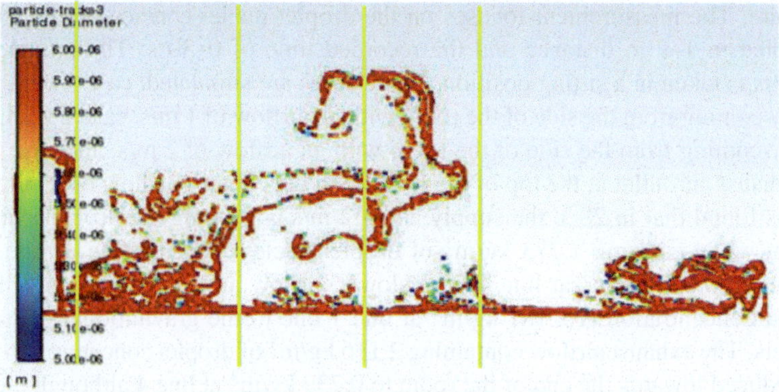

Fig. 7 The simulation result of injection droplets sized between 5–6 μm. The exhaust airflow rate is at 2 m/s. The droplet concentration is minimized by exhausting it to the top

Table 4 The flow measurement results of case 3 at 2 m/s of exhaust air by measuring the regions from line 1 to line 4

Measurement case 3	Streamline	Line 1	Line 2	Line 3	Line 4
DPM concentration (kg/m^3)	0.547	2.116	0.385	0.109	*0.029*
Droplet velocity (m/s)	−0.054	0.0136	0.029	0.093	0.0005
Air velocity (m/s)	0.270	0.105	0.197	0.449	0.239

6 Conclusion

A discrete phase model simulation study has been performed to measure the distance throw and the average length of droplet nuclei travel between participants in a classroom. An airflow of 1 and 2 m/s had been measured and simulated using CFD

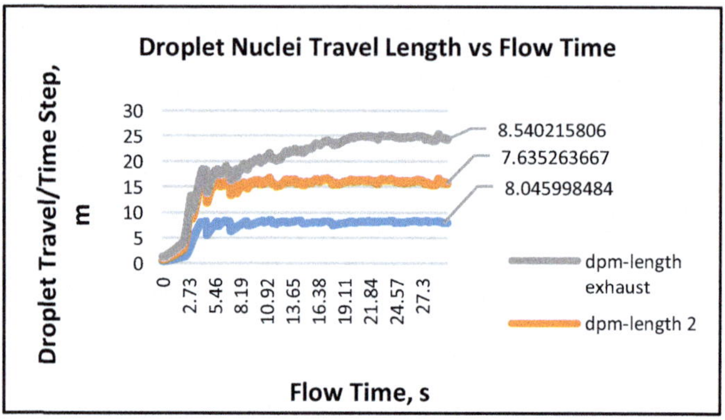

Fig. 8 The droplet nuclei travel length for 2 m/s supply air velocity is 7.6 and 8.5 m for exhaust

software. The measurement focuses on the droplet nuclei contacts to the students separated in 1–2 m distance and the recorded time of 0–30 s. The model of the students is taken in a sitting position. Three cases are simulated, case 1 with the air supply coming from the side of the room with an airflow of 1 m/s, case 2 with the air supply coming from the side of the room with an airflow of 2 m/s, and case 3 with the exhaust air outlet at the top of the room with the exhaust airflow of 2 m/s.

It is found that in 28 s, the supply air of 2 m/s is pushing the droplet nuclei up to 9 m while carrying 1.773 kg/m^3 of droplet nuclei concentration at line 1 and reduces to 0.029 kg/m^3 at line 4. The slower airflow at 1 m/s carries much more droplet concentration at 2.181 kg/m^3 at line 1 due to the gravitational pull on the droplets. The exhaust airflow containing 2.116 kg/m^3 of droplet concentration at line 1 is reduced towards the end of the room to 0.239 kg/m^3 at line 4 although there are still droplets moving around the 9 m room. Figure 8 shows the result of the studies. As shown in Fig. 9, the author concludes that an exhaust fan on top of the room is the best solution to reduce the droplet nuclei contacts between the airflow and the students.

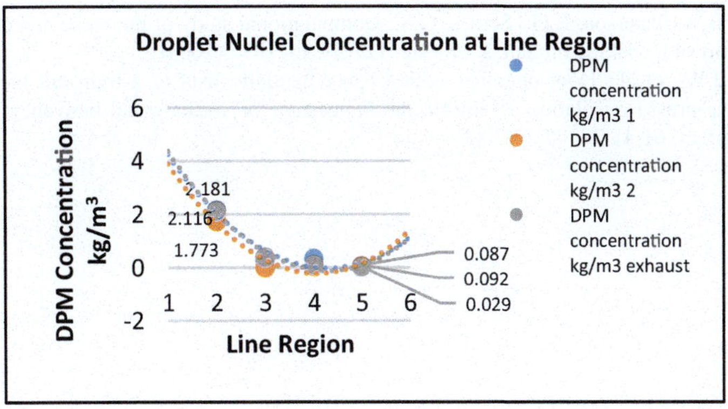

Fig. 9 The droplet nuclei travel length for 2 m/s supply air velocity and the polynomial forecast of the droplets, with the time flow increases

Acknowledgements The authors would like to acknowledge the material and financial support from the HVAC&R Department, UniKL Malaysia France Institute.

References

1. de Dear, R.J., Brager, G.S.: Thermal comfort in naturally ventilated buildings: revisions to ASHRAE Standard 55'. E. Build. **34**, 549–561 (2002)
2. Jamaluddin, A.A., Hussein, H., Mohd Ariffin, A.R., Keumala, N.: A study on different natural ventilation approaches at a residential college building with the internal courtyard arrangement. E. Build. **72**, 340–352 (2014)
3. Nicol, J.F., Humphreys, M.A.: Adaptive thermal comfort and sustainable thermal standards for buildings. E. Build. **34**, 563–572 (2002)
4. Su, X., Zhang, X., Gao, J.: Evaluation method of natural ventilation system based on thermal comfort in China. E. Build. **41**, 67–70 (2009)
5. Wong, N.H., Feriadi, H., Lim, P.Y., Tham, K.W., Sekhar, C., Cheong, K.W.: Thermal comfort evaluation of naturally ventilated public housing in Singapore. Build. Env. **37**, 1267–1277 (2002)
6. Wells, W.F.: Airborne Contagion, and Air Hygiene. Harvard University Press (1955)
7. Stetzenbach, L.D., Buttner, M.P., Cruz, P.: Detection and enumeration of airborne bio contaminants. Curr. Opin. Biotechnol. **15**(3), 170–174 (2004)
8. Tang, J.W., et al.: Factors involved in the aerosol transmission of infection and control of ventilation in healthcare premises. J. Hos. Infect. **64**(2), 100–114 (2006)
9. Lawrence, J.S.: Guidance for building operations during the Covid-19 pandemic (2020). https://www.ashrae.org/
10. Wong, K.C., Leung, K.S.: Transmission and prevention of occupational infections in orthopedic surgeons. J. Bone Joint Sur. Am. **86-A**(5), 1065–1076 (2004)
11. Priyadumkol, J., Kittichaikarn, C.: Application of the combined air-conditioning systems for energy conservation in data center. E. Build. **68**(A), 580–586 (2014)
12. Hayden, C.S., et al.: Air volume migration from negative pressure isolation rooms during entry/exit. Ap Occ. Env. Hygiene. **13**(7), 518–527 (1998)

13. Edge, B.A., Paterson, E.G., Settles, G.S.: Computational study of the wake and contaminant transport of a walking human. J. Fl. Eng. **127**(5), 967–977 (2005)
14. Tang, J.W., et al.: Door-opening motion can potentially lead to a transient breakdown in negative-pressure isolation condition: the importance of vorticity and buoyancy airflows. J. Hos. Infect. **61**(4), 283–286 (2005)

Algorithm for Optimization of Biomimick Vortex Generator—Airfoil Parameters

Mohamad Asmidzam Ahamat and Nur Faraihan Zulkefli

Abstract This paper proposed an algorithm or set of rules as guidance in the development of a biomimicry vortex generator for enhancement of airfoil performance. A biomimicry vortex generator is developed based on the features found in living organisms. An airfoil with higher performance increases the efficiency of engineering devices. For instance, the biomimicry vortex generator can be utilized in the production of a wind turbine with higher power coefficient or aircraft wing with higher lift-to-drag ratio. Thus, the objective of this paper is to develop an algorithm or set of rules for the development of a biomimicry vortex generator for airfoil. A complete cycle of development from the selection of organism, extraction of features from the selected organism, optimization of parameters for the combined vortex generator and airfoils and evaluation of the performance through simulation or experiment are included in the proposed algorithm. This algorithm is expected to contribute to a more systematic approach in development of engineering devices based on the biomimicry approach.

Keywords Algorithm · Vortex generator · Airfoil · Renewable energy · Aerospace

1 Introduction

This paper presents an algorithm or set of rules for guidance in the development of a nature inspired vortex generator for airfoil performance enhancement. Flow separation from the upper surface of an airfoil is not desirable since it causes significant reduction in lift force. One of the techniques to reduce the occurrences of flow separation from the airfoil is by adding vortex generators, a protruded structure on the upper surface of an airfoil. The function of a vortex generator is to energize the fluid

M. A. Ahamat (✉)
Universiti Kuala Lumpur Malaysia France Institute, Bangi, Malaysia
e-mail: asmidzam@unikl.edu.my

N. F. Zulkefli
Universiti Kuala Lumpur Malaysian Institute of Aviation Technology, Sepang, Malaysia
e-mail: nurfaraihan@unikl.edu.my

A. Ismail et al. (eds.), *Engineering Frontiers*, Advanced Structured Materials 208,
https://doi.org/10.1007/978-3-031-56844-2_17

within the boundary layer so that an inverse pressure gradient can be avoided. Many shapes of vortex generators have been reported in the literature. However, research related to nature inspired vortex generators are significantly less than the conventional shapes vortex generators. Thus, the objective of this paper is to propose an algorithm or set of rules for the development of biomimicry vortex generators.

Biomimicry is a technique of utilization of design or functionality from nature to create engineering devices. In aerospace applications, many engineering devices have been developed using the concept of biomimicry. A comprehensive review of biomimicry in aerospace applications has been recently reported [1]. The concept of biomimicry is not limited to aerospace engineering, this concept is also applicable to other sectors such as defense technology [2], construction industry [3] and power generation [4].

In engineering applications, airfoils are commonly used in aerospace and power generation. Vortex generators are usually installed on the upper surface of airfoils to reduce the risk of flow separation. There are many shapes of vortex generators that have been tested. Various types and shapes of vortex generators for wind turbines have been investigated [5]. The performance of the vortex generator-airfoil combination is determined by its shape and its location on the surface of the airfoil.

In term of biomimicry vortex generator, the shape of shark skin has been used. The shark skin vortex generator improves the lift coefficient of NACA 4415 airfoil [6]. Another different shape of vortex generator derived from the shark skin also increased the lift coefficient of NACA 4415 airfoil [7]. This shows one shape from nature can be adopted into various shapes in biomimicry applications. Thus, it is necessary to develop an algorithm or set of rules as guidance for the development of biomimicry inspired engineering devices.

Although many publications reported biomimicry in engineering applications, there is lack of publication discussing the method of adoption of the biomimicry approach in the development of engineering devices. The approach in developing an idea of biomimicry in the construction sector has been proposed [8]. Our recent publication proposed a new method in the biomimicry approach for developments in engineering inspired by marine animals [9]. Since the use of vortex generators is important for airfoils, an algorithm or set of rules for guidance in their development is essential.

In this paper, the algorithm or set of rules for the development of a biomimicry shape vortex generator is proposed. A complete cycle of development from the selection of organisms, selection and extraction of features, and performance evaluation through simulation or experiments in creating a satisfactory biomimicry shape has been proposed.

2 Selection of Organism and Extraction of Features

Before the shape from a living organism can be adopted to the vortex generator, it is required to select the suitable organism which has a similar principle of locomotion to the airfoil. Many organisms are immersed in fluid, whether in liquid or gas. During their motion, these organisms overcome resistances due to their interaction with the surrounding fluid. If this organism is living in water, its shape is usually optimized so that the higher resistance by water is compensated by a lower drag coefficient. The locomotion of fish such as its swimming speed can be found in Ref. [10].

For the same velocity and characteristic length, the Reynolds number of organisms living in water is approximately one order of magnitude higher than organisms that are living in air. This means that if a fish is swimming at 1 m/s, the equivalent velocity of a bird's flight is 10 m/s. By estimation of its Reynolds number, the type of flow around the organism could be determined. If it is comparable to the operating conditions of the vortex generator on the airfoil, the feature from that organism has a potential to be adapted.

A vortex generator is a protruded structure from the surface of the airfoil. To construct the analogy between organisms and vortex generator, the shape of any protrusion on organisms could be used as the shape of the vortex generator. For instance, the morphology of the surfaces of a fish like its fins, scale and any extended structures could be used to design the vortex generator.

Other than the surface morphology of the organism, other shapes such as fins of fish are also utilized in biomimicry for engineering purposes. Although the fins are mainly used for locomotion and stability control of fish, the shape the of shark caudal fin can be adopted to reduce wingtip vortices [11]. This means that there is no limit on the use of features from living organisms for engineering applications.

It is required to determine which feature needs to be extracted and used in the construction of the vortex generator. For illustration, Fig. 1 shows the shape of the dorsal fin of a dolphin. The shape of a vortex generator can replicate the whole shape of this dorsal fin, as shown in Fig. 1a. Another approach is to select a specific region from the fin such as in Fig. 1b or c. These three different shapes of vortex generators are expected to produce different effects on lift and drag coefficients of airfoils.

Another example is on the biomimicry of the caudal fin of a shark, shown in Fig. 2. In this example, instead of selecting a specific region from the caudal fin, the whole shape of fin can be rotated at 90° in anticlockwise direction, as in Fig. 2b. To obtain a flat surface at the interface between vortex generator and the surface of the airfoil, the caudal fin can be rotated by 135° in anticlockwise direction, as in Fig. 2c. To obtain another shape, Fig. 2c can be flipped vertically to create a mirror image, as in Fig. 2d.

Then, this example is further illustrated with the method to specify the dimensions of the vortex generator. The important parameters of vortex generators are height, length, and thickness [6]. It is required to define the length and height of the vortex generator. For vortex generators with conventional shapes, the measurement of height and length of the vortex generator is less complicated. A proposed method to define

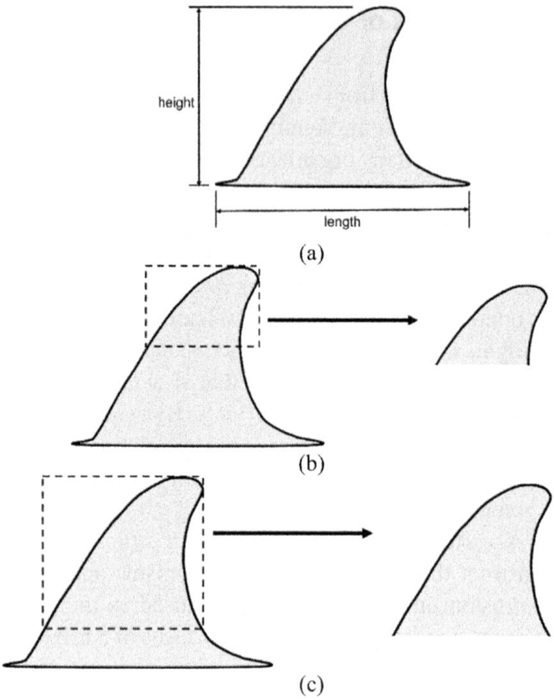

Fig. 1 **a** The whole shape of dorsal fin of dolphin. Possible extracted regions for vortex generator shapes are shown in (**b**) and (**c**)

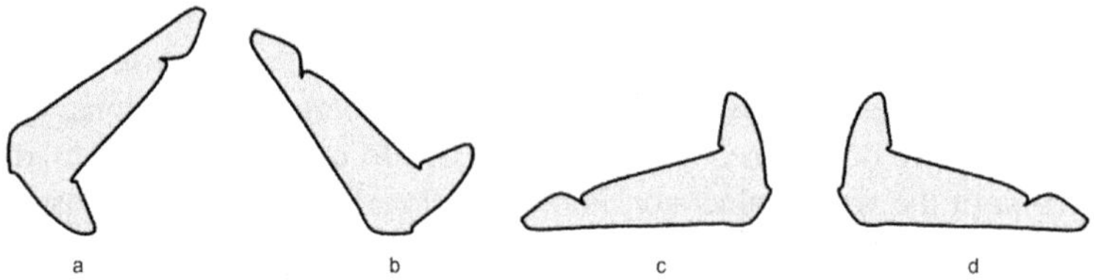

Fig. 2 **a** The shape of shark caudal fin at horizontal position, **b** the image **a** is rotated at 90° counterclockwise, **c** the image **a** is rotated at 135° counterclockwise and **d** is the mirror image of image **c**

the height and length of a vortex generator is shown in Fig. 1a. The thickness of the vortex generator can be defined at its maximum thickness.

3 Evaluation of Performance of Vortex Generator and Airfoil

The parameters of an airfoil can be defined as the chord length, thickness, maximum camber, and the position of the maximum camber. It is required to determine the location of the vortex generator on the upper surface of the airfoil. Two main references to pinpoint the position of the vortex generator on the surface of an airfoil are the locations along its chord and spanwise.

The vortex generators can have a single or paired configuration. For the single configuration, the angle of incidence with respect to the chord of airfoil needs to be specified. Then, the distance between two adjacent vortex generators is required. For a vortex generator with paired configuration, it can be set into counter rotating or co-rotating configurations.

The operating parameters for the vortex generator for airfoils concerned in this paper involve the Reynolds number and angles of attack. For a specified Reynolds number, the lift and drag coefficients of an airfoil with a vortex generator are usually measured or simulated at the range of angles of attack between θ_1 and θ_2. At different Reynolds numbers, the lift and drag coefficients curves may have different shapes. This justifies the usefulness of defining the operating conditions of an airfoil with a vortex generator at the beginning of design process.

4 The Proposed Algorithm

The proposed algorithm for the optimization of the biomimicry vortex generator is presented in Fig. 3. The first five steps are related to the selection and construction of the vortex generator using the biomimicry concept. The remaining steps are also applicable to the optimization of the non-biomimicry vortex generator. In this algorithm, the extraction of the feature from organism needs to be emphasized because it determines the effectiveness of the vortex generator.

There are two options for the loop of optimization. The first approach is to return to the extraction of features from the organism. This approach is applicable if the position of the vortex generator is well established. Otherwise, it is worth opting for the second approach where changes in the shape of the vortex generator is required. This approach is appropriate if the location of vortex generator on the upper surface of the airfoil is not yet known. If there is sufficient information available on the optimum location of the vortex generator for that airfoil, it is very likely that the shape of the vortex generator is not effective. Thus, it is more appropriate to review the extraction of the features from the organism.

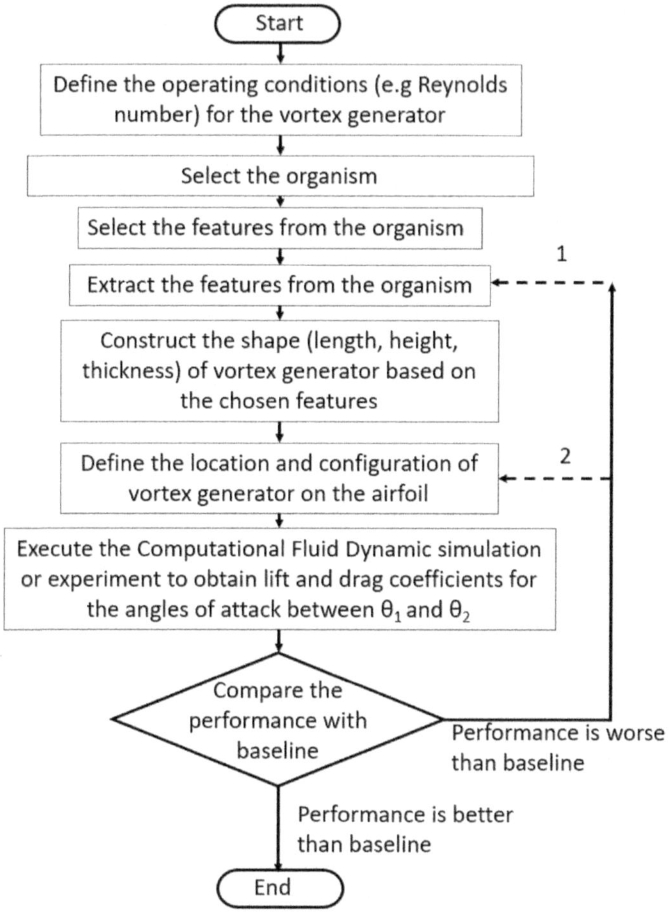

Fig. 3 The proposed algorithm for optimization of the biomimicry vortex generator for airfoils

5 Conclusion

In this paper, an algorithm or set of rules for optimization of the performance of a biomimicry vortex generator is proposed. As demonstrated in this paper, features from nature can be taken by cropping a section from the whole shape or changing the orientation of the shape to suit the requirement. The method to extract the features from organism requires further investigation to increase the probability of obtaining optimum parameters with less iterations in optimization loop. The proposed method in this paper is not only limited to the optimization of the vortex generator parameters for an airfoil, but also applicable to other applications such as the design of wingtip and any aerodynamics related devices.

Acknowledgements The authors would like to acknowledge the financial support from the Ministry of Higher Education (MOHE) through the Fundamental Research Grant Scheme (FRGS/1/2018/TK08/UNIKL/02/1).

References

1. Rose, J., Natarajan, S.G., Gopinathan, V.T.: Biomimetic flow control techniques for aerospace applications: a comprehensive review. Rev. Environ. Sci. Biotechnol. **20**(3), 645–677 (2021)
2. Wasilewski, T., Gębicki, J., Kamysz, W.: Bio-inspired approaches for explosives detection. Trends Anal. Chem. **142**, 116330 (2021)
3. Oguntona, O.A., Aigbavboa, C.O.: Biomimicry principles as evaluation criteria of sustainability in the construction industry. Energy Procedia. **142**, 2491–2497 (2017)
4. Carré, A., et al.: Extending the operating limits and performances of centimetre-scale wind turbines through biomimicry. Appl. Energy **326**, 119996 (2022)
5. Zhao, Z., et al.: Researches on vortex generators applied to wind turbines: a review. Ocean Eng. **253**, 111266 (2022)
6. Zulkefli N.F., et al.: Aerodynamic performance of shark skin shape vortex generator. In: Parvathy, R., et al. (eds.) Proceedings of International Conference of Aerospace and Mechanical Engineering 2019. Springer, Singapore (2020)
7. Arunvinthan, S., et al.: Aerodynamic characteristics of shark scale-based vortex generators upon symmetrical airfoil. Energies **14**(7), 1808 (2021)
8. Vanaga, R., Blumberga, A.: First steps to develop biomimicry ideas. Energy Procedia. **72**, 307–309 (2015)
9. Ahamat, M.A., et al.: Innovative approach for biomimicry of marine animals for development of engineering devices. In: Azman, I., Wardiah, M.D., Andreas, Ö. (eds.) Advanced Maritime Technologies and Applications. Advanced Structured Materials, pp. 301–310. Springer, Cham (2022)
10. Sfakiotakis, M., Lane, D.M., Davies, J.B.C.: Review of fish swimming modes for aquatic locomotion. IEEE J. Oceanic Eng. **24**(2), 237–252 (1999)
11. Zhongda, Y., Mingguang, L.: Maintenance differences between 737MAX and 320neo. In: 4th International Conference on Applied Materials and Manufacturing Technology. IOP Conference Series: Materials Science and Engineering, vol. 423, no. 1, p. 012059 (2018)

Comparison of Deposit Development on Heated Surface Using Neat Diesel Fuel and Blended Palm Oil Biodiesel (B10–B50)

Favian Jikol, Mohd Zaid Akop, Yusmady Mohamed Arifin, Mohd Azli Salim, and Safarudin Gazali Herawan

Abstract The purpose of this study is to investigate the deposition characteristics of different types of fuels by implementing the hot surface deposition test (HSDT) method as a replacement for the real engine deposition test. In this work, neat diesel fuel (DF) and blended palm oil biodiesel (B10–B50) were used to evaluate the deposit formation on the heated surface. Deposits formation relied on the type of fuels, surface temperature, and droplet interval (wet and dry conditions). In general, the results showed each fuel formed greater amount of deposit during wet condition test compared to that of dry condition test. For both wet and dry condition tests, DF produced the least amount of deposit followed by B10, B20, B30, B40, and B50, respectively.

Keywords Biodiesel · Deposit · Diesel · Dry/wet conditions · Hot surface

F. Jikol · M. Z. Akop (✉) · Y. M. Arifin · M. A. Salim
Fakulti Teknologi dan Kejuruteraan Mekanikal, Universiti Teknikal Malaysia Melaka, Hang Tuah Jaya, 76100 Durian Tunggal, Melaka, Malaysia
e-mail: zaid@utem.edu.my

Y. M. Arifin
e-mail: yusmady@utem.edu.my

M. A. Salim
e-mail: azli@utem.edu.my

S. G. Herawan
Industrial Engineering Department, Faculty of Engineering, Bina Nusantara University, West Jakarta, Indonesia
e-mail: safarudin.gazali@binus.edu

A. Ismail et al. (eds.), *Engineering Frontiers*, Advanced Structured Materials 208,
https://doi.org/10.1007/978-3-031-56844-2_18

1 Introduction

Given that it performs similarly to regular diesel fuel and does not require any engine modifications, biodiesel is frequently used in diesel engines, particularly in the transportation industry, and it is growing in popularity [1–4]. However, one of the biggest problems with utilizing biodiesel is the buildup of deposits inside the combustion chamber, which ultimately has an impact on the performance of the engine as a whole. Additionally, using biodiesel as diesel engine fuel on a continuous basis may cause serious engine issues and even engine failure [5–7]. Due to its incompatibility with various metallic and polymeric components, biodiesel has contributed to issues like metallic corrosion and is a serious concern that might shorten engine life [8].

As has been previously examined, biodiesel deposit formation in the combustion chamber is thought to be a complex phenomenon, and it is impossible to completely prevent deposits from forming inside the combustion chamber parts [9, 10]. Furthermore, long-term impacts of biodiesel on diesel engines are not widely known. Deposit building up on the diesel engine, especially near the combustion chamber, is one issue with using biodiesel. This is because of the dissimilarities between the characteristics of ordinary diesel fuel and biodiesel. These characteristics may have an impact on how deposits develop in the engine [11].

The wall temperature, the shape of the fuel injector nozzle, the use of lubricants, and the kind of fuel itself are just a few of the variables that may have a significant impact on deposit development inside the combustion chamber of real engines. Other than that, the properties of the fuel itself, especially for multi-component fuel such as biodiesel has a primary impact on the deposit development in the combustion chamber. Biodiesel has higher density, viscosity, carbon residue, flash point, and distillation temperature than neat diesel [12]. Compared to neat diesel, biodiesel possesses a higher boiling point components, thus the tendency of deposit formation is expected to be higher [13]. In addition, one of the elements that lead to fuel deposition on the wall surface is the liquid film formation [14] which could affect the evaporation behavior of the subsequent fuel droplet.

In this study, Malaysian palm oil biodiesel-diesel blend grades B10–B50 and neat DF were employed. For dry conditions and wet conditions, the droplet interval was set to 7 and 3 s, respectively. In dry conditions, the droplet interval of 7 s is longer than the fuel droplet lifespan. On the other hand, in wet conditions, the 3 s droplet interval is less than the fuel droplet lifespan. After every 1000 drops, pictures were taken and the mass of the collected deposits was reported. However, in this paper, the deposits formation was compared at droplet interval of $N_D = 5000$ and $N_D = 16{,}000$ to observe the differences at the early stage and end stage of the experiment.

Table 1 Fuel properties of neat diesel (DF) and Malaysian palm oil biodiesel-diesel blend (B10–B50) [25]

Properties	Fuel					
	DF	B10	B20	B30	B40	B50
Density (kg/m^3)	847	850	853	857	860	863
Kinematic viscosity (mm^2/s)	3.8	3.86	3.91	3.95	3.97	4.0
Heating value MJ/kg)	45.21	44.23	44.12	43.13	42.95	42.74
Acid value (mg KOH/g)	0.16	0.18	0.22	0.26	0.3	0.33

2 Methodology

The hot surface deposition test (HSDT) method was the one used in this investigation. Investigating the deposit development inside an engine's combustion chamber using this method is a simplified version of a genuine engine test. Additionally, the experimental HSDT concept was used in earlier research on deposit development and fuel evaporation characteristics on a heated surface [15–18].

In this work, neat diesel fuel (DF) and the Malaysian palm oil biodiesel-diesel fuel blends (B10–B50) were used to investigate the deposit formation on the heated aluminum alloy plate surface. Aluminum alloy was used to represent the piston, as most internal combustion (IC) engine pistons are made of aluminum alloy [19–22]. The fuel properties are presented in Table 1 while the schematic diagram of the HSDT experimental setup is illustrated in Fig. 1. Two conditions were tested in this study, which were the dry and wet conditions. For dry conditions and wet conditions, the droplet interval was set to 7 and 3 s, respectively. In dry conditions, the droplet interval of 7 s is longer than the fuel droplet lifetime. However, in wet conditions, the 3 s droplet interval is less than the fuel droplet lifetime. These droplet interval values were set based on the droplet lifetime obtained for each fuel in the evaporation test that was done before conducting the deposition test. The heated plate temperature was 315 °C and 340 °C for DF and B10–B50, respectively. Mass of the accumulated deposits were recorded and photographs were taken after every 1000 droplets.

3 Results and Discussion

The mass of accumulated deposits for the tested fuels for both wet and dry conditions at droplet interval N_D = 5000 and N_D = 16,000 are shown in Fig. 2 while the photographs of accumulated deposits were compared as presented in Table 2. In general, the deposits produced on the heated plate during the wet condition tests were heavier in mass compared to that of dry condition tests for all tested fuels. The fuel with higher blend ratio also generated more deposits, except for B40 fuel. As can be seen in Fig. 2, deposits formed by the B40 fuel at droplet interval N_D = 5000 were 6.9 and 25.1 mg for dry condition test and 17.4 and 62.4 mg for wet condition

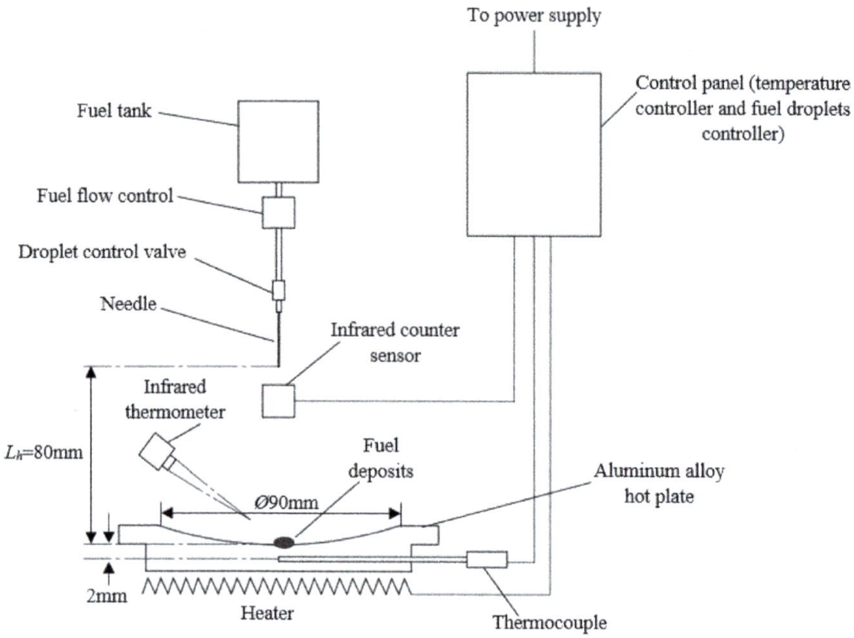

Fig. 1 Schematic diagram of HSDT experimental setup

test at droplet N_D = 16,000. These values exceed to that produced by the B50 fuel for both dry (N_D = 5000: 6.5 mg; N_D = 16,000: 28.8 mg) and wet (N_D = 5000: 11.0 mg; N_D = 16,000: 58.2 mg) condition test.

Fig. 2 Mass of accumulated deposits at droplet interval $N_D = 5000$ and $N_D = 16{,}000$

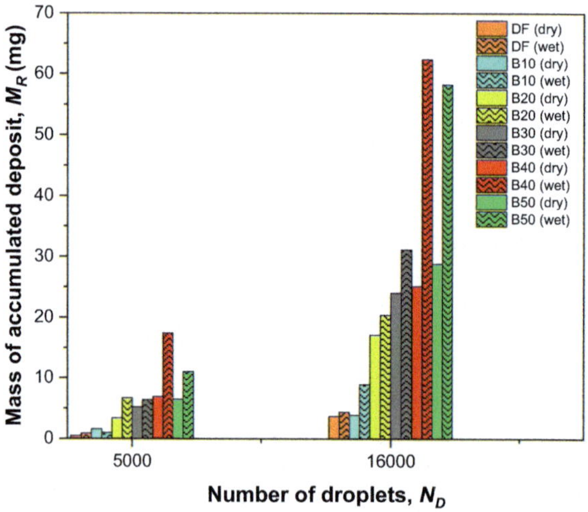

Table 2 Photographs of accumulated deposits at droplet interval $N_D = 5000$ and $N_D = 16{,}000$

Fuel	Condition	Number of droplets, N_D	
		5000	16,000
DF	Dry	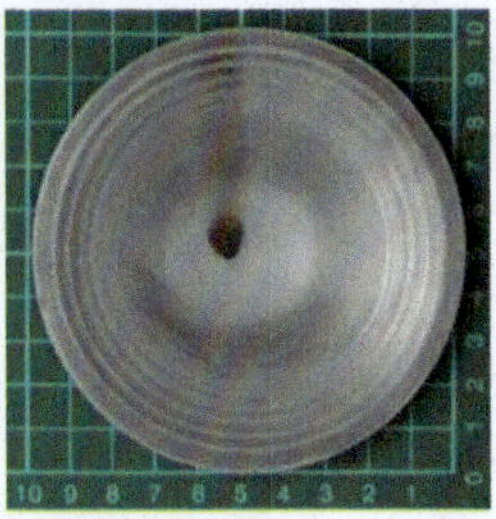 0.5 mg	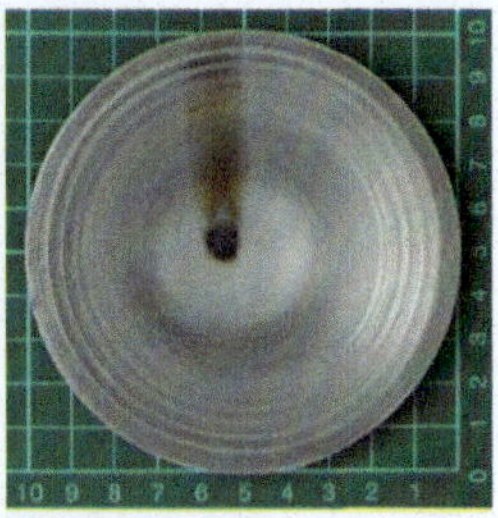 3.7 mg
	Wet	 0.9 mg	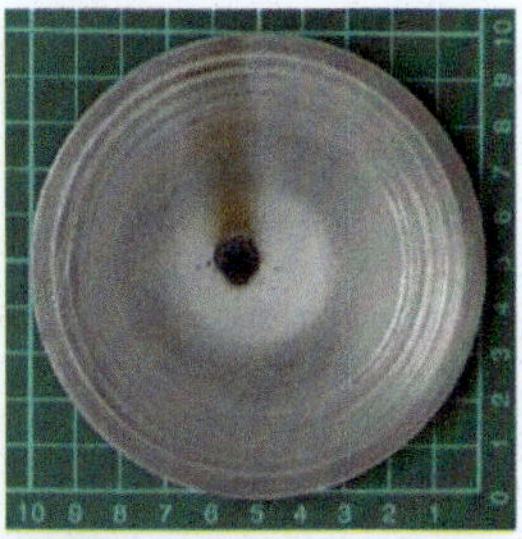 4.4 mg
B10	Dry	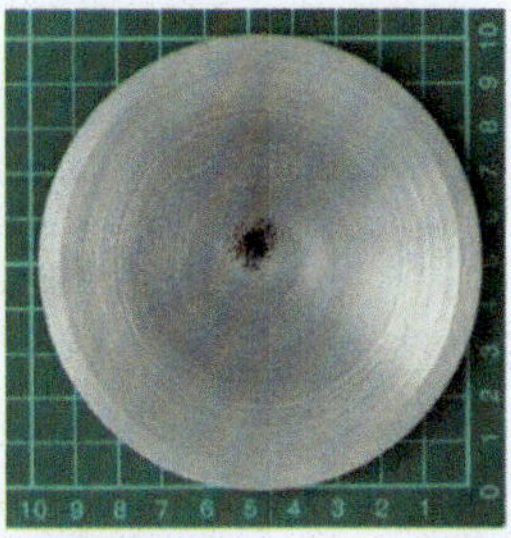 1.6 mg	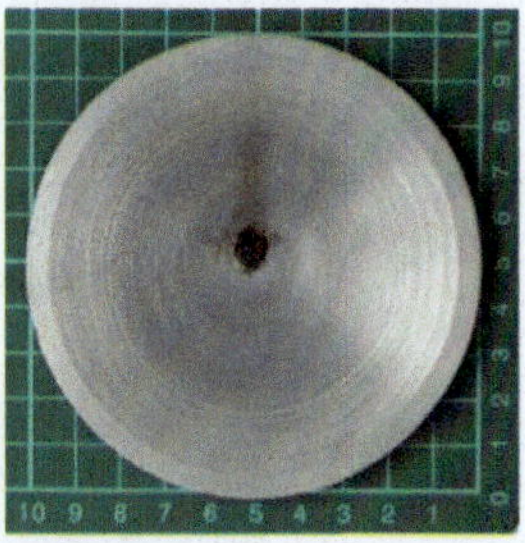 3.9 mg
	Wet	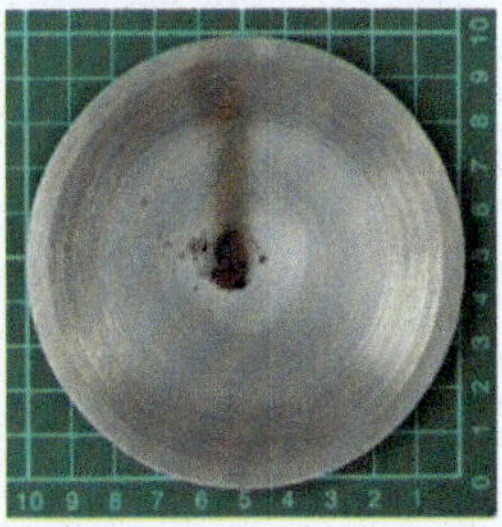 1.0 mg	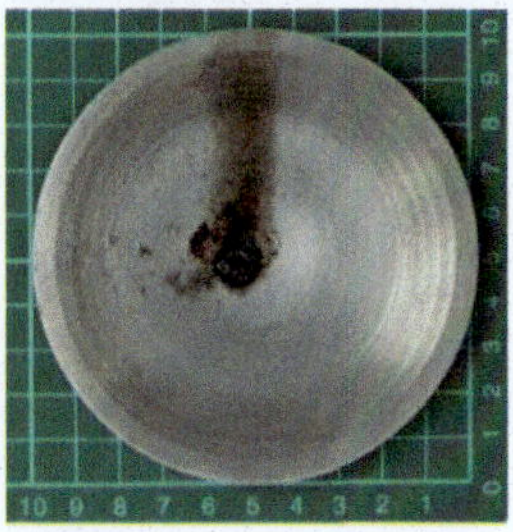 8.9 mg

(continued)

Table 2 (continued)

Fuel	Condition	Number of droplets, N_D	
		5000	16,000
B20	Dry	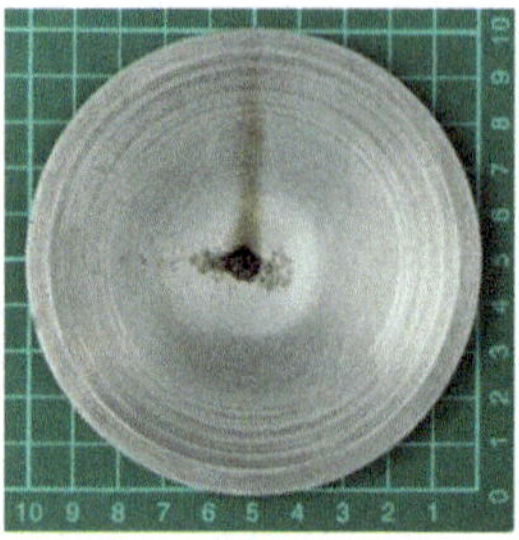 3.4 mg	 17.1 mg
	Wet	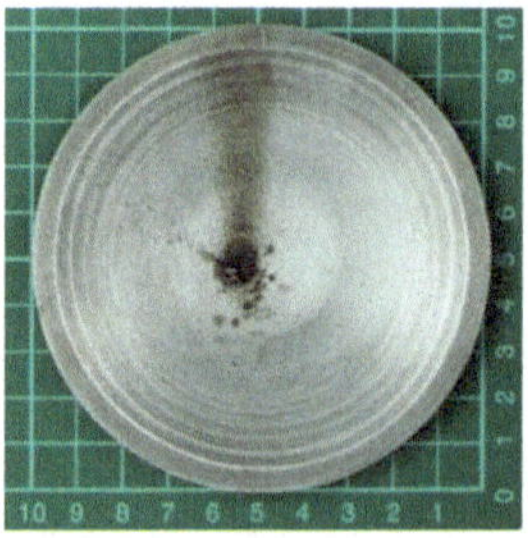 6.7 mg	 20.4 mg
B30	Dry	 5.2 mg	 24.0 mg

(continued)

Table 2 (continued)

Fuel	Condition	Number of droplets, N_D	
		5000	16,000
	Wet	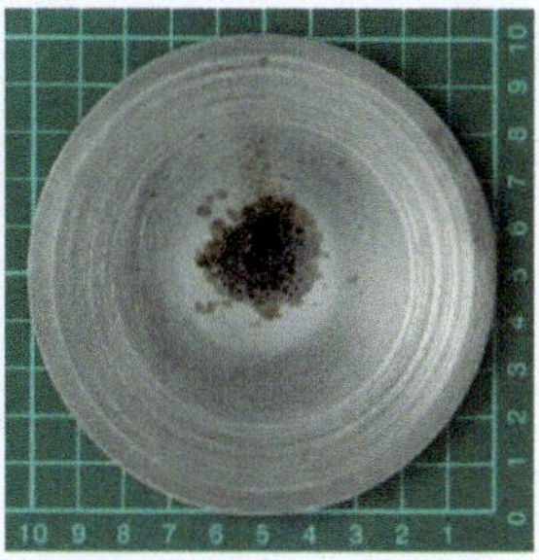 6.4 mg	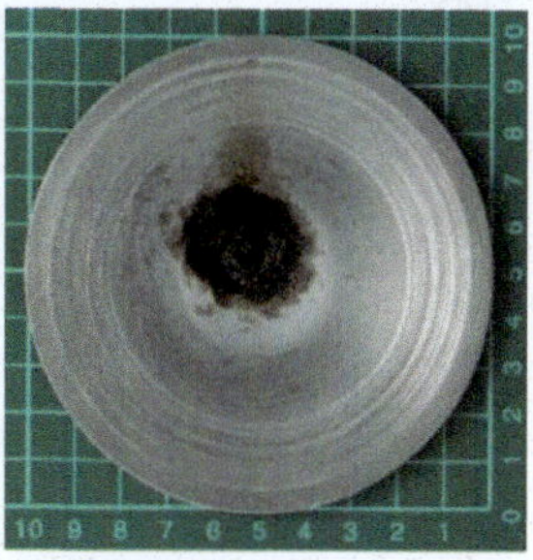 31.1 mg
B40	Dry	 6.9 mg	 25.1 mg
	Wet	 17.4 mg	 62.4 mg

(continued)

Table 2 (continued)

Fuel	Condition	Number of droplets, N_D	
		5000	16,000
B50	Dry	 6.5 mg	 28.8 mg
	Wet	 11.0 mg	 58.2 mg

The amount of deposit produced by the DF were the lowest among other tested fuels. During the dry condition test, DF only generated 0.5 and 3.7 mg of deposits at droplet interval $N_D = 5000$ and $N_D = 16{,}000$, respectively. Furthermore, there were only slight increases in mass of generated deposits during the DF wet condition test which were 0.9 and 4.4 mg at droplet interval $N_D = 5000$ and $N_D = 16{,}000$. This can be explained by the fact that the DF fuel used was neat diesel without any mixture of other fuel. For B10, B20, and B30 fuels, there were more significant increases in mass of deposits produced for the dry and wet condition test. At droplet interval $N_D = 5000$, the amount of accumulated deposit by B10 fuel was 1.6 mg during the dry condition test. However, during the wet condition test only 1.0 mg of deposits were formed on the heated plate. For droplet interval $N_D = 16{,}000$, B10 fuel produced 3.9 and 8.9 mg of deposits during the dry and wet condition test, respectively. For B20 fuel, the deposits produced were 3.4 mg ($N_D = 5000$) and 17.1 mg ($N_D = 16{,}000$) for dry condition test while for wet condition test, 6.7 mg ($N_D = 5000$) and 20.4 mg ($N_D = 16{,}000$) of deposits were accumulated. On the other hand, B30 fuel produced 5.2 mg ($N_D = 5000$) and 24.0 mg ($N_D = 16{,}000$) of deposits for dry condition test and 6.4 mg ($N_D = 5000$) and 31.1 mg ($N_D = 16{,}000$) of deposits for the wet condition test.

In terms of deposit appearances which can be observed in Table 2, it is obvious that the surface of the heated plated was a lot dirtier when the wet condition test was conducted compared to that of dry condition test. According to Rein [23], droplet's impact on a solid surface can cause it to stick to it, bounce off of it, spread out over the surface, or splash. Thus, it can be seen in Table 2 that fuels with higher content of palm oil biodiesel mixture produced more traces of splash deposits with larger radius. For DF, there was no obvious traces of splash deposit visible on the surface of the hot plate during the dry condition test for both droplet interval of $N_D = 5000$ and $N_D = 16{,}000$. In addition, the deposits produced were also thinner and easier to be cleaned off the hot plate. However, during the wet condition test, there were small amount but visible traces of splash deposit presented on the hot plate. This indicates that the wet condition has influenced the evaporation behavior of the impinged fuel droplets.

Visible traces of deposit splash also formed by the B10, B20, and B30 fuel for both dry and wet condition test. For each fuel, more deposits were scattered during the droplet interval $N_D = 16{,}000$. When the number of droplets increase, the evaporation lifetime of each subsequent droplets will be affected and this will alter the behavior of the droplet evaporation and magnitude of the droplet such as the bounce, splash, and rebound [24].

For B50 fuel, almost 30% of the hot plate surface was covered with traces of deposit splash during the wet condition test. This was more compared to the B40 fuel, even though B40 fuel formed more deposits in term of mass for both droplet interval of $N_D = 5000$ and $N_D = 16{,}000$. For B10, B20, and B30 fuel, the deposits were mostly scattered on the center region of the hot plate with moderate amount of traces of deposit splash compared to that of B40 and B50 fuel.

4 Conclusion

For all of the tested fuels, it was shown that the deposit development rate was faster under wet conditions than it was under dry conditions, ultimately leading to more deposits forming on the heated surface. Furthermore, the fuel with higher blend ratio produced more traces of deposit splash, which were resulted from the existence of wet condition when the number of droplets increase. This indicates that in the actual engine operation, deposits from B50 and B40 fuel might be scattered on larger surface area in the combustion chamber. For the lower blended fuel B10, B20, and B30, the deposits produced were moderate in term of traces of deposit splash. However, for an extended experiment with higher droplet number, the surface of the heated plate will be covered with more scattered deposits.

Acknowledgements This work was supported by Fundamental Research Grant Scheme (FRGS)—FRGS/1/2020/TK0/UTEM/02/20. The authors would like to thank the staff of the Faculty of Mechanical Engineering, Universiti Teknikal Malaysia Melaka for their support.

References

1. Agarwal, A.K., Bijwe, J., Das, L.M.: Effect of biodiesel utilization of wear of vital parts in compression ignition engine. J. Eng. Gas Turbines Power **125**, 604–611 (2003). https://doi.org/10.1115/1.1454114
2. Senthil, R., Silambarasan, R., Pranesh, G.: Exhaust emissions reduction from diesel engine using combined Annona-Eucalyptus oil blends and antioxidant additive. Heat and Mass Transfer/Waerme- und Stoffuebertragung. **53**, 1105–1112 (2017). https://doi.org/10.1007/s00231-016-1882-6
3. Amini, Z., Ong, H.C., Harrison, M.D., Kusumo, F., Mazaheri, H., Ilham, Z.: Biodiesel production by lipase-catalyzed transesterification of Ocimum basilicum L. (sweet basil) seed oil. Energy Conver. Manag. **132**, 82–90 (2017). https://doi.org/10.1016/j.enconman.2016.11.017
4. Tomić, M., Savin, L., Simikić, M., Kiss, F., Kešelj, K., Ivanišević, M., Ponjičan, O., Zoranović, M., Sedlar, A.: Effects of biodiesel on changes in IC engine performances: a long-term experiment with farm tractors. Fuel. **292** (2021). https://doi.org/10.1016/j.fuel.2021.120300
5. Cavalheiro, L.F., Misutsu, M.Y., Rial, R.C., Viana, L.H., Oliveira, L.C.S.: Characterization of residues and evaluation of the physico chemical properties of soybean biodiesel and biodiesel: diesel blends in different storage conditions. Renew. Energy **151**, 454–462 (2020). https://doi.org/10.1016/j.renene.2019.11.039
6. Hoang, A.T., Pham, V.V.: Impact of jatropha oil on engine performance, emission characteristics, deposit formation, and lubricating oil degradation. Combust. Sci. Technol. **191**, 504–519 (2019). https://doi.org/10.1080/00102202.2018.1504292
7. Sharma, A., Murugan, S.: Durability analysis of a single cylinder DI diesel engine operating with a non-petroleum fuel. Fuel **191**, 393–402 (2017). https://doi.org/10.1016/j.fuel.2016.11.086
8. Kugelmeier, C.L., Monteiro, M.R., da Silva, R., Kuri, S.E., Sordi, V.L., Della Rovere, C.A.: Corrosion behavior of carbon steel, stainless steel, aluminum and copper upon exposure to biodiesel blended with petrodiesel. Energy. **226** (2021). https://doi.org/10.1016/j.energy.2021.120344
9. Arifin, Y.M., Furuhata, T., Saito, M., Arai, M.: Diesel and bio-diesel fuel deposits on a hot surface. Fuel **87**, 1601–1609 (2008). https://doi.org/10.1016/j.fuel.2007.07.030
10. Feld, H., Oberender, N.: Characterization of damaging biodiesel deposits and biodiesel samples by infrared spectroscopy (ATR-FTIR) and mass spectrometry (TOF-SIMS). SAE Int. J. Fuels Lubr. **9**, 717–724 (2016). https://doi.org/10.4271/2016-01-9078
11. Hartono, Z.A., Cahyono, B.: Effect of using B30 palm oil biodiesel to deposit forming and wear metal of diesel engine components. Int. J. Marine Eng. Innov. Res. **5** (2020). ISSN: 2548–1479
12. El-Araby, R., Amin, A., El Morsi, A.K., El-Ibiari, N.N., El-Diwani, G.I.: Study on the characteristics of palm oil–biodiesel–diesel fuel blend. Egypt. J. Pet. **27**, 187–194 (2018). https://doi.org/10.1016/j.ejpe.2017.03.002
13. Lin, Y.S., Lin, H.P.: Spray characteristics of emulsified castor biodiesel on engine emissions and deposit formation. Renew. Energy **36**, 3507–3516 (2011). https://doi.org/10.1016/j.renene.2011.05.039
14. Furuhata, T., Ohmori, T., Arai, M.: Evaporation deposits of diesel and bio-diesel fuels on a hot surface. SAE Tech. Papers. 1–14 (2011). https://doi.org/10.4271/2011-01-1933
15. Pham, V.V.: Analyzing the effect of heated wall surface temperatures combustion chamber deposit formation. J. Mech. Eng. Res. Dev. **41**, 17–21 (2018). https://doi.org/10.26480/jmerd.04.2018.17.21
16. Suryantoro, M.T., Setiapraja, H., Yubaidah, S., Sugiarto, B., Mulyono, A.B., Attharik, M.I., Halomoan, T., Ariestiawan, M.R.: Effect of temperature to diesel (B0) and biodiesel (B100) fuel deposits forming. AIP Conf. Proc. **2062**, 1–6 (2019). https://doi.org/10.1063/1.5086591
17. Zare, M., Ghobadian, B., Hassan-beygi, S.R., Najafi, G.: Evaporation characteristics of diesel and biodiesel fuel droplets on hot surfaces. J. Renew. Energy Environ. **7**, 1–7 (2020)

18. Kompinsky, E., Dolan, G., Sher, E.: Experimental study on the dynamics of binary fuel droplet impacts on a heated surface. Chem. Eng. Sci. **98**, 186–194 (2013). https://doi.org/10.1016/j.ces.2013.04.047
19. Buyukkaya, E., Cerit, M.: Thermal analysis of a ceramic coating diesel engine piston using 3-D finite element method. Surf. Coat. Technol. **202**, 398–402 (2007). https://doi.org/10.1016/j.surfcoat.2007.06.006
20. Doetein, I., Muremi, L., Bokoro, P.: An analysis of geometry characteristics of metal oxide blocks on heat flow rate using finite element analysis. In: 2020 IEEE Electrical Insulation Conference, EIC 2020, pp. 163–166 (2020). https://doi.org/10.1109/EIC47619.2020.9158759
21. Ganapathy Pandian, S., Puzhuthivakkam Rengarajan, S., Babu, T.P., Natarajan, V., Kanagasabesan, H.: Thermal and structural analysis of functionally graded NiCrAlY/YSZ/Al2O3 coated piston. SAE Int. J. Mater. Manuf. **8**, 578–585 (2015). https://doi.org/10.4271/2015-01-9081
22. Srikanth, P.V., Rao, B.V.V.P., Laxmi, K.M., HariBabu, N.: Material coating optimization and thermal analysis of a four stroke C I engine piston. Int. J. Mech. Eng. Technol. **8**, 988–997 (2017)
23. Rein, M.: Introduction to drop-surface Interactions. Presented at the (2002)
24. Reis, M.W.F.: Study of boiling phenomenon using the droplet evaporation method (2019)
25. Ali, O.M., Mamat, R., Abdullah, N.R., Abdullah, A.A.: Analysis of blended fuel properties and engine performance with palm biodiesel-diesel blended fuel. Renewa. Energy **86**, 59–67 (2015). https://doi.org/10.1016/j.renene.2015.07.103

Methylene Blue in Coating Solutions for Steel Corrosion Protection

Wan Ahmad Heikal Azaha, Nurul Syazmeen Saiful Rijal, Mohd Al-Fatihhi Mohd Szali Januddi, Mahzan Johar, and Shahrulzaman Shaharuddin

Abstract Corrosion occurres in the plant industry and millions of dollars are spent for corrosion protection management per year. The project is aimed to evaluate the effect of Methylene Blue (MB) presence on coating solutions for corrosion inhibition. The dipping method was used to prepare the coated mild steel. The coated mild steel was examined with weight loss analysis, scanning electron microscope (SEM) observation and corrosion rate was calculated. The result showed a positive effect of the MB presence in the coating solution that reduced the weight loss and corrosion rate. SEM morphology supported the results with smoother surface of the MB coated steel was observed compared to other samples. These findings concluded the great potential of MB as healing agent in the corrosion protection coating solution. It could potentially improve the durability, reliability and safety of the steel structure, and a cost-effective approach in the corrosion management in industrial plants.Please check and confirm if the author names and initials are correct.Edited

Keywords Corrosion · Methylene blue · Coating · Inhibition · Healing agent

W. A. H. Azaha · N. S. Saiful Rijal · M. A.-F. Mohd Szali Januddi · M. Johar ·
S. Shaharuddin (✉)
Plant Engineering Technology Section, Advanced Facilities Engineering Technology Research Cluster, Universiti Kuala Lumpur Malaysian Institute of Industrial Technology, Jalan Persiaran Sinaran Ilmu, Bandar Seri Alam, 81750 Masai, Johor, Malaysia
e-mail: shahrulzaman@unikl.edu.my

W. A. H. Azaha
e-mail: heikal.azaha@s.unikl.edu.my

N. S. Saiful Rijal
e-mail: nurulsyazmeen@unikl.edu.my

M. A.-F. Mohd Szali Januddi
e-mail: mohdalfatihhi@unikl.edu.my

M. Johar
e-mail: mahzan@unikl.edu.my

A. Ismail et al. (eds.), *Engineering Frontiers*, Advanced Structured Materials 208,
https://doi.org/10.1007/978-3-031-56844-2_19

1 Introduction

Corrosion is a chemical reaction that causes metal materials to corrode due to the action of the surrounding environment and the entrance of corrosion ions [1]. Authors of Ref. [2] says that one of the identified primary causes for oil and gas infrastructure failures was corrosion. For instance, the transportation of crude oil and natural gas through pipelines from wellheads to processing plants. This continuous hazard is posed during installation and demolishment. According to Ref. [3], the total annual cost of corrosion in the oil and gas industry is estimated to be US$1372 billion, which includes $589 million in surface pipeline and facility costs, $463 million in downhole tube costs, and another $320 million in corrosion-related capital expenditures. The impacts of effective corrosion management shall significantly enhance and contribute to cost-effective planning, safety, health, and environmental requirement.

The catastrophic effects of the rusting process have become a global issue [4]. Corrosion has caused ten pipeline leaks and twelve explosions in Canada, and researching these phenomena and its management is significance for highly corrosive agent exposure in oil, gas, and petrochemical sectors [5]. According to accounts of corrosion-related failures, the cause of this phenomena is mostly due to catastrophic negligence in the manufacturing and installation of plumbing and equipment, which results in explosions, fires, and the spread of hazardous chemicals into the living environment. Aside from the costs of replacing corroded equipment, shutting down plants due to equipment replacement, process disruption due to equipment corrosion, impurity of processing products due to corrosion-related leakage, and waste of the products of those vessels attacked by corrosion are all significant costs and losses caused by corrosion.

A smart coating with self-healing property is one of current features that attracts attention of many researchers, since they can recover their protective characteristics after being harmed by the surrounding environment [6]. Self-healing coatings, which are inspired by biological processes, can repair physical damage and restore functionality with little or no intervention [7]. The use of nanotechnology in the creation of self-healing coatings is one area that has gotten a lot of interest. These stay dormant until the healing agent is released in response to injury caused by room temperature, producing a new polymer layer that will close the area for inhibitors. Factors such as pH changes or mechanical impact could initiate the self-healing process. For instance, a specific stimulus response would occur in the coating during micro or nano cracks that cause the release of a self-healing agent and recover the coating integrity [8]. The coating reactivity and longevity will be maintained with the direct addition of a self-healing agent. To our best knowledge, there is no study of methylene blue inclusion in anti-corrosion coating and potential as self-healing agent towards corrosion was reported. This study evaluated the corrosion rate and morphology of the encapsulated mild steel with methylene blue self-healing coating.

2 Methodology

2.1 *Preparation for Coating Solution*

The preparation started with filling 2 ml 5% ethyl methacrylate (EMA) into 100 ml distilled water. About 0.16 g of MB was added then. The mixture was stirred at 300 rpm stirring speed for 3 min. After that, 2.5 g urea and 0.25 g resorcinol were inserted, and the mixture was mixed for 3 min. About 6.35 g formaldehyde was poured into the solution and mixed for 3 min. The pH of the mixture solution was adjusted to 3.5. Then, about 2 drops of 1-octanol are filled to the solution and mixed.

2.2 *Preparation of Mild Steel*

The dimensions of the mild steel used were 20 × 20 × 2.0 mm. The mild steel was polished with sandpapers (P1000 and P2000), degreased with acetone and rinsed with distilled water.

2.3 *Coating Process*

The surface of the mild steel was dried for 3 min at 150°C. After that, the mild steel was coated through immersing the steel into a high gloss alkyd enamel paint for 7 days and dried for 15 min at 27 °C. Then, the steel was immersed into the coating solution for 1 week. Control samples were uncoated mild steel and coated mild steel without MB.

2.4 *Weight Loss and Corrosion Rate Determination*

The samples were weighed and recorded. Then, the samples were immersed into a 1.0 M NaOH solution for three distinct periods of time: 24, 72, and 120 h. In every period of time, the sample was withdrawn, dried and weighted. The analysis was performed in triplicate. The following formulae were used to calculate weight loss percentage and corrosion rate (mm/yr).

$$\text{Weight loss, W} = \frac{(\text{Initial weight} - \text{Final weight})}{(\text{Initial weight})} \quad (1)$$

$$\text{Corrosion rate(mm/yr)} = (534 \times \text{W})/\text{DAt} \quad (2)$$

where W is the weight loss, D is the density of the mild steel in g/cm^3, A is the area of mild steel plate in sq. inch and t is the exposure time of the mild steel plates in corroding media in hours.

2.5 Scanning Electron Microscope Analysis

The samples involved in this analysis were enamel paint mild steel (control), enamel paint mild steel with coating and enamel paint mild steel with methylene blue incorporated coating. The samples were fully immersed in 1 M sodium hydroxide for 120 h. Then, the surface morphology of the samples was analysed using SEM at magnification of 1000x without any further coating treatment.

2.6 Statistical Analysis

Data were analysed using IBM SPSS Statistics software (64-bit edition). The mean between treatment groups were compared using one-way-ANOVA, and the later the data were subjected to post-hoc Duncan Multiple Range Test to determine the significant values at $P < 0.05$.

3 Results and Discussion

3.1 Weight Loss and Corrosion Rate

There are three types of samples that were analysed in this analysis; uncoated mild steel, coated mild steel without MB and coated mild steel with MB. The weight loss and corrosion rate are indicated in Table 1. In every sampling point, the result shows clearly that the weight loss and corrosion rate were lowest for the sample of coated mild steel with MB. While the highest weight loss and corrosion rate were observed in the uncoated sample. During 120 h of exposure in 1 M NaOH solution, an increasing trend of weight loss in all samples was observed. The sample of coated mild steel with MB demonstrated the lowest weight loss (0.29 g $\pm$ 0.05) compared to other samples. The presence of MB was beneficial for corrosion protection of the steel. The potential inhibition and healing effects synergistically promoted the self-healing function of the coating [9]. It could improve durability and reliability of the steel structure subsequently.

Table 1 Weight loss and corrosion rate in NaOH exposure

Sample	Uncoated mild steel		Coated mild steel without MB		Coated mild steel with MB	
Immersion time (h)	Weight loss (g)	Corrosion rate (mm/yr)	Weight loss (g)	Corrosion rate (mm/yr)	Weight loss (g)	Corrosion rate (mm/yr)
24	0.31 ± 0.06^{b}	9.45	0.24 ± 0.03^{b}	7.14	0.12 ± 0.01^{a}	5.63
72	0.70 ± 0.03^{b}	4.12	0.32 ± 0.03^{a}	2.64	0.27 ± 0.05^{a}	3.38
120	0.98 ± 0.11^{c}	2.98	0.56 ± 0.06^{b}	2.03	0.29 ± 0.05^{a}	1.26

[abc] Different letters within a row indicate significantly different values ($p < 0.05$)

3.2 *Morphology Observation*

Figure 1 illustrates the morphology images of the sample surfaces after 120 h of NaOH exposure. For the uncoated sample, the image presented significant porosity and severe damage to the surface of the sample. This is due to a direct effect on the NaOH on the surface without any protection coating. Meanwhile, it could be seen that the coated mild steel shows less porosity and damage compared to the uncoated sample. The incorporation of MB improves the corrosion retention and a very low number of pores was observed on the surface of the coated mild steel with MB sample. The MB presence preserved smoothness of the surface compared to the samples. This result was in line with the values of the weight loss and corrosion rate obtained in Table 1. The result was similar to observation obtained in [10] that showed the release of self-healing agent recovers the protection layer of the coated steel. This shows the potential of MB to become an anti-corrosion substance as corrosion inhibitor or self-healing agent. The coated mild steel with MB demonstrated the best protection towards the corrosion among the samples and showed the potential of self-healing agent in corrosion management plan.

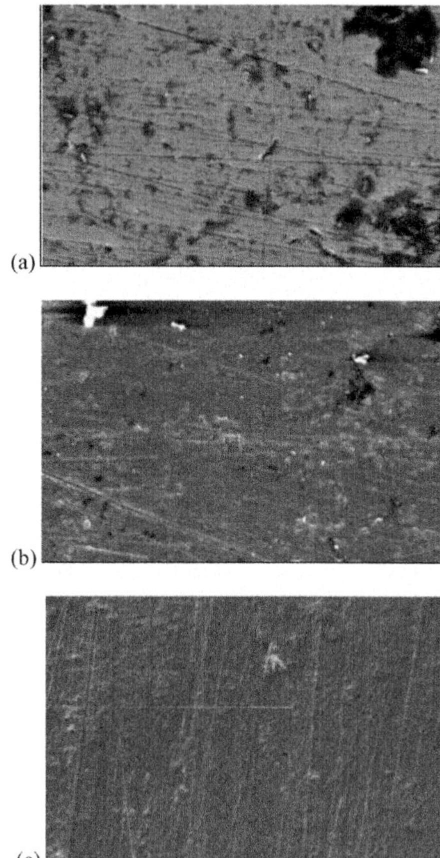

Fig. 1 SEM morphology of coated-mild steel after 120 h of NaOH exposure, **a** uncoated mild steel, **b** coated mild-steel without MB, and **c** coated mild-steel with MB

4 Conclusion

In this study, the coating solution with MB for mild steel was successfully developed. By measurement, the best weight loss and corrosion rate after 120 h of NaOH exposure was obtained by coated mild steel with MB (0.29 g $\pm$ 0.05 and 1.26, respectively). The SEM images supported the values with smooth surface and minimal pores were observed on the sample's image. These findings showed the potential function of MB as healing agent or inhibitor in a steel coating solution against the corrosion and improving the reliability of steel structure.

Acknowledgements The authors appreciated the facilities provided by Plant Engineering Technology Section, Universiti Kuala Lumpur Branch Campus Malaysian Institute of Industrial Technology.

References

1. Lyu, K., Liu, X., Liu, R., Yang, H., Qiao, Y., Shah, S.P., et al.: Evaluation of the anticorrosion performance of CeO_2-modified graphene oxide nanocomposite epoxy coating subjected to simulated saline-alkali solution. Polymers **14**, 1412 (2022)
2. Tamalmani, K., Husin, H.: Review on corrosion inhibitors for oil and gas corrosion issues. Appl. Sci. **10**(10), 3389 (2020)
3. Perez, T.E.: Corrosion in the oil and gas industry: an increasing challenge for materials. J. Min. Met. Mater. Soc. (TMS) **65**, 1033–1042 (2013)
4. Popoola, L.T., Grema, A.S., Latinwo, G., Gutti, B., Balogun, A., et al.: Corrosion problems during oil and gas production and its mitigation. Int. J. Ind. Chem. **4**, 35 (2013)
5. Samimi, A.: Study an analysis and suggest new mechanism of 3 layer polyethylene coating corrosion cooling water pipeline in oil refinery in Iran. Int. J. Innov. Appl. Stud. **1**(2), 216–225 (2012)
6. Yuwei, Y., Hao, C., Yangjun, Z., Ying, Y., Haichao, Z., et al.: Corrosion protective mechanism of smart graphene-based self-healing coating on carbon steel. Corros. Sci. **174**, 108825 (2021)
7. Zhang, F., Ju, P., Pan, M., Zhang, D., Huang, Y., Li, G., Li, X., et al.: Self-healing mechanisms in smart protective coatings: a review. Corros. Sci. **144**, 74–88 (2018)
8. Zhao, X., Jiang, D., Ma, L., Zeng, X., Li, Z., Huang, G. et al.: Special issue: corrosion effects and smart coatings of corrosion protection. Coatings **12**, 1378 (2022)
9. Lingwei, M., Chenhao, R., Jinke, W., Tong, L., Hao, Y., Yajie, W., Yao, H., Dawei, Z., et al.: Self-reporting coatings for autonomous detection of coating damage and metal corrosion: a review. Chem. Eng. J. **421**, 127854 (2021)
10. Minghan, H., Stefan, P., Yaowen, X., Diana, D., Lucas, C., Wolfgang, B., Michael, K., Markus, B., et al.: Monitoring crack appearance and healing in coatings with damage self-reporting nanocapsules. Mater. Horiz. **5**, 51–58 (2018)

Fuzzy Logic Control Strategy and Instrumentation of Shell and Tube Heat Exchangers

Mohd Ismail Yusof, Nuraisyhah Azhar, Mohd Aliff Afira, Azavitra Zainal, Ashraf Rohanim Asari, Mohd Aminuddin Mohd Rozi, Fatin Nabilah Mohd Yasin, Noor Azzah Abdul Manap, and Nafrizuan Yahya

Abstract The main control issue in a heat exchanger system is to regulate the outlet cold fluid temperature according to the setpoint. An appropriate control system is required in handling the heat exchange, whereby it involves the setpoint achievement of the system resulting in an excellent performance of the temperature control. In this paper, a comprehensive study regarding the standard PID controller and intelligent control such as Fuzzy logic control (FLC), FPID controller has been made.

M. I. Yusof · M. A. Afira · A. Zainal · A. R. Asari · M. A. M. Rozi · F. N. M. Yasin · N. A. A. Manap
Instrumentation and Control Engineering Section, Universiti Kuala Lumpur Malaysian Institute of Industrial Technology, Persiaran Sinaran Ilmu, 81750 Bandar Seri Alam, Johor Bahru, Malaysia
e-mail: mohdismaily@unikl.edu.my

M. A. Afira
e-mail: mohdaliff@unikl.edu.my

A. Zainal
e-mail: azavitra@unikl.edu.my

A. R. Asari
e-mail: ashraf@unikl.edu.my

M. A. M. Rozi
e-mail: aminuddin.rozi@s.unikl.edu.my

F. N. M. Yasin
e-mail: fatin.yasin@s.unikl.edu.my

N. A. A. Manap
e-mail: azzah.manap@s.unikl.edu.my

N. Azhar (✉)
Institut Kemahiran MARA Bintulu, Jalan TVET, KM20, Jalan Bintulu Sibu, 97000 Bintulu, Sarawak, Malaysia
e-mail: nuraisyhah@mara.gov.my

N. Yahya
Faculty of Manufacturing and Mechatronic Engineering Technology, Universiti Malaysia Pahang, 26600 Pekan, Pahang, Malaysia
e-mail: nafrizuanmy@ump.edu.my

A. Ismail et al. (eds.), *Engineering Frontiers*, Advanced Structured Materials 208,
https://doi.org/10.1007/978-3-031-56844-2_20

This study aims to develop a control system and to find the most effective optimization technique for the temperature control system of STHE. The comparison between controllers' performance based upon the transient response concerning the time domain specification has been carried out. As the findings from the data analysis, the FPID controller is one of the effective control techniques in controlling the outlet temperature of cold water in STHE as an improvement to the PID controller. From the transient response functionality, it is proven that the combination of FLC and PID controller delivered a good output response with minimal overshoot, rise time, and settling time.

Keywords Heat exchanger · PID · Fuzzy logic control

1 Introduction

In the case of nonlinear processes such as heat exchanger processes, the classical PID controller becomes insufficient when the operating condition is not fixed to the demanding performance [1]. Usually, due to the linear behavior of conventional PID controllers, it fails to perform satisfactorily for complex systems with nonlinear characteristics and structural uncertainties like temperature control systems. One of the reasons is that the linear controller is not appropriate for nonlinear process operation [2]. In addition, the availability of a precise mathematical model of the system to be controlled is also essential. Many processes being handled by human operators in businesses cannot be automated using standard control approaches as the performance of these controllers is often inferior to the operators [3]. For this reason, the system with nonlinear behavior should be tuned by a nonlinear optimization method like an intelligent controller, for example, FLC in order to regulate the system performance [4]. The FLC is derived from Type 1 Fuzzy sets and due to their capabilities have been used in different applications [5]. Also proven that previous researcher has attempted to use a Fuzzy controller as the desired degree of nonlinearity that can be easily incorporated in the controller behaviour to overcome the multi-characteristics of nonlinearity constraints such as delay time, time lag, and time-varying of control objects without any issues related to knowledge of the mathematical model [6]. Zadeh has presented Fuzzy logic as an alternative to address the erroneous information available in the human language as expression such as 'almost', 'a little', and 'too' is not part of the domain of crisp set since they cannot interpret in the form of 'all' or 'nothing' [7]. In most research, the intelligent controller which is FLC is also used to replace the PID controller for better system control [8–10]. In fact, the FLC has demonstrated better control accuracy and faster response [11]. The Fuzzy self-tuning technique gives stable performance compared with PID since Fuzzy can reduce the overshooting [12]. FLC has been demonstrated to be an effective control technique to many complicated nonlinear systems since it was first developed as a model-free controller schemes approach [13].

2 Experimental Setup of Shell and Tube Heat Exchanger

Figure 1a illustrates the STHE plant used in this study. This research focused on the heating phase as the aim of the research is to control the outlet fluid temperature of the STHE. Generally, a heat exchanger with shell and tube type is built-up with two main parts which are the shell and tube. Two media, whether liquid or gas, are allowed to flow individually into shell side and tube side. All of the data measurement of the medium used throughout this experiment are crucial and are done in dynamic way. Figure 1b illustrates a simplified schematic diagram of a STHE to represent the fluid flow for heating operation.

Apart from the mechanical design point of view, the software used in this research work included MATLAB/Simulink, TIA Portal for SIEMENS, Programmable Logic Control (PLC), and Open Platform Communication (OPC) configuration (shown in Fig. 1c). An extensive experiment was carried out to evaluate the control performance

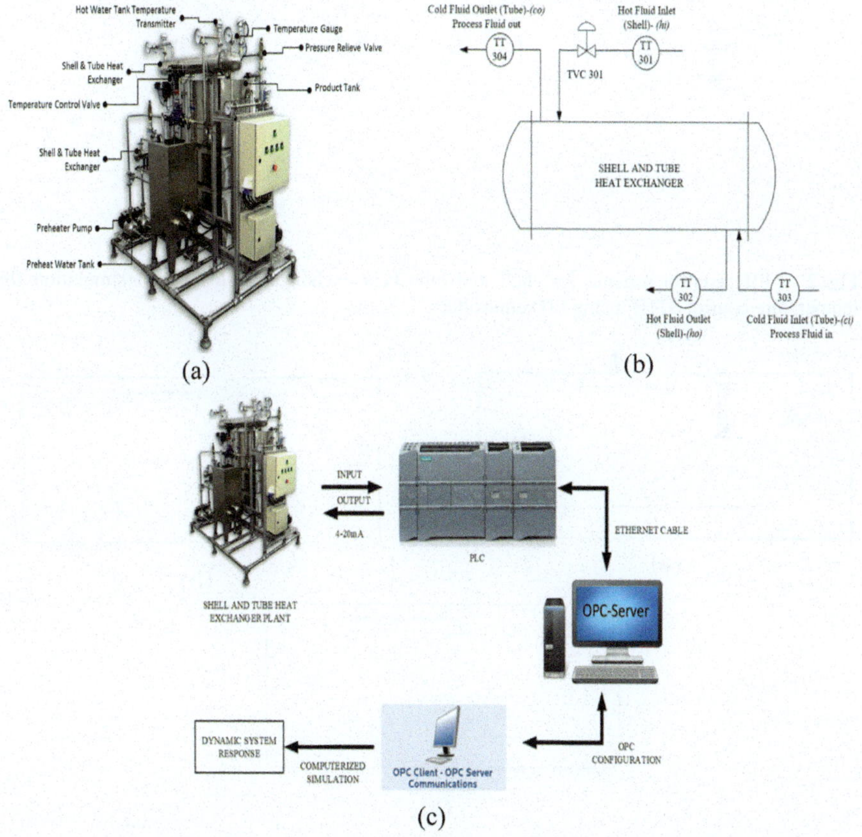

Fig. 1 **a** The downscale industrial shell and tube heat exchanger. **b** The schematic layout of shell and tube heat exchanger with all instruments tags, **c** software configuration in the STHE plant.

by applying the PID controller in the STHE system as illustrated in Fig. 2a. The behavior of the PID controller is observed under three different values of gains which is k_p, τ_i, and τ_d. These three parameters of the PID controller must be tuned according to the processing system to obtain a satisfactory closed-loop process performance. Figure 2b indicates the model of the PID controller constructed in Simulink for MATLAB simulation.

Implementation of the membership function is necessary in the Fuzzy system to assess non-Fuzzy input values. The membership function used is a triangular membership function for both input and output variables. For instance, the membership function for linguistic terms of temperature variables has been plotted on the scale of −20 to 60 °C according to the adaptability of theSTHE plant. Figure 3a and b illustrate the membership function of inputs FPID controller. The phases include:

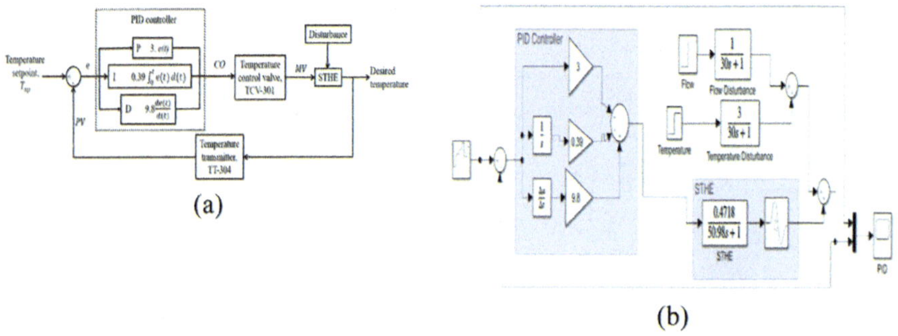

Fig. 2 **a** PID control scheme for shell and tube heat exchanger. **b** MATLAB simulation for temperature control STHE using PID controller

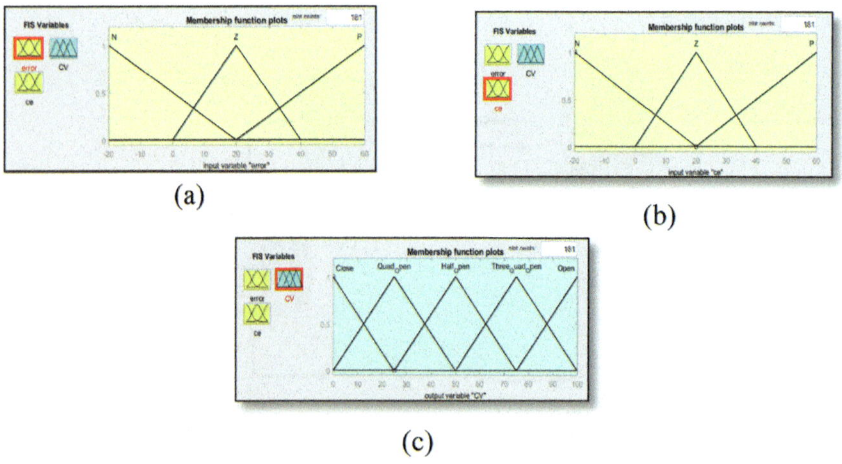

Fig. 3 Membership function of **a** FPID controller for error, **b** inputs of FPID controller for change of error, **c** the output of FPID controller for control valve action

- "Negative" (N) phase is when the error or changes of error is between −20 and 20,
- "Zero" (Z) phase is when the error or changes of error is between 0 and 40, and
- "Positive" (P) phase is when the error or changes of error is between 20 and 60.

The membership function for the output variables has been considered on a scale of 0–100%. Figure 3c represents the membership function for the output of FPID controller for the control valve, TCV301 action. The membership function for output variables indicates five phase where,

- Close (C) phase indicates the control valve opening between 0 and 25%,
- Quadrant open (QO) phase indicates the control valve opening between 0 and 50%,
- Half open (HO) phase indicates the control valve opening between 25 and 75%,
- Three quadrant open (TQO) phase indicates the control valve opening between 50 and 100%, and
- Open (O) phase indicates the control valve opening between 75 and 100%.

Importantly, the rules-based is constructed to control the output variables. The rules describe all the relationships between the input and output variables. The IF–THEN rules are simple, generated based on the combination of inputs condition and output conditions (conclusion). Figure 4a describes the Fuzzy rules arrangement in the Fuzzy system.

The rules can be read as:

- IF "e(t)" is "negative" (N) and "Δe(t)" is "negative" (N) THEN "u(t)" is close (C),
- IF "e(t)" is "negative" (N) and "Δe(t)" is "zero" (Z) THEN "u(t)" is quadrant open (QO),
- IF "e(t)" is "negative" (N) and "Δe(t)" is "positive" (P) THEN "u(t)" is half-open (HO),
- IF "e(t)" is "zero" (Z) and "Δe(t)" is "negative" (N) THEN "u(t)" is quadrant open (QO),

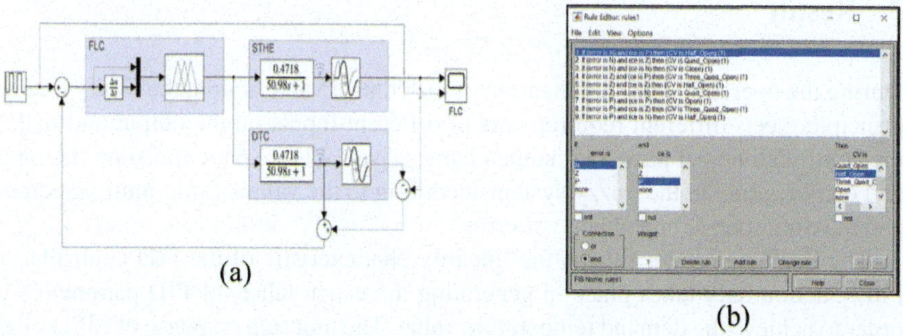

Fig. 4 **a** Block diagram of temperature control using FLC. **b** The Fuzzy rules applied FPID controller

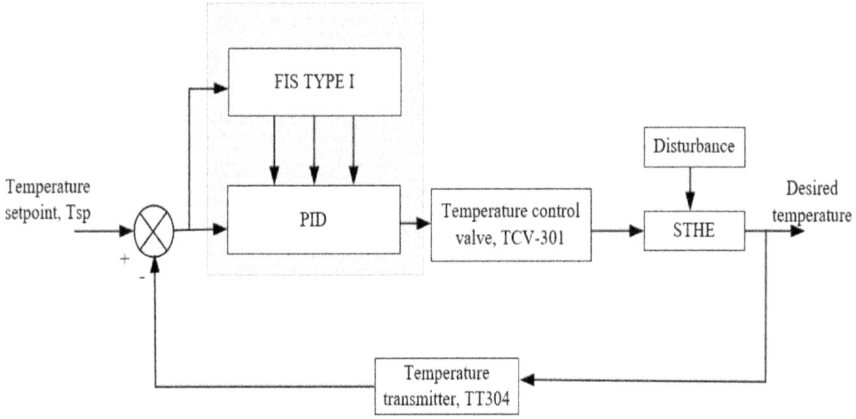

Fig. 5 Block diagram of FPID controller

- IF "e(t)" is "zero" (Z) and "Δe(t)" is "zero" (Z) THEN "u(t)" is half open (HO),
- IF "e(t)" is "zero" (Z) and "Δe(t)" is "positive" (P) THEN "u(t)" is three quadrant open (TQO),
- IF "e(t)" is "positive" (P) and "Δe(t)" is "negative" (N) THEN "u(t)" is half-open (HO),
- IF "e(t)" is "positive" (P) and "Δe(t)" is "zero" (Z) THEN "u(t)" is three quadrant open (TQO), and
- IF "e(t)" is "positive" (P) and "Δe(t)" is "positive" (P) THEN "u(t)" is open (O)

Figure 4b represents the control scheme of the temperature control system by conventional FLC in MATLAB/Simulink simulation. The general control strategy of the Fuzzy logic control for both type 1 is applied as the optimization tuning method for the PID controller in controlling the control valve opening, TCV301 to obtain the desired temperature setpoint at TT304 as shown in Fig. 5.

3 Result

During the operation and modification of the control valve action, the Fuzzy control principle meets different requirements of different inputs on the output and makes the controlled object have a dynamic characteristic. Figure 6a–c illustrate the input and output action in the Fuzzy system according to the value of minimum, medium, and maximum error and change of error.

The FPID controller retains the linearity characteristic of the PID controller as Fuzzy action only takes place in generating the exact values of PID parameters in order to achieve the demand temperature value. The unit step response of FPID gives near-zero peak overshoot with the shortest delay time, faster response, and shortest settling time.

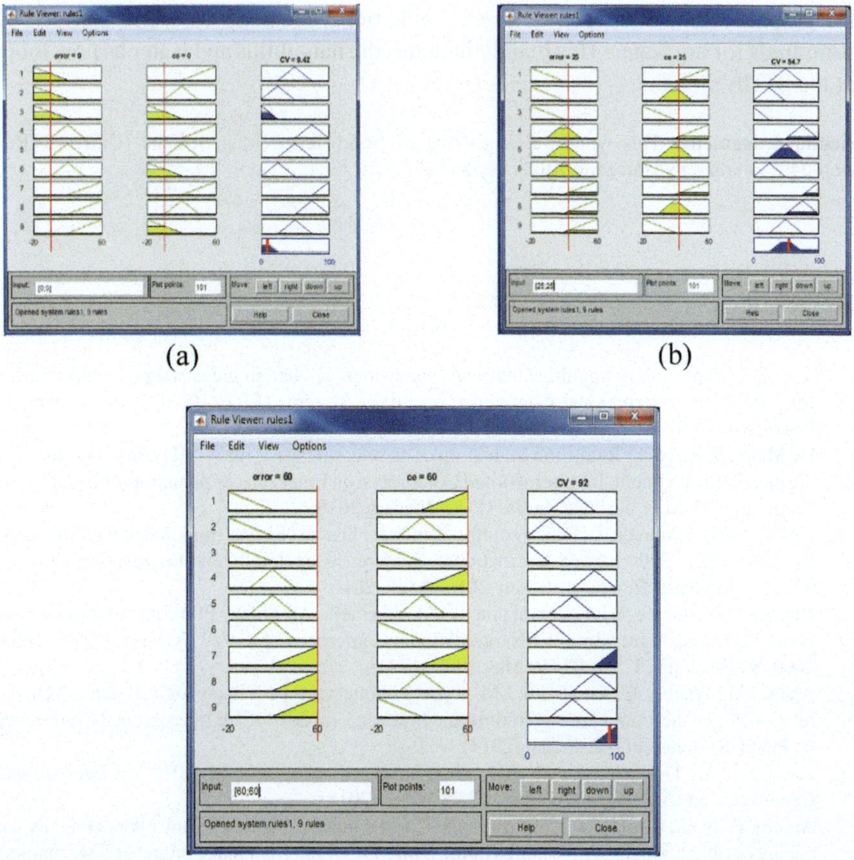

Fig. 6 **a** Control valve open about 10% if the error and change of error are 0. **b** Control valve open about 50% if the error and change of error are 25. **c** Control valve open about 100% if the error and change or error are 60

4 Conclusion and Future Research

In this study, the FPID has been introduced to overcome the drawbacks of conventional PID controller without considering uncertainties in output produce. This development is done to support the objective of this research study which is to design robust intelligent controllers for STHE. The overall simulation results also proved that the FPID controller has better performance in terms of rise time with (1.2 s) compared to the conventional PID controller (32 s).

Future research should also contribute to various sizes of the heat exchanger that might usefully focus on a particular performance. It may be advantageous to be able to assess how the size of the heat exchanger impacts the effectiveness of the tuning rules. In order to investigate the heat energy transfer performance in the STHE,

the graphite element should be used as additional elements in water that acts as nanofluids for hot water. This mainly includes the nanofluids and water as flow loops in the STHE system.

Acknowledgements This work was supported by Short Term Grant Scheme (UniKL/STRG/str18040) awarded by Universiti Kuala Lumpur.

References

1. Souza, L., et al.: Fuzzy cognitive maps and fuzzy logic applied in industrial processes control. In: 2018 IEEE International Conference on Fuzzy Systems (FUZZ-IEEE), Rio de Janeiro, Brazil, 8–13 July 2018
2. De Maity, R.R., et al.: Real-time evaluation of an interval type-2 fuzzy PID controller on servo position control system. In: International Conference on Emerging Applications of Information Technology (EAIT), Kolkata, India, 12–13 January 2018
3. Nesticky, M., Skulavik, T.: Fuzzy control of indirect heat exchanger implemented in Simulink. In: 2016 IEEE 20th Jubilee International Conference on Intelligent Engineering Systems (INES), Budapest, Hungary, 30 June 2016–2 July 2016
4. Huang, Y., Yasunobu, S.: A general practical design method for fuzzy PID control from conventional PID control. In: Ninth IEEE International Conference on Fuzzy Systems. FUZZ- IEEE 2000, San Antonio, USA, 07–10 May 2000
5. Aladi, J.H., Wagner, C., Garibaldi, J.M.: Type-1 or interval type-2 fuzzy logic systems—On the relationship of the amount of uncertainty and FOU size. In: 2014 IEEE International Conference on Fuzzy Systems (FUZZ-IEEE) (2014)
6. Xie, G., et al.: Design of Fuzzy PID temperature control system. In: 2018 3rd International Conference on Circuits and Systems (CAS 2018) (2018)
7. Araújo, F., et al.: Comparison of fuzzy type-2 and conventional fuzzy controllers tuned by ant colony optimization. In: Proceedings of the 24th ABCM International Congress of Mechanical Engineering, Curitiba, Brazil (2017)
8. Mondal, B., et al.: Performance comparison of conventional PID and fuzzy logic controller in the field of over headed water level control system. Int. J. Comput. Sci. Eng. **4**, 76–81 (2016)
9. Dhakad, S., Dandotiya, D., Pandey, P., Singh, D.: An intelligent control strategies implemented on heat exchanger system: a case study. Int. J. Emerg. Technol. Comput. Appl. Sci. **6**, 195–200 (2013)
10. Nour, M.I.H., Ooi, J., Chan, K.Y.: Fuzzy logic control versus conventional PID control of an inverted pendulum robot. In: 2007 International Conference on Intelligent and Advanced Systems (2007)
11. Gaurav, A.K., Kaur, A.: Comparison between conventional PID and fuzzy logic controller for liquid flow control: performance evaluation of fuzzy logic and PID controller by using MATLAB/Simulink. Int. J. Innov. Technol. Explor. Eng. (IJITEE) **1**, 84–88 (2012)
12. Sivakumar, P., Prabhakaran, D., Kannadasan, T.: Temperature control of shell and tube heat exchanger by using intelligent controllers-case study. **2**, 285–291 (2012)
13. Vasičkaninová, A., Bakošová, M., Meszaros, A., Oravec, J.: Fuzzy controller design for a heat exchanger. In: 2015 IEEE 19th International Conference on Intelligent Engineering Systems (INES), Bratislava, Slovakia, 03–05 September 2015

The Development of a Railway Route Crack Detection System Using an Arduino and an Ultrasonic Sensor

Nur Dania Ajrina Rahim, Mazhaniza Mazumil, Ashraf Rohanim Asari, and Mohd Aliff Afira Sani

Abstract The development of a railway route crack detection system using an Arduino and an ultrasonic sensor was based on the forms of railway tracks that exist today. The problem statement arises from the rising of train incidents that are caused by cracks and obstacles that obstruct along the railway route. Thus, this may result in significant losses of people's huge losses to the people's lives and property. Through the development of technology, the existing railway route crack detection system is available only at selected countries due to its expensive cost of production. To reduce train incidents, the installation of a railway route crack detection system is essential. The approach of railway crack detection system is by using two sensors which are an ultrasonic sensors to detect crack and an infrared sensor to identify any obstacle that blocks the railway track such as in KTM train area. The ultrasonic and infrared sensors are then integrated with an Arduino itself and send messages to the maintenance team with the aid of GPS and GSM module so that the maintenance could fix it faster. As the result of implementing railway route crack detection system, the inspection robot was successfully developed that is equipped with all the sensors to identify any defects on the railroad. The accuracy of the ultrasonic sensor to detect crack was recorded at 66.67% for the left side and 83.33% for the right side of inspection robot. When the inspection robot identified a crack or obstacle on the railway track, a message is sent to the control room as the monitoring system that is comprised in the development of the railway route crack detection system. Last but not least, all the characteristics associated with this system serve to quantify and

N. D. A. Rahim · M. Mazumil · A. R. Asari (✉) · M. A. A. Sani
Instrumentation and Control Engineering, Universiti Kuala Lumpur Malaysian Institute of Industrial Technology, 81750 Masai, Johor, Malaysia
e-mail: ashraf@unikl.edu.my

N. D. A. Rahim
e-mail: daniarahim@s.unikl.edu.my

M. Mazumil
e-mail: mazhaniza@unikl.edu.my

M. A. A. Sani
e-mail: mohdaliff@unikl.edu.my

A. Ismail et al. (eds.), *Engineering Frontiers*, Advanced Structured Materials 208,
https://doi.org/10.1007/978-3-031-56844-2_21

direct the optimal possibilities. For future recommendations, next researcher may use better specification for GPS antenna to track the location of defects.

Keywords Ultrasonic sensor · Arduino · GSM and GPS module · Crack detection system · Railway track

1 Introduction

Mobility is a critical requirement in the logistics area since it allows for the processing and production of goods to take place in various areas. Hence, rail transport plays an important role in Malaysia in terms of supplying the required transportation structure to sustain and satisfy the always increasing requirements of progressively rising industry [1]. During the period when trains were being developed as a mode of transportation, one tragic crash was reported in Berita Harian where there were two passengers dead in a railway wreck. Another incident was recorded by Mstar Blog on 20 August 2012. As the Express Wau heading to Tumpat from Kuala Lumpur in the voyage. An average of 11 people were hurt, eight of them were seriously hurt, along with a cut in the face, whereas the remaining two suffered a fractured hip and broken arms. Railroad cracks have just been recognized as the primary source of fatalities in earlier, but there have been no low-cost automated testing options available.

Annually, faults in railway routes create numerous collisions, resulting in injuries, accidents, maintenance expenses, environmental degradation, and a loss of usage, among other things. As a result, there is a continuous need for route investigation and maintenance, which is done by human inspectors. Because of the major repercussions of this difficulty, it is necessary to create an optimal and premium solution appropriate for huge applications. In the long run, the structure will be used to encourage greater protection procedures and create an effective evaluation framework in order to supply superior results in the presence.

2 Methodology

Project Overview. Development of this project is to design a crack detection system application an Arduino and ultrasonic sensor. This project is using an ultrasonic and infrared sensor to detect crack and obstacle respectively. The inspection robot is a vehicle that is operated wirelessly outfitted with a sensor that serves as the "eye" for monitoring any defects in the railway route. The data collected from the sensor than will be sent to the control room with the aids of GPS and GSM module to transmit message using SMS. This features act as the monitoring system of this project; therefore, the defect may be identified. The capacity to recognize the quality or obstruction of a railway route and notify the status in a single system is the basis of this railway route crack detection system. As a result, this method has a good chance of eliminating or lowering the number of deaths caused by railway disasters. Once implemented, the expenditure is lowered opposed to the current system, resulting in wider application throughout every train terminal.

The main objectives of this project are:

- To design a prototype system of railway route crack detection using an ultrasonic sensor with an Arduino.
- To monitor the data of railway track condition using an ultrasonic and infrared (IR) sensors.
- To transmit a message to the control room by using the GPS and GSM module.

Flowchart. Figure 1 explains the process flow for the inspection robot. It begins with activating all the required components such as the Arduino Uno, ultrasonic sensor, infrared sensor, GPS, and GSM module. These are all the important packages for the inspection robot to detect a crack and obstacle at the railway track. The inspection robot will move along the railway route and check for two conditions which are cracks and obstacles on the route. If cracks are found, then the inspection robot will stop, and the GPS module starts to search for the exact location. The virtual GSM module sends the message of the location of crack to the control room by using the SMS. At the same time, the inspection robot will check for any obstacle by using the IR sensor. When obstacles are captured by the IR sensor, the inspection robot will stop, and the GPS modules start searching for the current location and the virtual GSM module sends that location to control room by the aids of SMS. The GSM module will continuously send the message until the obstacle has been removed.

Circuit Diagram Design. The hardware components are presented by the schematic circuit diagram which includes the Arduino Uno, GSM, GPS module, DC motor, and sensors as stated in Fig. 2. The circuit has two input signals; ultrasonic sensor and infrared sensor, and two output signals; DC motor and GSM together with the GPS module. The infrared sensor is connected at the digital input port while the ultrasonic sensors is connected at the analog input port. Digi sim card is placed at the GSM module to transmit the data as text SMS to a control room. The inspection robot is connected to the external battery to power up the prototype.

Prototype Design. Based on Fig. 3, the prototype consists of three parts which are (1) the ultrasonic and infrared sensor mounted at front, (2) GSM and GPS module placed at the back and (3) Arduino Uno with the L298N Driver Shield. While Fig. 4 displays the railway that is already design with a crack on the route for demonstration. Lastly, the 4 wheels inspection robot car is mounted on the base of the prototype to simulate the change in speed of the DC motor during the operation and it is attached with the L298N Driver Shield on top of it.

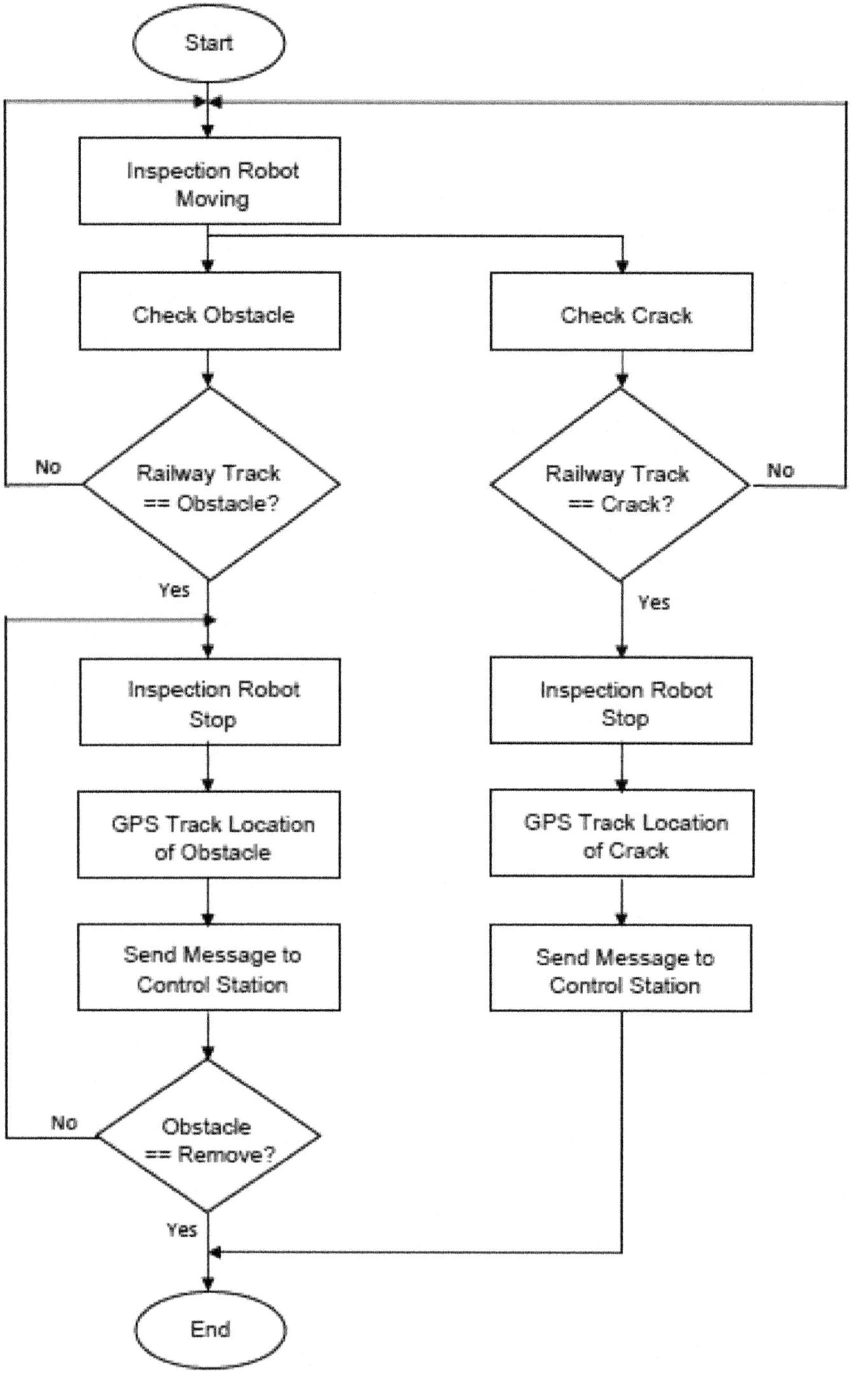

Fig. 1 Flow chart of the project

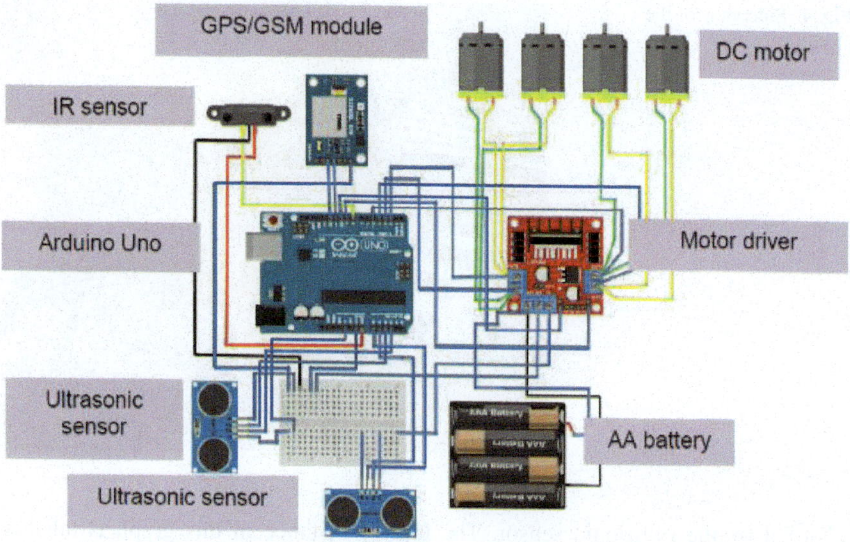

Fig. 2 Circuit diagram of the project

Fig. 3 Front view without cover

3 Results and Discussion

Data Accuracy. In this project, there are two tests that have been simulated and recorded for the ultrasonic sensor. The sensors are mounted at right- and left- hand side of the inspection robot and were used to examine the depth of the railway route crack. Figure 5 shows the data accuracy taken for a left-hand side of the ultrasonic sensor. There are two parameters involve in this graph which are the actual and measured distance of the railway route crack. The actual distance is the distance that is evaluated manually using a ruler while the measured distance is the distance

Fig. 4 Side view of the prototype

recorded by the ultrasonic sensor. The unit of distance in this graph is taken in centimeter. For distance 10 cm, the ultrasonic sensor measured a slightly lower value which at 9 cm only and at distance 24 cm, the ultrasonic sensor measured a different value which exceed 2 cm from the actual distance which is 26 cm. The different value are recorded because during the movement of the inspection robot, there is a vibration created by the motor. This vibration is considered as the disturbance of this project. As the vibration happened, the ultrasonic sensor will measure an error value in the range of ±2–4 cm. Meanwhile, other than that distance the ultrasonic sensor recorded the same value as the actual distance which are 7, 15, 20, and 30 cm. The accuracy for the left-hand side of ultrasonic sensor is about 66.67%.

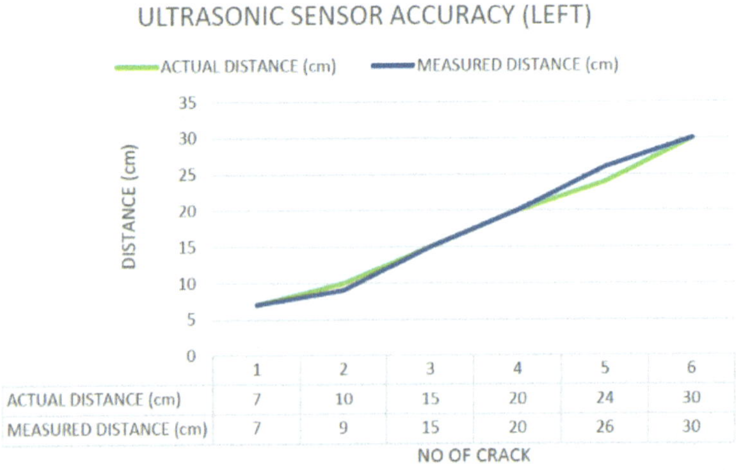

	1	2	3	4	5	6
ACTUAL DISTANCE (cm)	7	10	15	20	24	30
MEASURED DISTANCE (cm)	7	9	15	20	26	30

Fig. 5 Graph data accuracy of ultrasonic sensor (left)

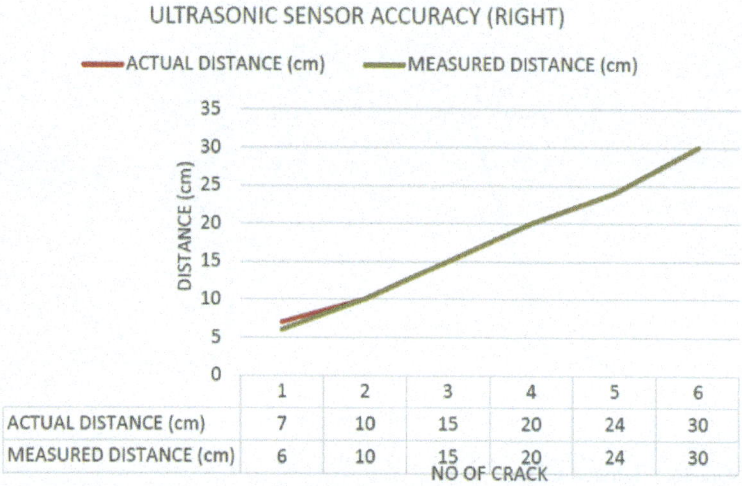

Fig. 6 Graph data accuracy of ultrasonic sensor (right)

Figure 6 demonstrates the graph of ultrasonic sensor accuracy for the right-hand side of the inspection robot. This graph includes two parameters: the actual and measured distance of the railroad crack and the unit of measurement is centimeters. For distance 7 cm, the ultrasonic sensor only measured 6 cm which is slightly lower than the actual distance. The lower value was obtained due to the vibration during the inspection robot's movement. While at the other distance, the ultrasonic sensor measured the same value as actual distance which at 10, 15, 20, 24, and 30 cm. The right side of the ultrasonic sensor has an accuracy of around 83.33%.

Sensor Behavior. Infrared sensor is specifically utilized to detect any obstacles that obstruct along the railway route. The test has been simulated and recorded with a duration of 8 s. Figure 7 shows the infrared sensor reading that appears once an obstacle is detected. When an object comes close to the sensor, the infrared light from the LED reflects off the object and is detected by the receiver and the sensing range is set at 5 cm. If the sensor detects no objects, the digital value will be 0; otherwise, it will be 1. The conversion of digital output to voltage is shown in Table 1 and Fig. 8 is demonstrated the action of the inspection robot when it detects obstacles.

Monitoring System. The GSM and GPS modules were used as a monitoring system for the inspection robots. This system is essentially developed to monitor the conditions of railway track and send message to the nearby control room by using SMS. Figure 9 displays the message send include the conditions of the railway track and the current locations of the defect. When the left side of ultrasonic sensor detect crack, the inspection robot will stop moving for 1 min and a message of "left rail crack" is display on the screen while "right rail crack" message for the right-side crack before the inspection robot continue examine the railway track.

Besides, if the infrared sensor detects any obstacles, a message of "object detected on rail" will appear and continue notify the user until the obstacles being removed.

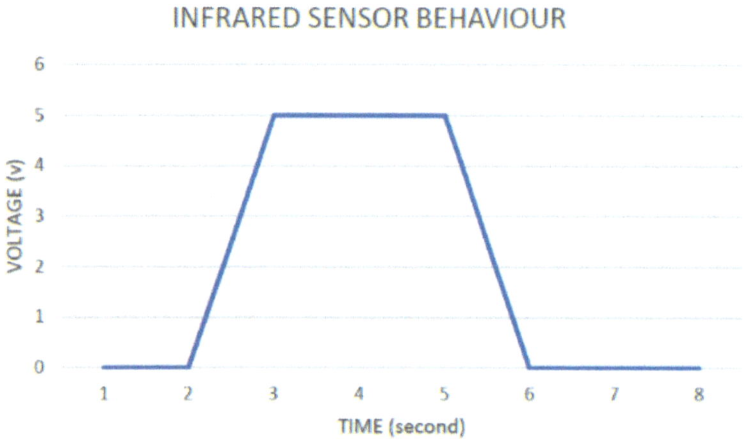

Fig. 7 Graph for infrared sensor

Table 1 Conversion digital output to voltage

Time (second)	Digital output	Voltage (v)
1	0	0
2	0	0
3	1	5
4	1	5
5	1	5
6	0	0
7	0	0
8	0	0

These features were created as a safety highlights. Based on Fig. 10, the testing is done at UniKL MITEC which is located at Jln Persiaran Sinaran Ilmu, Bandar Seri Alam, 81750 Masai, Johor. A detail location of the cracks or obstacles can be identified by using the Google Map as the message send is included the latitude and longitude of the site.

Fig. 8 Obstacle detected by the inspection robot

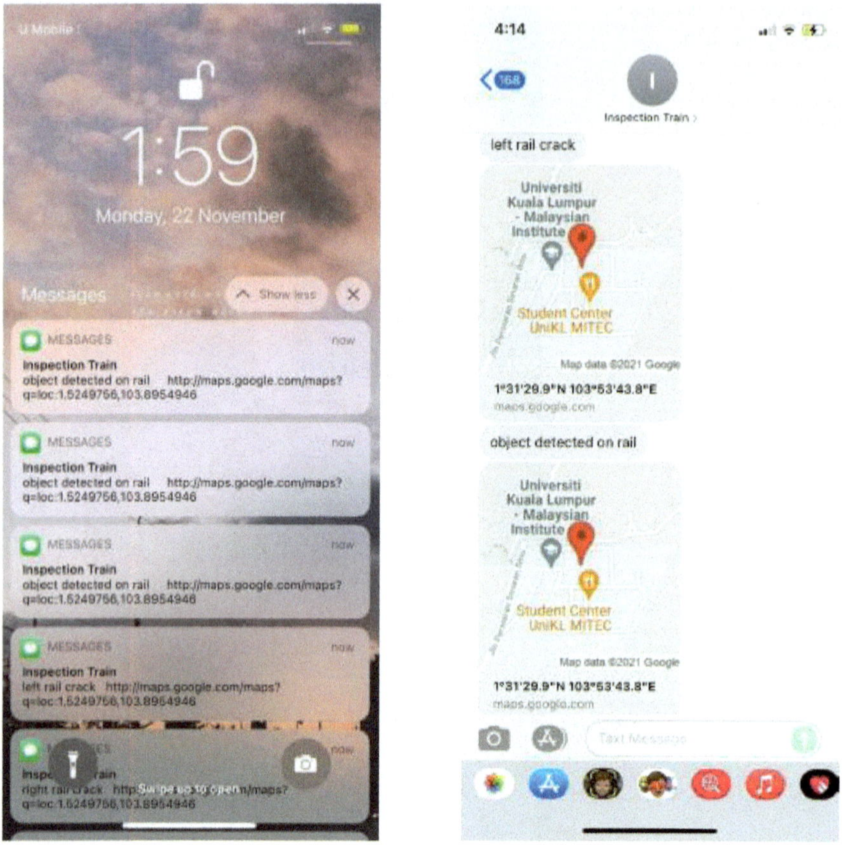

Fig. 9 Message notifications and detail

Fig. 10 Locations on Google Maps

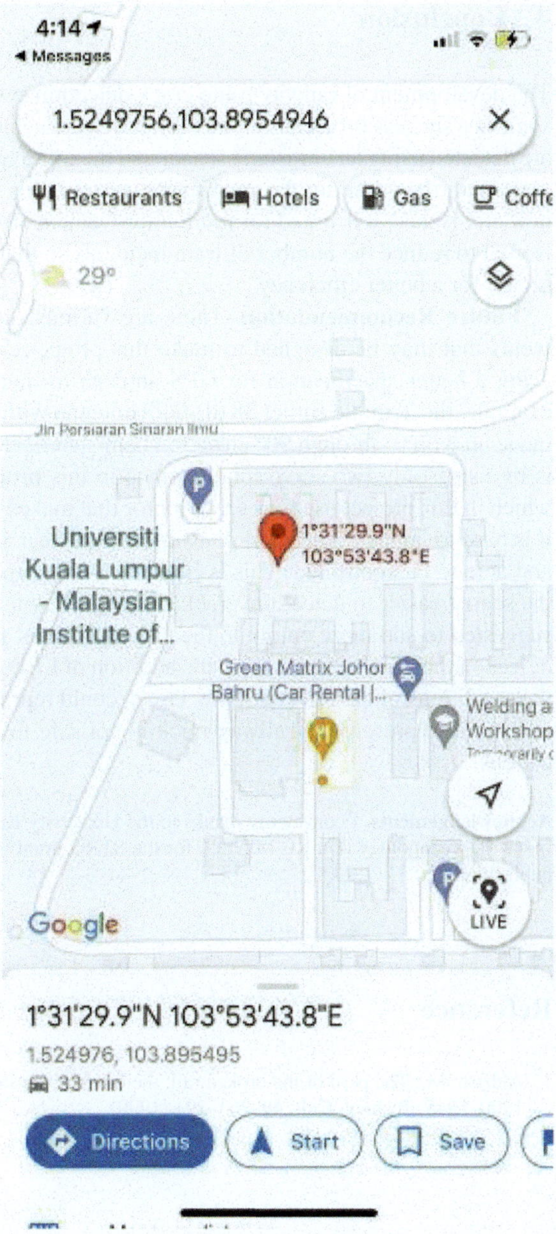

4 Conclusion

The development of railway route crack detection system using Arduino and Ultrasonic is a success project as all the objectives are achieved. The railway route cracks are detected using an ultrasonic sensor and the infrared sensor is used to detect obstacles. With the addition of a monitoring system, this project is more reliable in the new era. Hence, it is urged for more improvement about this project to be done as it is vital to reduce the number of train incidents so that a safety transportation can be secure for a better efficiency.

Future Recommendation. There are various recommendations and enhancements that may be designed to make this project even more effective. Firstly, by using a better specification for GPS antenna to track the location of defects. For example, the Trimble Bullet 38 dB GPS antenna with 25 MHz (min) bandwidth and made up with 0° through 90° elevation hemispherical of elevation coverage. Second is by using only two servo motors since in this project we use four servo motors which in this project use four servo motor that makes the inspection move faster and it is hard to control. The most plausible and perfect servo motor must be low speed and able to be controlled. This is because a lower speed is more secure and makes the sensor easier to detect the crack and obstacle on the railway track. Finally, it is suggested to add more output to the project such as a buzzer to alert the user of the defect on the railway track. With the addition of LED, it could ease the user to notify the conditions of the railway route. Green could represent that the conditions is safe and red to represent the railway track is not safe to be used and needs immediate actions.

Acknowledgements I convey my thanks to the University Kuala Lumpur Malaysian Institute of Industrial Technology (UniKL MITEC) for the STRG grant to my co-author which leads to this publication.

Reference

1. Aldrich, M.: The peril of the broken rail: the carriers, the steel companies, and rail technology, 1900–1945. Technol. Cult. **40**, 263–291 (1999)

Geometric Error Assessment of CNC Turning Machine Tools with a Case Study on Gildemeister CTX 310 Eco

Herman Budi Harja, Nandang Rusmana, Elan Suherlan, Mohammad Nurdin, Dini Hadiani, and Mohd Amran Mohd Daril

Abstract This paper studies the measurement and assessment of circular geometric errors on CNC lathes under loading and unloading conditions. Hence it could generate certain information about maintenance activities. The research method was conducted by measuring the machine tools' angular geometric error using a ballbar following ISO 230-4:2005 and cutting tests referring to ISO 13041-6:2009. The angular geometric error status of the machine was determined by comparing the measurement result with standard tolerances according to ISO 10791-2:2001, ISO 10791-4:1998, and ISO 13041-4:2004. The object of this machine study is a Gildemeister CTX 310 Eco with travel axis of X, and Z, about 160 and 450 mm. The ballbar measurements were tested by measuring the position of the cylindrical test, flatness test, circular test, and modification workpiece test. Geometric error in the results of the modification workpiece test, flatness test, and circular test that ate the machine tools are still in tolerance in the no-load condition test and the in-load condition test. The measurement result value of the bidirectional circular deviation of the modification workpiece test, flatness test, and circular test is 15.24, 16.64, and 16.28 μm. But the geometric error of the cylindrical test is 24.748 μm, and it exceeds the tolerance in the load condition test. Based on the assessment, the recommended maintenance activities are inspecting and repairing the mechanical driving components, such as servo motors and couplings, ballscrews, and linear guideway components.

H. B. Harja (✉) · N. Rusmana · E. Suherlan · M. Nurdin · D. Hadiani
Politeknik Manufaktur Bandung, Bandung, Indonesia
e-mail: herman@polman-bandung.ac.id

N. Rusmana
e-mail: nandangr@polman-bandung.ac.id

M. Nurdin
e-mail: mdnd@polman-bandung.ac.id

D. Hadiani
e-mail: dinibhs@polman-bandung.ac.id

M. A. M. Daril
Universiti Kuala Lumpur Malaysian Institute of Industrial Technology, Masai, Johor, Malaysia
e-mail: mamran@unikl.edu.my

A. Ismail et al. (eds.), *Engineering Frontiers*, Advanced Structured Materials 208,
https://doi.org/10.1007/978-3-031-56844-2_22

Keywords CNC turning machine · Circular geometric error · Ballbar measurement · Cutting test

1 Introduction

A computer numerical control (CNC) machine tool is an automatic machine whose operation method is according to digital information. The operation process is through G-code programmed commands to control the axial movement of the machine. The quality of the machine will decrease during the usage phase. Machine errors are caused by deviations in machine components, such as geometric errors, servo errors, and thermal errors. These geometric errors have the most significant influence on the accuracy and precision of the machine. The accuracy and precision of machine tools, especially CNC machine tools, are essential because they will affect the quality of the products produced. Then the geometric deviation of the CNC machine tool should be checked to ensure the machine is suitable for operation.

Several topical studies about performance assessment of machine tools have been carried out. Those topics are the efficiency of power consumption, machine tool geometric errors, and evaluation of cutting results. The study of machine tools quality could be started from the power consumption level and geometric error of machine tools itself, and geometric error of machining workpiece as a cutting test result. Harja evaluated a lathe's energy efficiency on the condition of the aligned and misaligned machine pulley-belt transmission system. This study assesses the lathe machining process's specific energy consumption (SEC) indications. It was stated that the mechanical loss aspect of the misalignment condition factor in the engine pulley-belt transmission system must be considered because it decreases the lathe's energy efficiency performance by more than 5% [1]. Also, Harja explained an assessment of linear positional error for CNC milling machines using a laser interferometer based on ISO 230-2, the measurement standard of positioning accuracy and rapid ability [2]. Similarly, Sabahudin measured the machine error using a QC-10 ballbar with a transducer length of 100 mm and analyzed the measurement result to find the errors in the machine [3]. Moreover, Holub explained the possibility of using the QC20-W ballbar as a diagnostic tool for monitoring the geometric accuracy of CNC machine tools. The measurement data can be statistically calculated to determine the value of the deviation [4]. Likewise, Monika and Majda researched the reproducibility of the circularity test of the machine following ISO 230-4. They explain that the ballbar is very well used to evaluate the condition of machine geometry deviations [5]. Also, Robert's diagnostic machine tools use the Renishaw ballbar with measurement four times on the same parameters. The measurement principle compares the actual deviation with the ideal circle. The calculation carried out is to calculate the arithmetic mean to be compared with the standard deviation [6].

This study aims to determine the deviation value of the CNC lathe machine tool and whether the geometric deviation is still within the ISO tolerance standard or has exceeded the permitted limit. It is also intended to find out the recommendations for action on the CNC lathe, which refers to the measurement results of geometric deviations.

2 Methodology

The research method consists of two stages; firstly, it measures the geometry error of the machine tools using a ballbar and analyses the measurement data. Secondly is the cutting test with standard product features of machining workpieces and measuring the geometric error of machining workpieces using a CMM machine.

The object of this machine study is a CNC turning machine with a case Gildemeister CTX 310 Eco. It is equipped with a Siemens 810D machine controller and has a distance travel axis of X and Z of about 160 and 450 mm. Figure 1 shows a visualisation of the working area of the CNC turning machine Gildemeister CTX 310 Eco.

2.1 Measuring the Machine Tools' Geometric Error

The geometric error assessment of the CNC turning machine in unloading conditions is conducted by measuring the geometric error using a ballbar following ISO 230-4:2005 [7].

Ballbars are used to evaluate machine performance, which was adopted by ISO 230-1 to perform circular tests on machine tools [8]. The ballbar consists of a telescopic bar with a magnetic ball and socket joint at the end and has a displacement sensor inside. While the machine moves circular interpolation, it measures the magnitude of the deviation of the machine spindle from the specified program path. The circular path results in significant deviations from the contours of the machine tool's

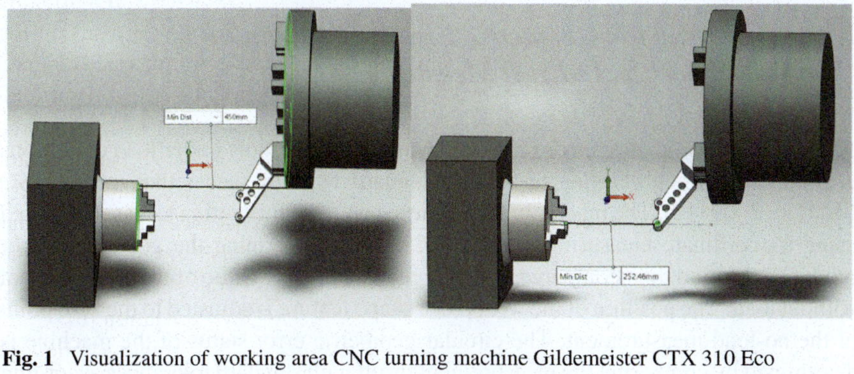

Fig. 1 Visualization of working area CNC turning machine Gildemeister CTX 310 Eco

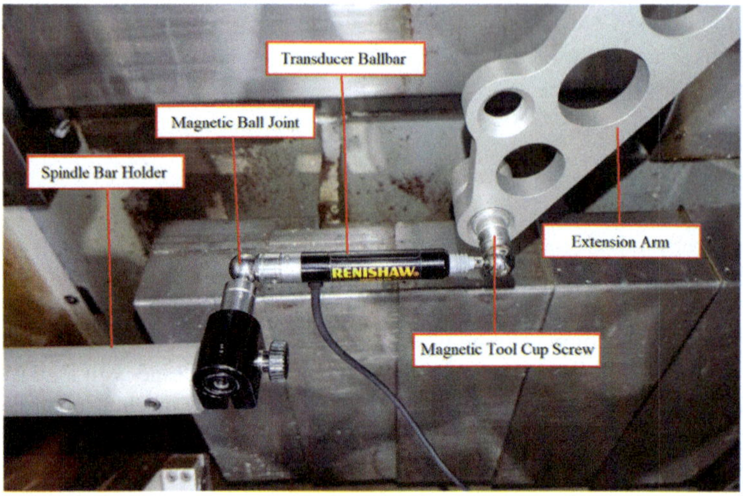

Fig. 2 Measurement configuration of circular interpolation using ballbar QC10

performance. If the machine tool is tested in the direction clockwise (CW) and counterclockwise (CCW), then the data analysis can provide detailed information about the machine tool's performance and the source of the deviation. Circular interpolation changes the translational position of the linear axis to the circular axis motion. The circular interpolation accuracy is one of the program parameters that can significantly affect the shape geometry and tolerance of the machining workpiece being processed on the machine [9]. The ballbar is a measurement tool for measuring the geometric deviations under no-load conditions. The ballbar is used to determine the accuracy of CNC machine tool geometry through circular interpolation motion [10, 11]. This study used the Renishaw ballbar QC10 type. Figure 2 shows the measurement configuration of the circular interpolation using the ballbar QC10.

2.2 Measuring the Geometric Error of Machining Workpieces as Indirect Measurement

Measuring the geometric error of CNC turning machine tools under load conditions is conducted through cutting tests with standard working pieces that refer to ISO 13041-6: 2009 [12]. The machining workpiece results of the cutting test are measured using a coordinate measuring machine (CMM) to determine the geometric error product value and then compared with the geometric error value of the machine tool without load. The position of the workpiece shape feature is adjusted to the zero point of the no-load measurement. The circular geometric error status of the machine is determined by comparing the measurement result with standard tolerances according to ISO 10791-2:2001, ISO 10791-4:1998, and ISO 13041-4:2004 [13–15].

The cutting test referred to ISO 13041-6 is about the test conditional for numerical controlled turning machines and turning centers-accuracy of a finished test piece. It explains several standard feature products tested, such as the cylindrical test piece (Fig. 3), flatness test piece (Fig. 4), and circular test piece (Fig. 5). In addition, the cutting test was conducted with modified test working piece objects representing three standard product-tested features, as shown in Fig. 6.

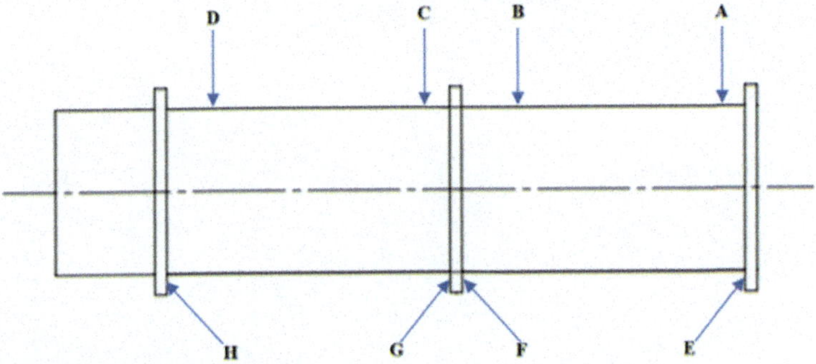

Fig. 3 Cylindrical test piece

Fig. 4 Flatness test piece

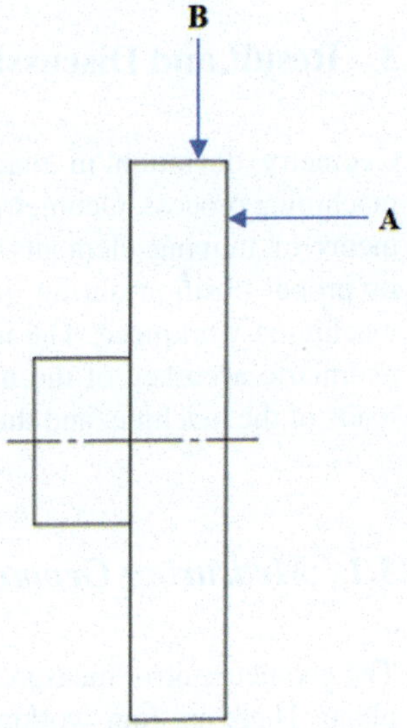

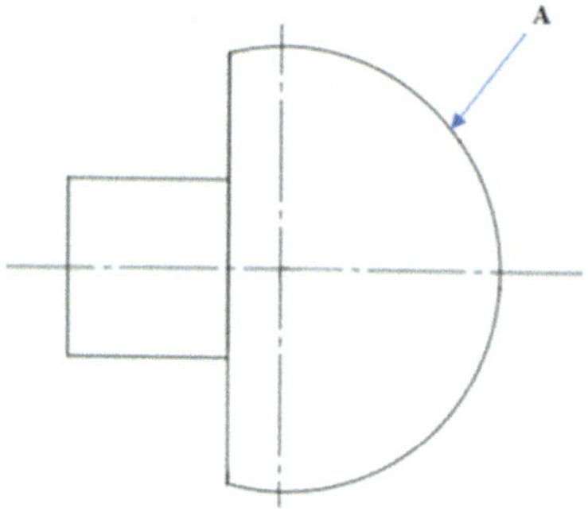

Fig. 5 Circular test piece

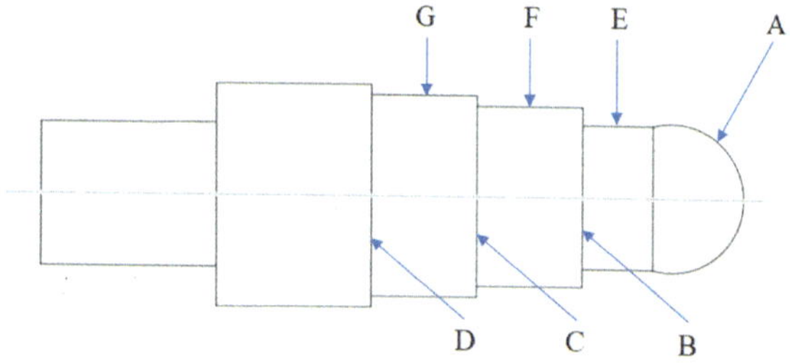

Fig. 6 Modified test piece

3 Result and Discussion

Geometry deviation in machine tools changes accuracy conditions during the machining process. Geometric deviations are caused by imperfections and misalignments of moving elements. The position of the cutting edge will change from its proper position during the machining process, resulting in inaccuracies in the machining workpiece. The inaccuracies magnitude depends on three main factors: geometric accuracy of the machine tool, the deflection due to static and dynamic loads of the machine, and thermal drift [16].

3.1 Measuring Geometric Error Results Using a Ballbar

The geometric error measurement on the CNC turning object machine is on the *Z–X* plane. There are four workpiece object types of measures, modification test piece, cylindrical test piece, flatness test piece, and circular test piece. Figure 6 shows the measuring position of the modification test piece. There are five measuring coordinate positions, which (MCS) machine coordinate system of each position on the *X* and *Z* axis are position 1 (167, 279), position 2 (167, 259), position 3 (167, 229), position 4 (167, 199) and position 5 (167, 284).

Measurements with circular interpolation motion using a ballbar were conducted in 5 positions for five repetitions. The following is the result of measuring the ballbar at position 1 (Figs. 7 and 8).

The most considerable radial deviation value occurs at the point of reversal of the direction of rotation, namely at an angle of 270°–360° the direction of CCW movement.

Table 1 shows the average value of the measurement result of circular interpolation on the modification workpiece test. The mean value of the counterclockwise circular deviation (G) in the *XZ* plane is 12.952 μm. The mean value of the clockwise circular deviation (G) in the *XZ* plane is 12.656 μm. The mean value of the maximum counterclockwise radial deviation (F) in the *XZ* plane is −2.392 μm. The mean

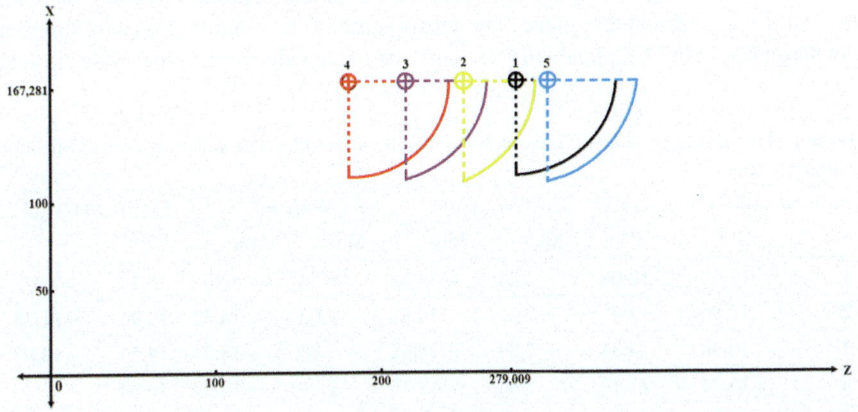

Fig. 7 Measuring the position of the modification test piece

Fig. 8 Ballbar measurement plot of a modification test piece on position 1

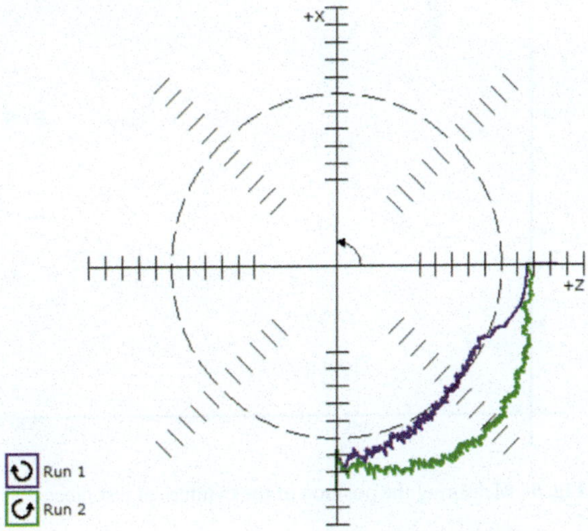

value of the minimum counterclockwise radial deviation (F) in the *XZ* plane is − 14.936 μm. The mean value of the maximum clockwise radial deviation (F) in the *XZ* plane is −0.256 μm. The mean value of the minimum clockwise radial (F) deviation in the *XZ* plane is −12.916 μm. The mean value of the bidirectional circular deviation (G(b)) in the *XZ* plane is 14.916 μm. The mean value of the mean bidirectional radial deviation in the *XZ* plane is −10.756 μm.

Figure 9 shows the measuring position of the cylindrical test piece. There are five measuring coordinate positions, which (MCS) machine coordinate system of each position on the *X* and *Z* axis are position 1 (167, 304), position 2 (167, 301), position 3 (167, 230), position 4 (167, 227), and position 5 (167,157) (Fig. 10).

The most considerable radial deviation value occurs at the starting point of the rotation direction, which is at an angle of 360°–270° the direction of CW movement.

Table 2 shows the average value of the measurement result of circular interpolation on the cylindrical test piece. The mean value of the counterclockwise circular deviation (G) in the *XZ* plane is 15.16 μm. The mean value of the clockwise circular

Table 1 The average value of measurement result of circular interpolation on modification workpiece test

Position	G CCW (μm)	G CW (μm)	F CCW (μm)		F CW (μm)		G (b) (μm)	D (μm)
			Max	Min	Max	Min		
1	11.6	12.08	− 4.24	− 13.84	6.24	− 5.86	15.24	− 7.02
2	12.56	11.3	− 2.32	− 14.88	− 3.02	− 14.32	12.96	− 11.48
3	10.78	14.34	− 7.84	− 18.62	− 3.88	− 18.24	14.54	− 14.98
4	16.24	11.42	2.86	− 13.36	− 1.7	− 13.22	16.64	− 9.7
5	13.58	14.14	− 0.42	− 13.98	1.08	− 13.04	15.2	− 10.6
Average	12.95	12.65	− 2.39	− 14.93	− 0.25	− 12.93	14.91	− 10.75

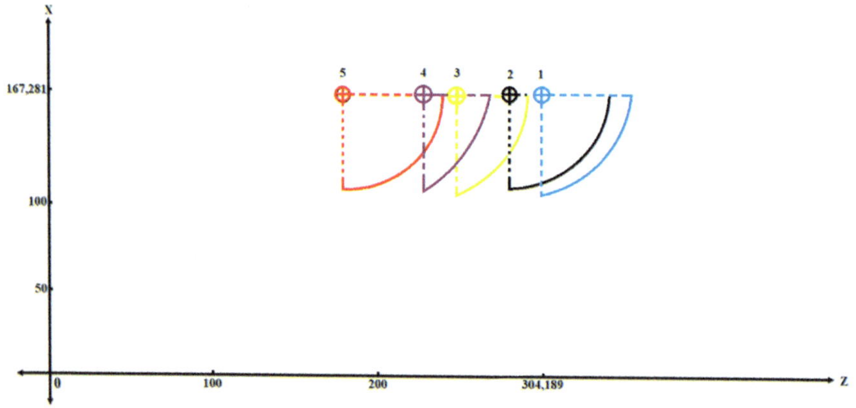

Fig. 9 Measuring the position of the cylindrical test piece

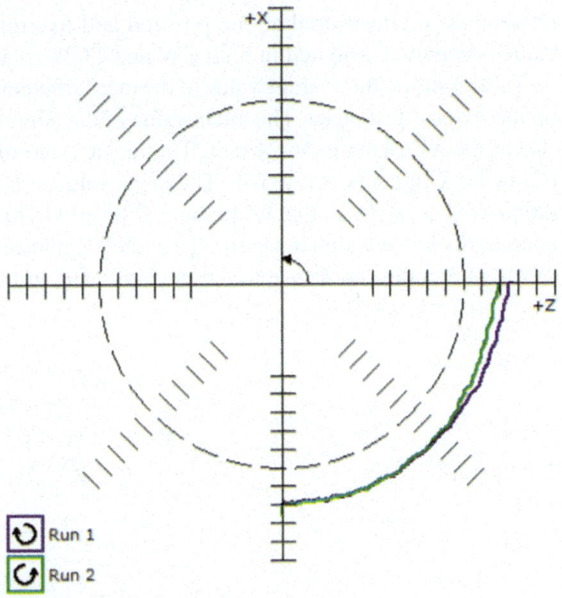

Fig. 10 Ballbar measurement plot of a cylindrical test piece on 1St position

deviation (G) in the *XZ* plane is 14.356 μm. The mean value of the maximum counterclockwise radial deviation (F) in the *XZ* plane is 74.356 μm. The mean value of the minimum counterclockwise radial deviation (F) in the *XZ* plane is 59.184 μm. The average value of the maximum clockwise radial deviation (F) in the *XZ* plane is 80.016 μm. The mean value of the minimum radial deviation (F) clockwise in the *XZ* plane is 65.664 μm. The mean value of the bidirectional circular deviation (G(b)) in the *XZ* plane is 24.748 μm. The mean value of the mean bi-directional radial deviation in the *XZ* plane is 64.44 μm.

Figure 11 shows the measuring position of the flatness test piece. The measuring coordinate position on the *X* and *Z* axis is 167.281 and 194.129. Figure 12 shows the Balbar measurement plot of the flatness test piece. The circular interpolation

Table 2 The average value of measurement result of circular interpolation on the cylindrical test piece

Position	G CCW (μm)	G CW (μm)	F CCW (μm)		F CW (μm)		G (b) (μm)	D (μm)
			Max	Min	Max	Min		
1	16.18	13.02	56.12	39.92	52.82	39.84	22	44.14
2	11.34	14	– 2.24	– 13.6	2.8	– 11.24	15.1	– 8.82
3	11.84	20.4	204.94	193.12	233.74	213.34	41.78	207.58
4	21.8	12.18	98.64	76.82	96.36	84.2	29.46	84.26
5	14.64	12.18	14.32	– 0.34	14.36	2.18	15.4	5.04
Average	15.16	14.35	74.35	59.18	80.01	65.66	24.74	66.44

movements are measured at one position and five repetitions. The radial deviation value is evenly distributed in both CW and CCW rotations.

Table 3 shows the average value of the measurement result of circular interpolation on the flatness test piece. The mean value of the counterclockwise circular deviation (G) in the *XZ* plane is 16.24 μm. The mean value of clockwise circular deviation (G) in the *XZ* plane is 11.42 μm. The mean value of the maximum counterclockwise radial deviation (F) in the *XZ* plane is 2.86 μm. The mean value of the minimum counterclockwise radial deviation (F) in the *XZ* plane is −13.36 μm. The mean value of the maximum clockwise radial deviation (F) in the *XZ* plane is −1.7 μm. The

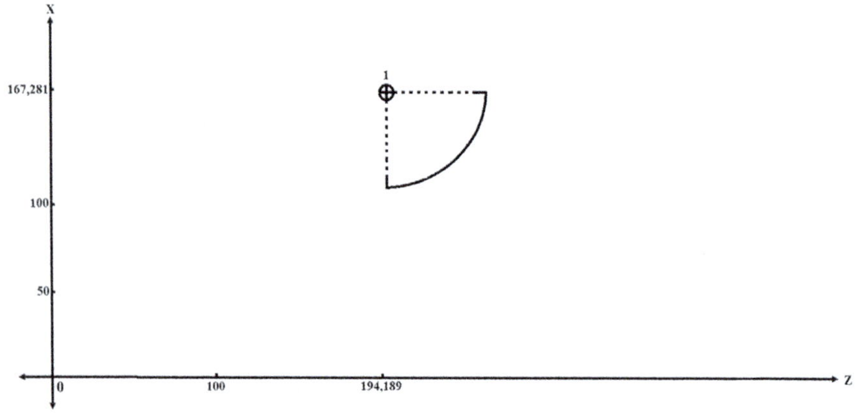

Fig. 11 Measuring the position of the flatness test piece

Fig. 12 Ballbar measurement plot of the flatness test piece

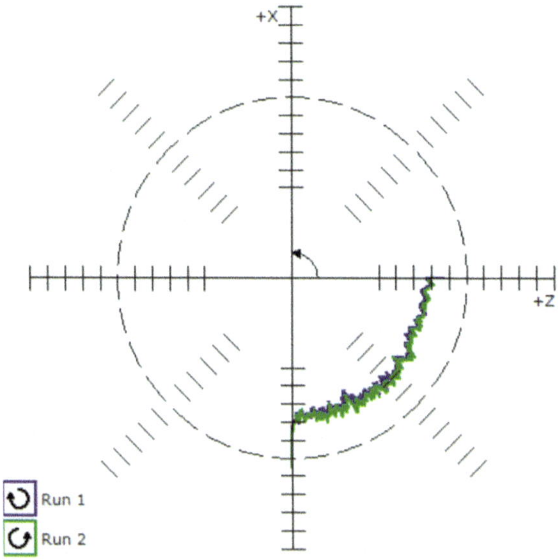

mean value of the minimum clockwise radial deviation (F) in the *XZ* plane is −13.22 μm. The mean value of the bidirectional circular deviation (G(b)) in the *XZ* plane is 16.64 μm. The mean value of the mean bidirectional radial deviation in the *XZ* plane is −9.7 μm.

Figure 13 shows the measuring of the position of the circular test piece. The measuring coordinate position on the *X* and *Z* axis is 167.281 and 204.189. Figure 14 shows the ballbar measurement plot of the circular test piece. Its measuring is conducted at one position and five repetitions. The largest radial deviation value occurs at the starting point of the rotation direction, which is at an angle of 360°–270° the movement direction is CW.

Table 4 shows the average value of measurement result of the circular interpolation on the circular test piece. The mean value of the counterclockwise circular deviation (G) in the *XZ* plane is 14.22 μm. The mean value of the clockwise circular deviation (G) in the *XZ* plane is 13.3 μm. The mean value of the maximum counterclockwise radial deviation (F) in the *XZ* plane is −2.42 μm. The mean value of the minimum counterclockwise radial deviation (F) in the *XZ* plane is −16.64 μm. The mean value of the maximum clockwise radial deviation (F) in the *XZ* plane is −9.12 μm. The mean value of the minimum clockwise radial deviation (F) in the *XZ* plane is −16.8 μm. The mean value of the bidirectional circular deviation (G(b)) in the *XZ* plane is 16.28 μm. The mean value of the mean bidirectional radial deviation in the *XZ* plane is 15.86 μm.

Table 3 The average value of measurement result of circular interpolation on the flatness test piece

Position	G CCW (μm)	G CW (μm)	F CCW (μm)		F CW (μm)		G (b) (μm)	D (μm)
			Max	Min	Max	Min		
1	16.24	11.42	2.86	− 13.36	− 1.7	− 13.22	16.24	− 9.7

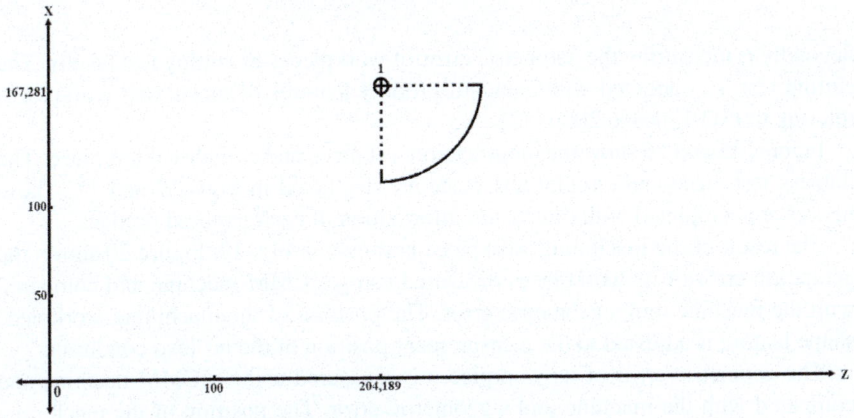

Fig. 13 Measuring the position of the circular test piece

Fig. 14 Ballbar measurement plot of the circular test piece

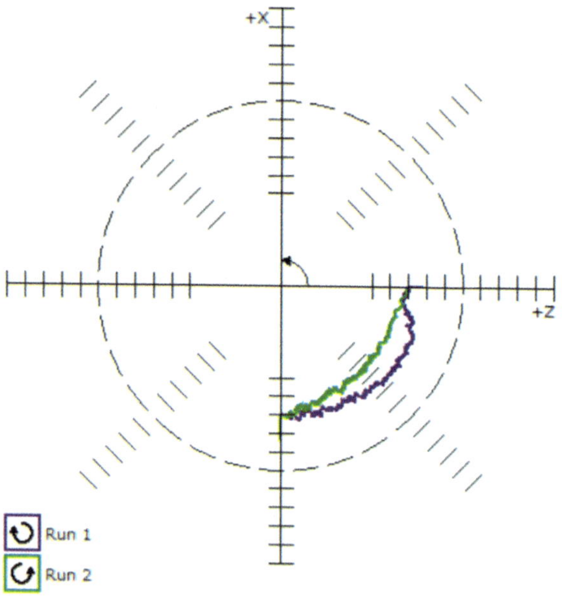

Table 4 The average value of measurement result of circular interpolation on the circular test piece

Position	G CCW (μm)	G CW (μm)	F CCW (μm)		F CW (μm)		G (b) (μm)	D (μm)
			Max	Min	Max	Min		
1	14.22	13.3	− 2.42	− 16.64	− 9.12	− 16.82	16.28	− 15.86

3.2 *Measuring the Geometric Error of Workpieces as Cutting Tests Result*

Secondly is measuring the geometric error of workpieces as cutting test results. The cutting test is conducted with standard product features of machining workpieces refering to ISO 13041-6:2009 [12].

Figures 15 and 16 show the modification test piece and cylindrical test piece. The flatness test piece and circular test piece are displayed in Figs. 17 and 18. Those figures are completed with dimension information of each featured product.

The result of the machining workpiece is shown in Fig. 19. Figure 20 shows the geometric error of its workpieces measured using a CMM machine and compared with the machine tool's geometric error. The position of the machining workpiece shape feature is adjusted to the measurement position of the no-load condition.

The geometric error of its workpieces is measured using a CMM machine and compared with the machine tool's geometric error. The position of the machining workpiece shape feature is adjusted to the measurement position of the no-load condition.

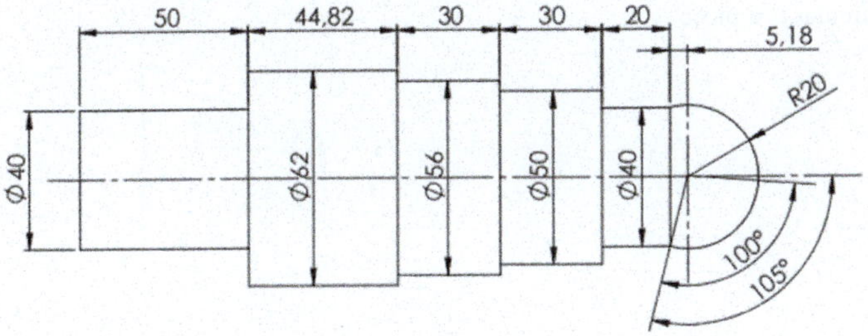

Fig. 15 Modification test piece

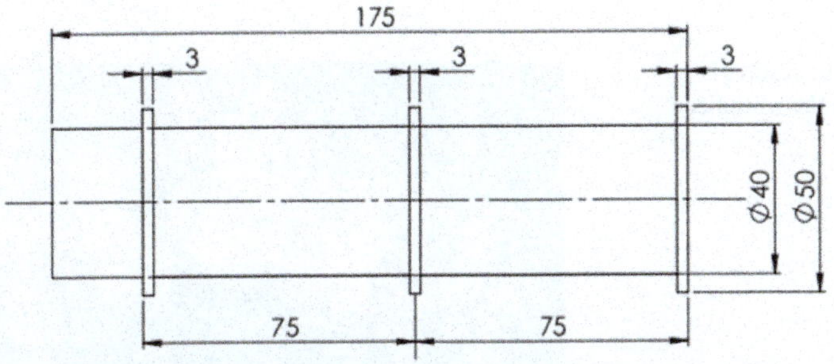

Fig. 16 Cylindrical test piece

Fig. 17 Flatness test piece

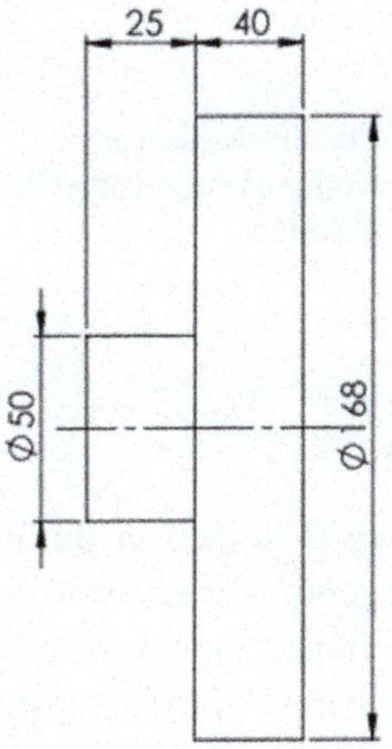

The measurement result of geometric deviations is obtained by comparing the value between the results of the ballbar measurement with the value of the measurement results of the cutting test object. The value compared is the deviation magnitude that occurs at the same machine coordinates. The tolerance standards used are ISO 10791 and ISO 13041. The geometric error in the results of the modification workpiece test, flatness test, and circular test state that the machine tools are still

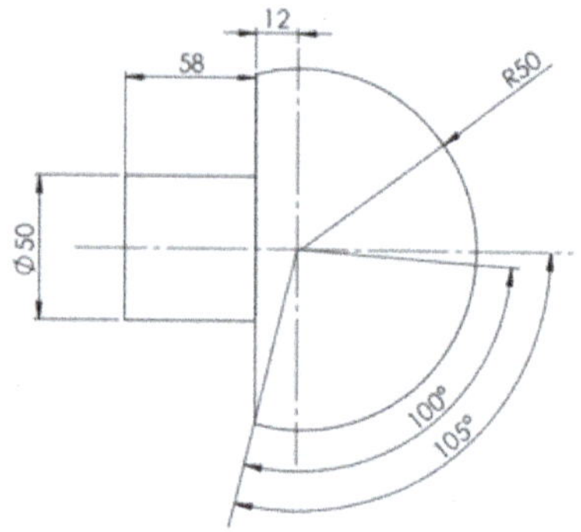

Fig. 18 Circular test piece

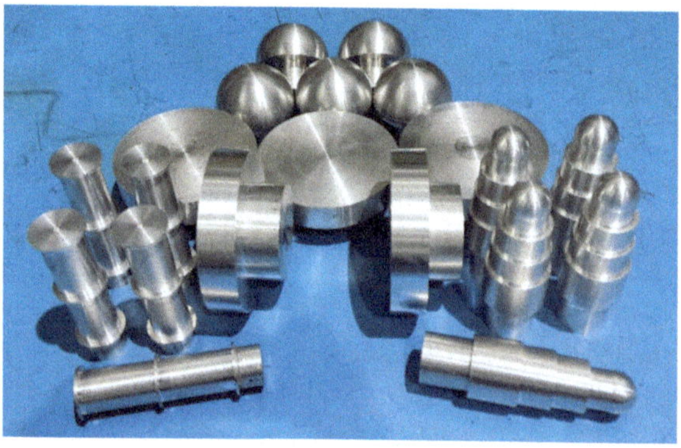

Fig. 19 The result of machining the workpiece

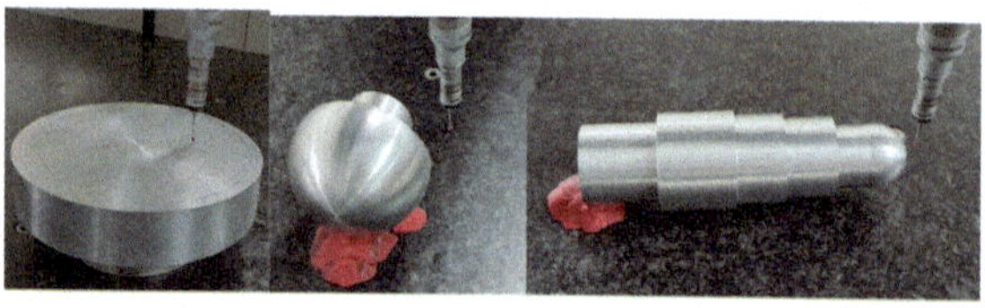

Fig. 20 Measuring workpiece test using CMM machine

in tolerance in the no-load condition test and the in-load condition test. But the geometric error of the cylindrical test exceeds tolerance in the load condition test. Table 5 shows the measurement result of the bidirectional circular deviation for each featured product tested. The geometric error in machining products is caused by the machine tools' geometric error, cutting tools deflection, thermal drift, or other factors. Hence the geometric error on the product should be studied in future research.

Table 5 Measurement result of bidirectional circular deviation

	No-load condition test (μm)	Load condition test (μm)
Cylindrical test	22	24.748
Flatness test	16.64	16.64
Circular test	16.28	16.28
Modification workpiece test	15.24	14.916

4 Conclusion

The circular geometric error assessment of CNC machine tools with a case Gildemeister CTX 310 Eco has been conducted. The ballbar measurements are tested by measuring the position of the cylindrical test, flatness test, circular test and modification workpiece test. The geometric error in the results of the modification workpiece test, flatness test, and circular test that ate the machine tools are still in tolerance in the no-load condition test and the in-load condition test. The measurement result value of the bidirectional circular deviation of the modification workpiece test, flatness test, and circular test is 15.24, 16.64, and 16.28 μm. Yet, the geometric error of the cylindrical test is 24.748 μm, and it exceeds the tolerance of the in-load condition test. Based on the assessment, the recommendations for maintenance activities include the inspection and repair of active mechanical driving components, such as servo motors and couplings, ballscrews, and linear guideway components.

Acknowledgements This work was supported by the Manufacturing Engineering Department of POLMAN Bandung, Indonesia. We would like to thank them for their support.

References

1. Harja, H.B., Fauzi, M., Fathan, A.: Evaluasi kinerja efisiensi energi mesin bubut melalui penilaian indikasi specific energy consumption. J. ELKOMIKA. **9**, 389 (2021)
2. Harja, H. B., Candra, A.: Assessment of positional error CNC machine tools using laser interferometer. In: Proceedings of the 4th International Conference on Applied Science and Technology on Engineering Science—ICAST-ES (2023). https://doi.org/10.5220/0010965300003260
3. Ekinovic, S., Begovi, E., Hekic, E.: Ballbar QC10 measuring device: fast experimental check of machine tools accuracy. J. Trends Dev. Mach. Assoc. Technol. **19**(1), 29–32 (2015)
4. Holub, M., Bradac, F., Pokorny, Z., Jelinek, A.: Application of a ballbar for diagnostics of CNC machine tools. MM Sci. J. **12**, 8032 (2018)
5. Monika, W., Paweł, M.: Reproducibility of machine tools' circularity test according to ISO 230–4 with respect to testing position. Arch. Mech. Technol. Mater. **37**(1), 22–26 (2017)
6. Čep, R., Malotová, Š., Kratochvíl, J., Stančeková, D., Czán, A.: Diagnosis of machine tool with using Renishaw ballbar. 01006, 1–7 (2018)
7. International Organization for Standardization ISO 230-4: Circular Test for Numerically Controlled Machine Tools. International Organization for Standardization, Switzerland (2022)

8. International Organization for Standardization ISO 230-1: Tests Code for Machine Tools—Part 1. Geometric Accuracy of Machines Operating Under No-Load or Quasi-Static Conditions. International Organization for Standardization, Switzerland (2012)
9. Beňo, M., Zvončan, M., Kováč, M., Peterka, J.: Circular interpolation and positioning accuracy deviation measurement on five axis machine tools with different structures. Teh. Vjesn. **20**(3), 479–484 (2013)
10. Renishaw: QC-10 Ballbar Training Course Manual 9 Part 1 and 2 (2001)
11. Renishaw: QC-20 Ballbar Training Course Manual 9 Part 1 and 2 (2010)
12. International Organization for Standardization ISO 13041-6: Test Conditions for Numerically Controlled Turning Machines and Turning Centres—Part 6 . Accuracy of a Finished Test Piece. International Organization for Standardization, Switzerland (2009)
13. International Organization for Standardization ISO 10791-2: Test Conditions for Machining Centres—Part 2. Geometric Tests for Machines with Vertical Spindle or Universal Heads with Vertical Rotary Axis (Vertical Z-Axis). International Organization for Standardization, Switzerland (2001)
14. International Organization for Standardization ISO 10791-4: Test Conditions for Machining Centres-Accuracy and Repeatability of Positioning of Linear and Rotary Axes. International Organization for Standardization, Switzerland (1998)
15. International Organization for Standardization ISO 13041-4 : Test Conditions for Numerically Controlled Turning Machines and Turning Centres—Part 4 : Accuracy and Repeatability of Positioning of Linear and Rotary Axes. International Organization for Standardization, Switzerland (2004)
16. Prakosa, T., Wibowo, A., Yuwana, Y., Nurhadi, I.: Pengujian Ketelitian Geometrik Mesin Perkakas CNC Milling Vertikal Buatan Dalam Negeri. Seminar Nasional Teknik Mesin IX, Palembang (2010)

Assessment of Geometric Error for a CNC Vertical Milling Machine at No-Load Condition

Haris Setiawan, Herman Budi Harja, Julianti Putri Sa'adiyah, Risky Ayu Febriani, Andri Pratama, Addonis Candra, and Mohd Amran Mohd Daril

Abstract CNC machine tools are used for a long time and high-frequency results in various types of machine components experiencing damage and wear. It makes deviations from the original positioning accuracy of the CNC machine tool occur. This paper is carried out to identify geometric errors that occur in the DMTG XD-40A CNC vertical milling machine by measuring at the no-load condition with a ballbar and laser interferometer to ensure optimal machine performance. The results of the measurement with a ballbar show that the measurement values met the assumption of normality and the values of circular deviation, radial deviation, bi-directional circular deviation, and mean bi-directional radial deviation in μm of the *YZ* plane are 9.822; 12.074; 31.944; and 11.972, while the *ZX* plane is 9.832; 16.899; 31.112; 17.192. The results of the measurement and analysis of the *XY* plane show that the type of deviation (error) that contributes greatly to circular error is scaling mismatch error with 15%. The results of the laser interferometer measurements are compared with ISO 10791-2:2001 and ISO 10791-4:1998, hence the geometric deviation conditions are obtained. The results of the no-load measurement give the recommendations for maintenance activities in the form of compensation values for machine parameters as well as recommendations for maintenance and repair activities on

H. Setiawan (✉) · H. B. Harja · J. P. Sa'adiyah · R. A. Febriani · A. Pratama · A. Candra
Politeknik Manufaktur Bandung, Jl. Kanayakan No. 21, Bandung 40135, Indonesia
e-mail: haris@polman-bandung.ac.id

H. B. Harja
e-mail: herman@polman-bandung.ac.id

R. A. Febriani
e-mail: riskyayuf@polman-bandung.ac.id

A. Pratama
e-mail: pratama@polman-bandung.ac.id

A. Candra
e-mail: addonisc@polman-bandung.ac.id

M. A. M. Daril
Universiti Kuala Lumpur Malaysian Institute of Industrial Technology, Masai, Johor, Malaysia
e-mail: mamran@unikl.edu.my

A. Ismail et al. (eds.), *Engineering Frontiers*, Advanced Structured Materials 208,
https://doi.org/10.1007/978-3-031-56844-2_23

the machine's linear guideway as a form of maintenance related to the geometric deviation conditions of the CNC DMTG XD-40A machine tool in the X-axis and Y-axis.

Keywords Geometric error measurement · CNC milling machine · Ballbar · Laser interferometer · Linear guideway

1 Introduction

Long-term usage of CNC machine tools will decrease the quality of machine tools, which is related to deviations of machine tools. The amount of deviation should not exceed the standard allowable limit. Its deviation can influence the quality of the machined workpiece. The usage intensity of the machine tool also affects the decline in the ability and quality of the machine. Its usage intensity of CNC machine tools will have an impact on the decrease of machine performance, especially causing damage and wear of various types of components that can cause deviations to the machine's accuracy.

More than 50% of the total machining process errors are caused by machine tools fault sources, such as geometric errors, servo errors, and thermal errors [1]. Geometric errors are the biggest cause of machine tool accuracy and precision. Hence, the machine tool geometric error contributes significantly to the dimensional and geometric errors of machine workpieces. Machine tool geometric errors that occur can be reduced in two ways (i) design and manufacture of precise machine tools, and (ii) adjust error compensation parameters [2].

The geometric error in machine tools is a condition of moving accuracy during the machining process. The geometry accuracy affects the machine tool performance and directly affects the geometry accuracy of the workpiece [3]. Geometric errors of one-axis machine tools have six geometric errors. Those are linear positional geometric error, linear straightness vertical error, linear straightness horizontal error, angular error to the *X* axis (rolling motion), angular error to the *Y* axis (pitching motion), and angular error to the *Z* axis (yawing motion) [4].

The machine tools with three axes (*X*, *Y*, and *Z*) have eighteen geometric errors from three axes *X*–*Y*–*Z* and three perpendicular geometric errors between each axis [4]. Machine tools with capabilities of more than three axes, it has an additional error type in the form of rotational errors on the rotating motion axis. The table of machine tool geometric error types is shown in Table 1. Furthermore, a study of the linear position error and straightness geometric error assessment was carried out on a CNC milling machine using a laser interferometer [5, 6].

One of the advantages of CNC machine tools compared to conventional machine tools is the ability to produce interpolation movements that are realized by moving more than one axis simultaneously [7]. The measurement of geometric accuracy of this study was to measure the 21 geometric errors contained in Table 1, and also to

Table 1 Table components of geometric error in machine tools [4]

No	Item		Notation
1	Linear displacement error		$\delta x(x)$, $\delta y(y)$, $\delta z(z)$
2	Straightness error	Vertical straightness errors	$\delta z(x)$, $\delta z(y)$, $\delta y(z)$
		Horizontal straightness errors	$\delta y(x)$, $\delta x(y)$, $\delta x(z)$
3	Angular error	Pitch angular error	$\varepsilon y(x)$, $\varepsilon x(y)$, $\varepsilon x(z)$
		Yaw angular error	$\varepsilon z(x)$, $\varepsilon z(y)$, $\varepsilon y(z)$
		Roll angular error	$\varepsilon x(x)$, $\varepsilon y(y)$, $\varepsilon z(z)$
4	Squareness error		$\phi\, x(y)$, $\phi\, x(z)$, $\phi y\,(z)$,
5	Rotational error		θx, θy, θz

measure the circular error on the CNC machine tool where measurement and analysis refer to applicable ISO standards.

2 Methodology

The machine tool used as the object of this research is a CNC milling DMTG XD-40A machine with a travel axis on the *X*-axis, *Y*-axis, and *Z*-axis are 600 mm × 420 mm × 520 mm as shown in Fig. 1.

There are several stages for geometric error assessment activities on machine tools: measuring geometric errors on machine objects, analyzing measurement data results, evaluating deviation status by comparing deviation item information values with ISO standard tolerances related to geometric errors, calculating compensation

Fig. 1 CNC vertical milling machine dmtg xd 40-a

parameter values, determining factors that the main causes of deviations, and provide recommendations for maintenance and repairs on the machine tool test object based on the identification of the causative factors. The activity of identifying the magnitude of the deviation values and the status of the deviations of the machine geometry is based on ISO 10791-2:2001 and ISO 10791-4:1998. The tools that used in this test are:

a. Ballbar: Renishaw ballbar QC-10
b. Laser interferometer: Renishaw laser interferometer XL-80

Measurements using a laser interferometer are carried out based on the standard determination and measurement of accuracy and repeatability of the numerically controlled axis position of the machine tool at no-load conditions by means of geometric measurements and machining tools ISO 230-2, while the circular interpolation test is adjusted to the testing standards and methods of evaluating the precision of the machine tool geometry through the circular error and radial error of the circular path generated by the interpolation motion of the CNC machine, ISO 230-4 CNC.

Circular interpolation motion can be defined as a movement of more than one axis that moves simultaneously where the movement is carried out in a circular state [8]. Geometric accuracy testing using a laser interferometer cannot be used for circular interpolation accuracy because measurements are made for each axis, while measurements between axis only produce squareness parameters. In this test, the ballbar is used to test the accuracy of circular interpolation (Fig. 2).

The trajectory of the measurements can be seen in Fig. 3. Measurements with the ballbar that were carried out are in the *XY*, *YZ*, and *ZX* planes. In the *XY* plane measurement, the g-code moves the test object machine in a circular path with a

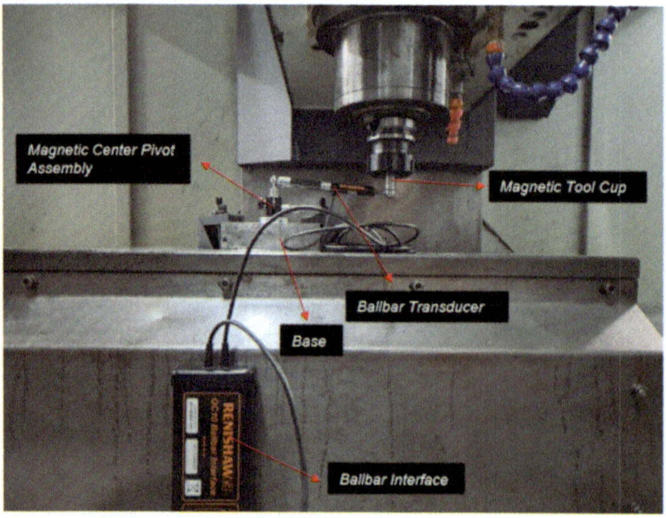

Fig. 2 Measurement configuration using ballbar

ballbar movement circle angle of 360° where the radius is the same as the selected ballbar length, 150 mm. In the measurement of the *YZ* plane and *ZX* plane, the g-code moves the test object machine in a circular path with a ballbar movement circle angle of 180° where the radius is equal to the length of the ballbar radius, 150 mm [9] (Fig. 4).

Measurements were made at five points of rotation center or five measurement positions where in each position five measurements were repeated. Determination of the chosen position is determined because the position represents a place that is often used for the work area and represents the maximum position of the axis travel that can be reached by the ballbar [9]. A trajectory carried out on measurements is shown in Fig. 5.

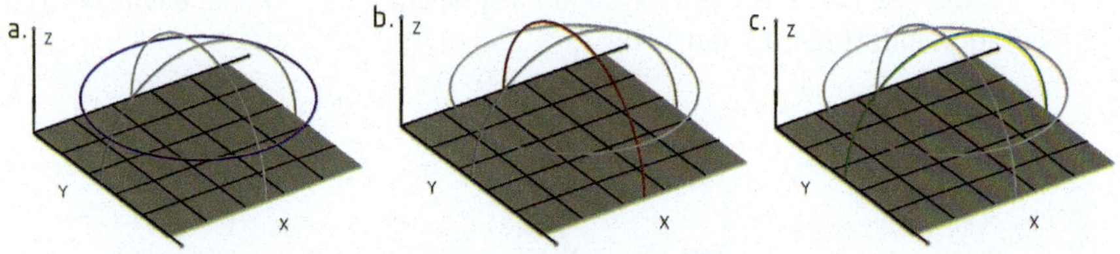

Fig. 3 Ballbar measurement path in **a** *xy*, **b** *yz*, and **c** *zx* planes

Fig. 4 Ballbar measurement features based on modified circle diamond square test piece measurement features

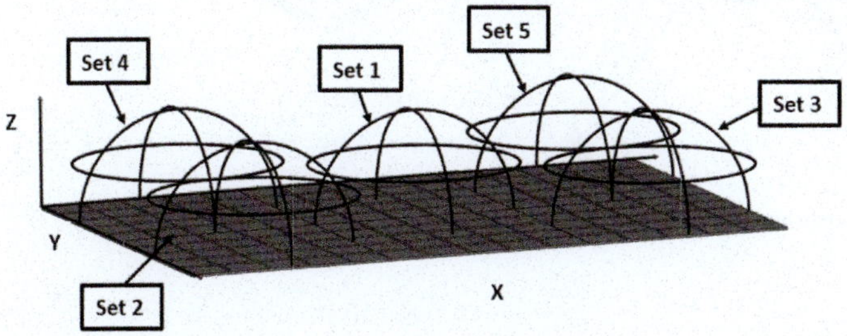

Fig. 5 Trajectory of measurement with ballbar at five positions

Figure 6 shows the setup measurement using a laser interferometer. Interference can be formed when the laser beam is separated so it is divided into two parts, a test beam, and a reference beam. In the measurement process using a laser interferometer, there is a phase difference between the reference beam and the measurement beam due to an error in the geometry of the machine and the laser system will calculate the magnitude of the phase difference. The geometrics error that can be measured by using a laser interferometer in this research is positioning accuracy of linear axes, horizontal and vertical straightness, angular pitch and angular yaw, and squareness errors between two axes as shown from Figs. 7, 8, 9, 10, 11 and 12 [7, 10].

Environmental and thermal conditions of machine tools are two factors that must be considered in measuring operations with a laser interferometer. Environmental condition factors are air temperature, air pressure, and relative humidity [11]. Therefore, the laser interferometer system must be equipped with an environmental compensation unit (Figs. 13 and 14).

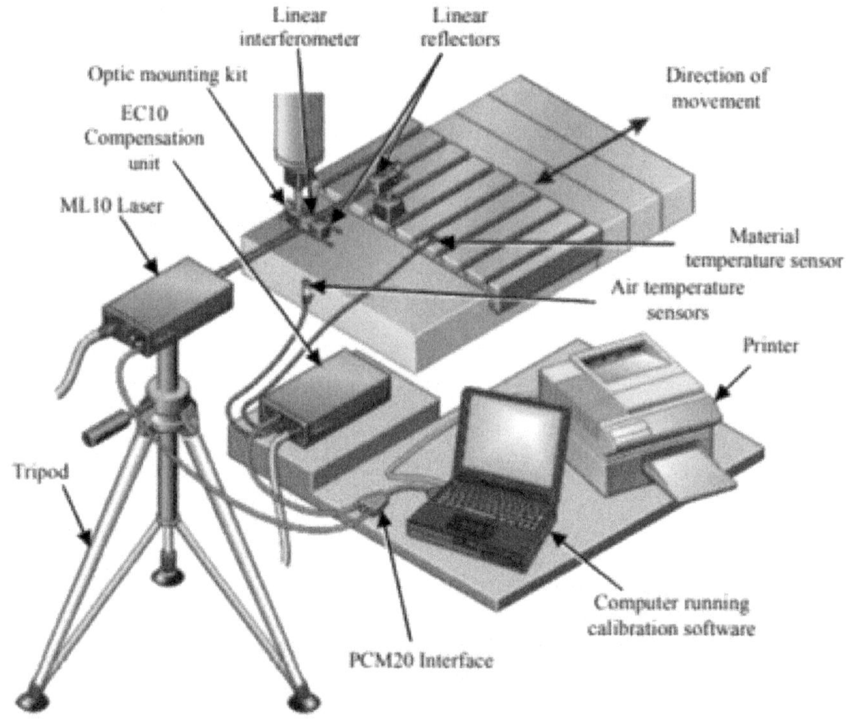

Fig. 6 Measurement set up using a laser interferometer

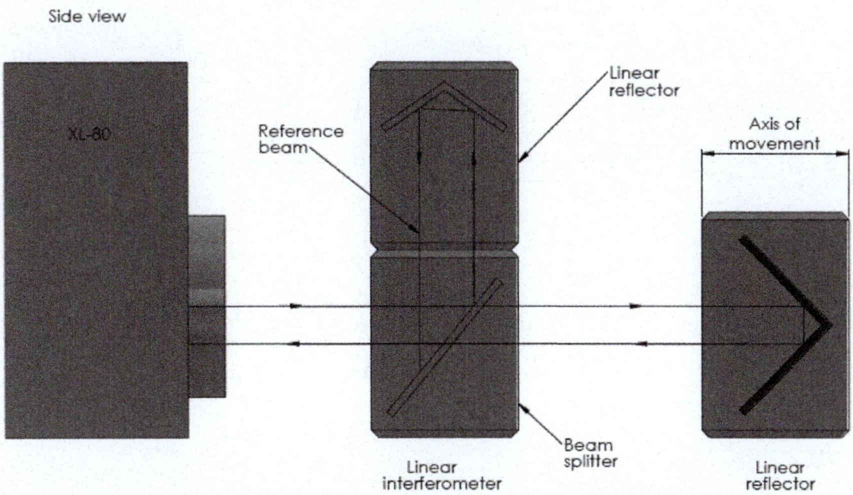

Fig. 7 Principle of positioning error measurement with laser interferometer

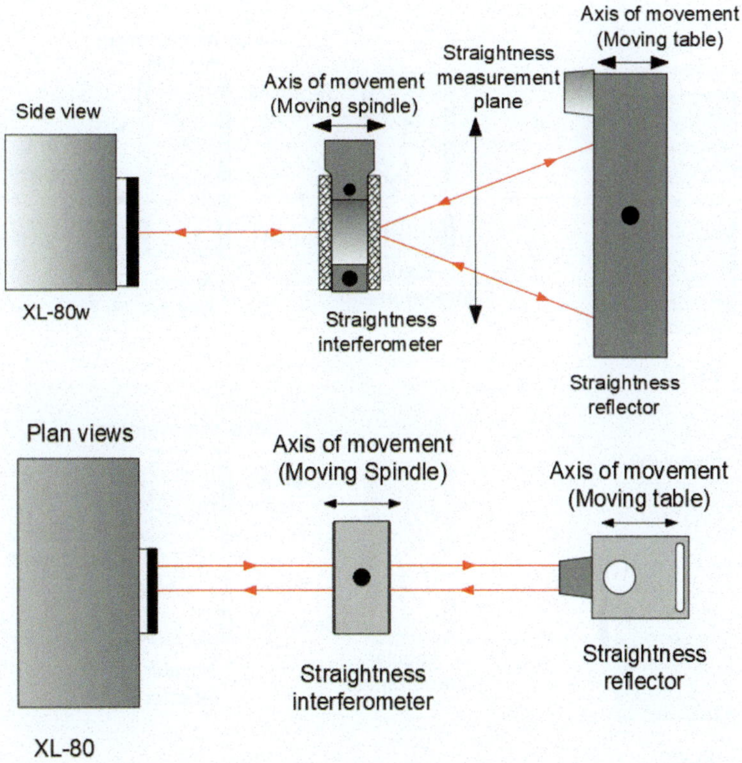

Fig. 8 Principle of horizontal axis straightness error measurement with laser interferometer

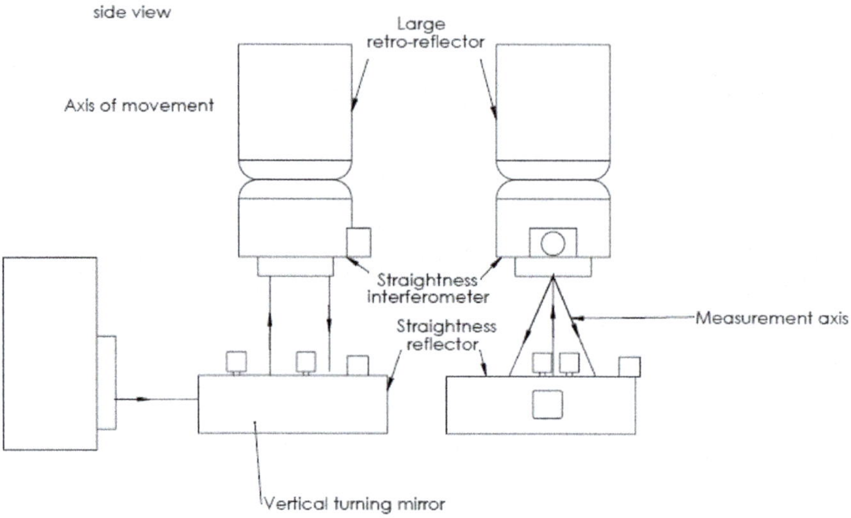

Fig. 9 Principle of vertical axis straightness error measurement with laser interferometer

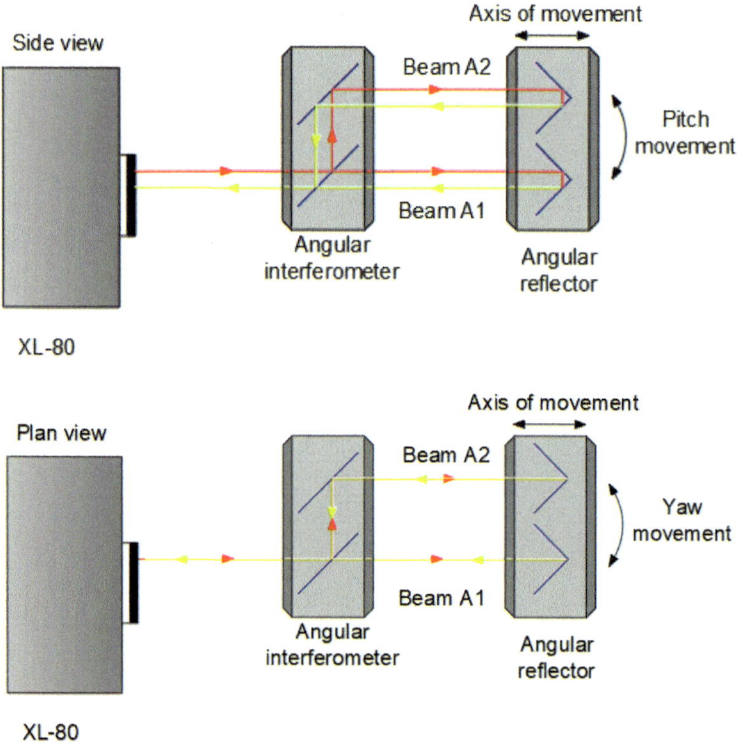

Fig. 10 Principle of angular error measurement with laser interferometer

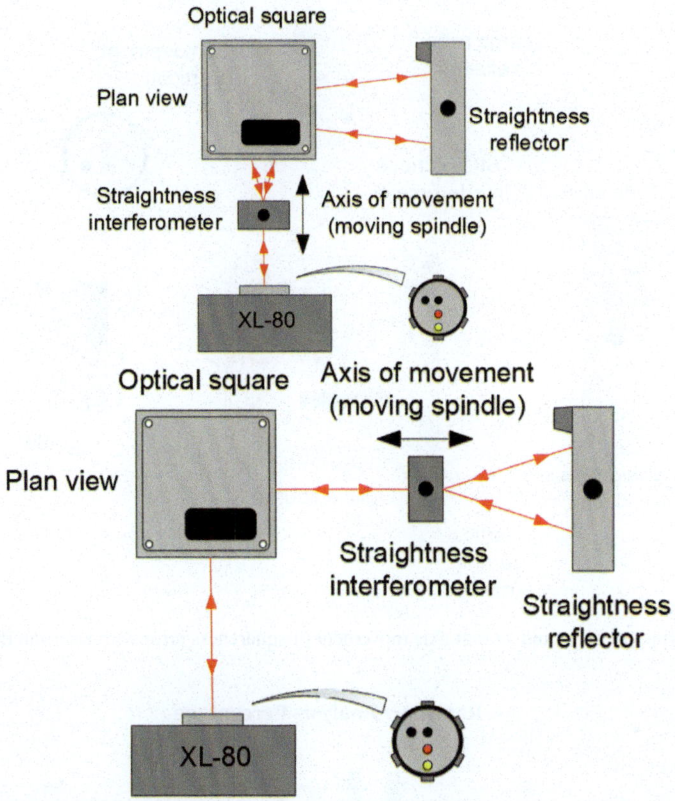

Fig. 11 Measurement of xy-squareness error with a laser interferometer

3 Results and Discussion

Measurements using a ballbar were carried out five times on the same plane with the same measurement parameters. The results of measurements in the *XY* plane are shown in Table 2 and Fig. 13. From the histogram graph, it can be seen that the deviation item that contributes the highest to the occurrence of circular error is the scaling mismatch error with 15%. Then, in the pareto diagram, the cumulative value of 80% is selected as the limit for taking deviations that contribute to the circular error. The types of deviations included in the 80% limit value include scaling mismatch error, lateral play *X* and *Y* error, squareness error, backlash *X* and *Y* error, cyclic error *X* and *Y*, and straightness error on the *X* axis. The main factors that cause geometric deviations are determined according to the previous analysis where the determination of the main factors is based on fishbone diagrams of geometric and kinematic error points on machine tools as shown in Fig. 15.

Based on the results of the analysis, the main factors that cause geometric deviations in the *XY* plane are caused by damage and errors that occur in the machine's

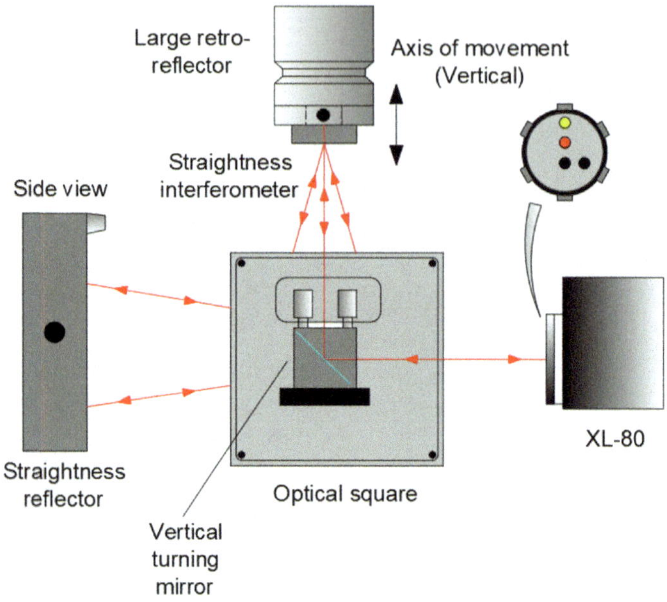

Fig. 12 Principle of yz and xz first axis movement of squareness error with laser interferometer

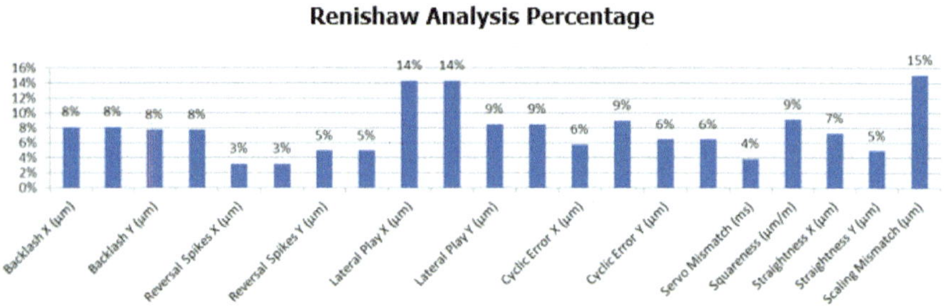

Fig. 13 Histogram graph of percentage contribution of deviations from renishaw plot analysis results on *xy* plane

linear guideway. Damage and errors that occur in the linear guideway can be caused by several things such as misalignment in the linear guideway, wear on the linear guideway, linear guideway that does not move freely, bent linear guideway, and motion errors that occur in the linear guideway.

Deformation will occur between the raceway surfaces and rolling elements when the linear guideway is subjected to loads during movement (in machining activities). Under such conditions, the rolling balls on both sides of the raceway will experience different levels of contact loading, which causes a different normal stiffness at the rolling interface and therefore affects the vibration of the linear guideway and causes uneven loading of the guideway system. In measurements using a ballbar in the *YZ*

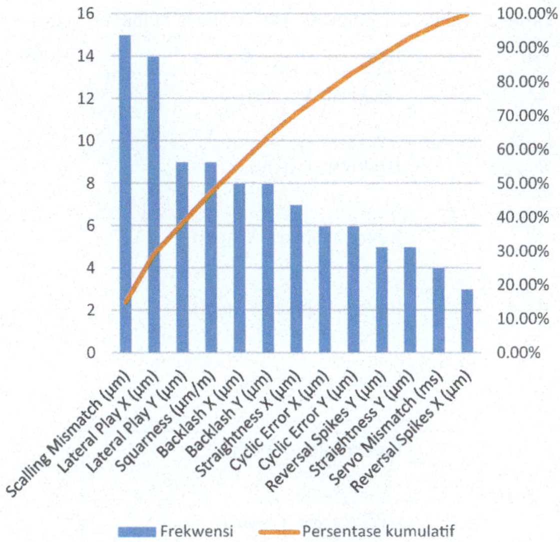

Fig. 14 Pareto diagram of percentage contribution of deviation result of renishaw plot analysis on *xy* plane

and *ZX* plane with a measurement angle of 180°, Renishaw ballbar plot analysis cannot be produced as in measurements with a measurement angle of 360°. The information output becomes less because the algorithm on the Renishaw ballbar requires a path of 2π so that the analysis results that can be seen are in accordance with the ISO-230-4:2005 standard.

An analysis was performed with a normality test to check whether the value of the measurement results for each position with five repetitions of measurements had a normal distribution or distribution of data and whether the values of the repeated measurements at each position had the same population or not. The calculation of the normal distribution was carried out using the Kolmogorov–Smirnov normality test formulation. The results of the normality test indicated that the measurement values carried out met the normality assumption with the significance value of the normality test resulting greater than 0.05 (significance determined). The average deviation in the *YZ* and *ZX* planes can be seen in Table 3.

The laser interferometer cannot be used to measure the angular roll error. Thus, the errors measured when testing include the accuracy of the positioning of the linear axes, horizontal and vertical straightness error, angular pitch and angular yaw error, and squareness error between axes. To ensure that the number of measurements that are carried out is sufficient, a data adequacy test is carried out by comparing the value of N measured with N' the results of manual calculations. An example of the adequacy test performed on the measurement of the linear positioning error on the Y-axis is shown in Fig. 16. Table 4 shows the calculation of the data adequacy test carried out on each axis in each deviation item was measured using a laser interferometer with the same calculation formula as in the above.

Measurements were performed using a laser interferometer, based on the ISO 10791-4:1998 standard on the measurement of positioning and repeatability of

Table 2 Table of average deviation values in the *XY* plane

No	Item	Value	
1	Backlash *X* (μm)	<	− 2.004
		>	− 2.148
2	Backlash *Y* (μm)	∨	− 1.6
		∧	− 0.972
3	Reversal spikes *X* (μm)	<	0.808
		>	0.44
4	Reversal spikes *Y* (μm)	∨	− 1.576
		∧	1.184
5	Lateral play *X* (μm)	<	− 1.928
		>	− 3.204
6	Lateral play *Y* (μm)	∨	2.252
		∧	1.208
7	Cyclic error *X* (μm)	↑	2.148
		↓	2.96
8	Cyclic error *Y* (μm)	↑	3.22
		↓	2.408
9	Servo mismatch (ms)		0.0384
10	Squareness (μm)		27.82
11	Straightness *X* (μm)		− 7.152
12	Straightness *Y* (μm)		− 2.788
13	Scalling error *X* (ppm)		143.036
14	Scalling error *Y* (ppm)		93.218
15	Circurality (μm)		27.744

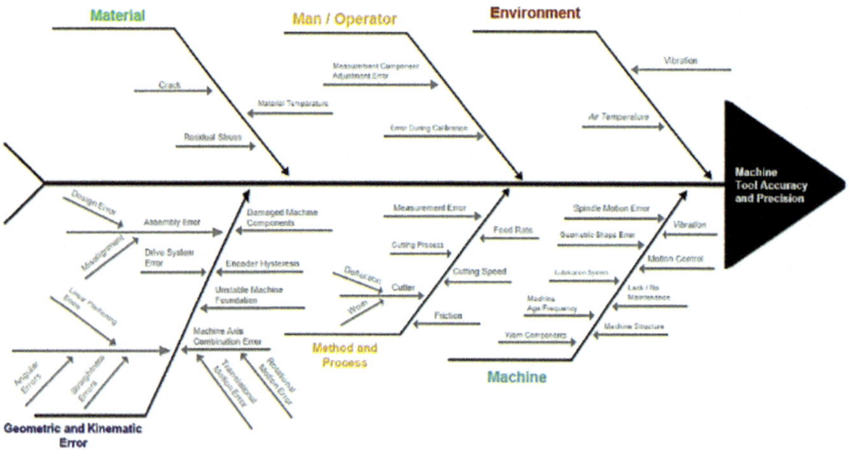

Fig. 15 Fishbone diagram of the factors causing geometric deviations on machine tools

Table 3 Table of average deviation values on *YZ* and *XZ* plane

Position	Circular deviation CCW (μm)	Circular deviation CW (μm)	Radial deviation CCW (μm)		Radial deviation CW (μm)		Bidirectional circular deviation (μm)	Bidirectional radial deviation (μm)
			Max	Min	Max	Min		
1	9.92	8.4	19.28	9.34	16.04	7.66	30.84	12.38
2	10.74	11.12	15.9	5.16	17.28	6.2	32.2	10.96
3	14.16	7.1	19.06	4.94	16.4	8.96	33.04	12.64
4	11.02	8.48	17.78	6.74	14.32	6.06	33.22	10.92
5	9.72	7.44	18.52	8.76	15.24	7.84	30.42	12.52
Mean	11.112	8.532	18.108	6.988	15.856	7.344	31.792	11.972

XZ plane

Position	Circular deviation CCW (μm)	Circular deviation CW (μm)	Radial deviation CCW (μm)		Radial deviation CW (μm)		Bidirectional circular deviation (μm)	Bidirectional radial deviation (μm)
			Max	Min	Max	Min		
1	9.98	8.36	18.8	8.82	15.54	7.2	30.8	12.38
2	10.74	11.12	15.9	5.16	17.28	6.2	32.2	10.96
3	9.56	7.1	21.46	11.9	18.48	11.38	31.94	15.58
4	10.94	8.48	17.86	6.92	14.66	6.14	33.6	11.18
5	9.7	7.44	18.94	9.18	15.68	8.26	30.42	12.86
Mean	10.184	8.5	18.592	8.396	16.328	7.836	31.792	12.592

Fig. 16 An example of the adequacy test performed on the measurement of the linear positioning error on the *y*-axis

$$N' = \left(\frac{\frac{C}{a} \cdot \sqrt{N \sum_{i=1}^{N} x_i^2 - \left(\sum_{i=1}^{N} x_i\right)^2}}{\sum_{i=1}^{N} x_i} \right)^2$$

$$N' = \left(\frac{\frac{2}{0.05} \cdot \sqrt{331.236 - (1655.864)^2}}{-40.692} \right)^2$$

$$N' = 0.305$$
$$N' = 5$$
$$(N' < N)$$

machine tool geometries. The positioning deviation and repeatability on the *X*-axis is 24.8 μm, the *Y*-axis is 11.4 μm, and the *Z*-axis is 17.7 μm. They are already past the permissible tolerances according to ISO standards where the tolerance values for the *X*-axis, *Y*-axis, and *Z*-axis are 10.0 μm. These values are shown in Table 5.

According to the ISO 10791-2:2001 standard, on angular measurements, the accuracy of the angular pitch on the *Y*-axis is 23.8 μrads and on the *Z*-axis is 16.1 μrads, it shows the accuracy of the angular pitch on the *Y*-axis and *Z*-axis exceeded the limit.

Table 4 Table of data adequacy test results

Item	N′	N	Note
Linear positioning *X*	4.063	5	Ok
Linear positioning *Y*	0.305	5	Ok
Linear positioning *Z*	0.325	5	Ok
Horizontal straightness *X*	1.392	5	Ok
Horizontal straightness *Y*	4.654	5	Ok
Horizontal straightness *Z*	4.624	5	Ok
Vertical straightness *X*	4.206	5	Ok
Vertical straightness *Y*	3.585	5	Ok
Vertical straightness *Z*	1.339	5	Ok
Angular pitch *X*	1.049	5	Ok
Angular pitch *Y*	1.459	5	Ok
Angular pitch *Z*	2.850	3	Ok
Angular yaw *X*	4.857	5	Ok
Angular yaw *Y*	1.598	5	Ok
Angular yaw *Z*	0.245	3	Ok

The repeatability of the angular pitch on the *Y*-axis is 20.7 μrads and this value has exceeded the permissible limit. The average deviation value in the angular pitch is still within the tolerance value determined by the ISO 10791-2:2001 standard. These values are shown in Table 6.

According to the ISO 10791-2:2001 standard, on angular measurements, it shows that the accuracy of angular yaw on the *Y*-axis and *Z*-axis has passed the permissible limit, which is 16.5 μrads for the *Y*-axis and 15.0 μrads for the *Z*-axis. The repeatability of angular yaw on the *Y*-axis is 17.1 μrads, where this value has exceeded the permissible limit of 16.5 μrads. The average deviation value of the angular yaw on the *Y*-axis and *Z*-axis is 17.3 and 20.5 μrads, respectively, shows that the average value of the angular yaw deviation on the *Y*-axis and *Z*-axis has crossed the specified limit permitted which is 16.5 μrads for the *Y*-axis and 15.0 μrads for the *Z*-axis. These values are shown in Table 7.

The measurement results of the straightness errors, both horizontal and vertical, all measurement items are still within the tolerance value limits according to the ISO 10791-2:2001 standard as shown in Tables 8 and 9.

The results of squareness error measurement on both the *XY*-plane, *YZ*-plane, and *XZ*-plane, all measurement items are still within the tolerance value limits according to the ISO 10791-2:2001 standard as shown in Table 10. Table 11 is an example of the value of the compensation parameter on linear positioning error on the *X*-axis machine. The value of the compensation parameter can be used as a reference for resetting the error compensation parameter on the machine control to correct the linear positioning geometric error of the machine tool.

Table 5 Software results of positioning and repeatability of linear axis measurement

X-axis (measurement travel = 500 mm)				Tolerance	
Item	Forward (μrads)	Reverse (μrads)	Bidir (μrads)	Unidir (μrads)	Bidir (μrads)
Accuracy	27.5 (F)	28.5 (F)	32.3 (F)	16	22
Repeatability	4.20 (O)	4.70 (O)	9.0 (O)	6	12
Systematic deviation	25.1 (F)	24.5 (F)	28.8 (F)	10	15
Reversal			5.6 (O)		10
Mean reversal			4.2 (O)		6
MeanDeviation			24.8 (F)		10
Y-axis (measurement travel = 300 mm)				Tolerance	
Item	Forward (μrads)	Reverse (μrads)	Bidir (μrads)	Unidir (μrads)	Bidir (μrads)
Accuracy	13.5 (O)	13.0 (O)	16.5 (O)	16	22
Repeatability	3.0 (O)	2.5 (O)	6.6 (O)	6	12
Systematic deviation	11.5 (F)	11.2 (F)	14.7 (O)	10	15
Reversal			5.0 (O)		10
Mean reversal			3.3 (O)		6
MeanDeviation			11.4 (F)		10
Z-axis (measurement travel = 300 mm)				Tolerance	
Item	Forward (μrads)	Reverse (μrads)	Bidir (μrads)	Unidir (μrads)	Bidir (μrads)
Accuracy	18.5 (F)	18.0 (F)	20.6 (O)	16	22
Repeatability	1.6 (O)	1.5 (O)	3.7 (O)	6	12
Systematic deviation	17.8 (F)	17.7 (F)	19.8 (F)	10	15
Reversal			2.3 (O)		10
Mean reversal			2.2 (O)		6
MeanDeviation			17.7 (F)		10

Note O = Ok; F = Failed

Table 6 Table of software results of angular pitch error measurement

X-axis (measurement travel = 500 mm)				Tolerance	
Item	Forward (μrads)	Reverse (μrads)	Bidir (μrads)	Unidir (μrads)	Bidir (μrads)
Accuracy	19.4 (O)	17.9 (O)	19.8 (O)	30	30
Repeatability	6.8 (O)	7.7 (O)	8.4 (O)		
Systematic deviation	15.5 (O)	14.5 (O)	15.6 (O)		
Reversal			2.3 (O)		
Mean reversal			− 0.3 (O)		
MeanDeviation			15.0 (O)		
Y-axis (measurement travel = 300 mm)				Tolerance	
Item	Forward (μrads)	Reverse (μrads)	Bidir (μrads)	Unidir (μrads)	Bidir (μrads)
Accuracy	23.0 (F)	23.3 (F)	23.8 (O)	16.5	16.5
Repeatability	20.7 (F)	18.3 (F)	20.7 (O)		
Systematic deviation	5.2 (O)	5.9 (O)	5.9 (O)		
Reversal			2.6 (O)		
Mean reversal			− 0.8 (O)		
MeanDeviation			4.8 (O)		
Z-axis (measurement travel = 300 mm)				Tolerance	
Item	Forward (μrads)	Reverse (μrads)	Bidir (μrads)	Unidir (μrads)	Bidir (μrads)
Accuracy	15.1 (F)	16.1 (F)	16.1 (F)	15	15
Repeatability	9.4 (O)	3.9 (O)	9.4 (O)		
Systematic deviation	10.1 (O)	12.2 (F)	12.2 (O)		
Reversal			1.1 (O)		
Mean reversal			0.1 (O)		
MeanDeviation			11.2 (O)		

Note O = Ok; F = Failed

Table 7 Table of software results of angular yaw error measurement

X-axis (measurement travel = 500 mm)				Tolerance	
Item	Forward (μrads)	Reverse (μrads)	Bidir (μrads)	Unidir (μrads)	Bidir (μrads)
Accuracy	18.3 (O)	18.8 (O)	20.7 (O)	30	30
Repeatability	5.1 (O)	6.4 (O)	8.3 (O)		
Systematic deviation	14.1 (O)	14.7 (O)	16.3 (O)		
Reversal			4.2 (O)		
Mean reversal			1.7 (O)		
MeanDeviation			14.4 (O)		
Y-axis (measurement travel = 300 mm)				Tolerance	
Item	Forward (μrads)	Reverse (μrads)	Bidir (μrads)	Unidir (μrads)	Bidir (μrads)
Accuracy	31.8 (F)	30.9 (F)	31.8 (F)	16.5	16.5
Repeatability	16.9 (F)	15.3 (O)	17.1 (F)		
Systematic deviation	16.7 (F)	17.9 (F)	17.9 (F)		
Reversal			1.7 (O)		
Mean reversal			– 0.4 (O)		
MeanDeviation			17.3 (F)		
Z-axis (measurement travel = 300 mm)				Tolerance	
Item	Forward (μrads)	Reverse (μrads)	Bidir (μrads)	Unidir (μrads)	Bidir (μrads)
Accuracy	22.1 (F)	23.1 (F)	23.5 (F)	15	15
Repeatability	2.4 (O)	4.0 (O)	4.0 (O)		
Systematic deviation	20.5 (F)	20.4 (F)	20.5 (F)		
Reversal			1.0 (O)		
Mean reversal			– 0.4 (O)		
MeanDeviation			20.5 (F)		

Note O = Ok; F = Failed

Table 8 Table of software results of vertical straightness error measurement

X-axis (measurement travel = 500 mm)				Tolerance	
Item	Forward (µrads)	Reverse (µrads)	Bidir (µrads)	Unidir (µrads)	Bidir (µrads)
Accuracy	–	–	7.7 (O)	10	10
Repeatability	2.1 (Ok)	1.5 (Ok)	2.2 (O)	10	10
Y-axis (measurement travel = 275 mm)				Tolerance	
Item	Forward (µrads)	Reverse (µrads)	Bidir (µrads)	Unidir (µrads)	Bidir (µrads)
Accuracy	–	–	2.8 (Ok)	10	10
Repeatability	1.4 (Ok)	2.2 (Ok)	2.2 (Ok)	10	10
Z-axis (measurement travel = 300 mm)				Tolerance	
Item	Forward (µrads)	Reverse (µrads)	Bidir (µrads)	Unidir (µrads)	Bidir (µrads)
Accuracy	–	–	4.2 (Ok)	10	10
Repeatability	0.8 (Ok)	0.8 (Ok)	0.8 (Ok)	10	10

Table 9 Table of software results of horizontal straightness error measurement

X-axis (measurement travel = 500 mm)				Tolerance	
Item	Forward (µrads)	Reverse (µrads)	Bidir (µrads)	Unidir (µrads)	Bidir (µrads)
Accuracy	–	–	6.3 (Ok)	10	10
Repeatability	2.5 (Ok)	2.9 (Ok)	3.0 (Ok)	10	10
Y-axis (measurement travel = 275 mm)				Tolerance	
Item	Forward (µrads)	Reverse (µrads)	Bidir (µrads)	Unidir (µrads)	Bidir (µrads)
Accuracy	–	–	2.5 (Ok)	10	10
Repeatability	0.7 (Ok)	0.8 (Ok)	1.1 (Ok)	10	10
Z-axis (measurement travel = 300 mm)				Tolerance	
Item	Forward (µrads)	Reverse (µrads)	Bidir (µrads)	Unidir (µrads)	Bidir (µrads)
Accuracy	–	–	3.4 (Ok)	10	10
Repeatability	2.4 (Ok)	2.3 (Ok)	2.5 (Ok)	10	10

Table 10 Table of software results of squareness error measurement

X–Y		Tolerance
Item	Bidirectional (μm/mm)	Bidirectional (μm/mm)
Squareness error	2.06 (μm/mm)	8 (μm/mm)
Y–Z		Tolerance
Squareness error	1.2134 (μm/mm)	12 (μm/mm)
X–Z		Tolerance
Squareness error	− 0.5142 (μm/mm)	12 (μm/mm)

Table 11 Table of compensation parameter error for linear positioning error of *X*-axis

Index	Position (mm)	Forward direction (μm)	Reverse direction (μm)
1	0	0	3
2	25	3	2
3	50	0	1
4	75	2	2
5	100	1	1
6	125	1	1
7	150	− 2	− 2
8	175	− 2	− 2
9	200	− 1	0
10	225	1	1
11	250	− 1	0
12	275	− 1	− 2
13	300	− 2	− 1
14	325	0	− 1
15	350	− 1	− 1
16	375	− 2	0
17	400	− 3	− 4
18	425	− 3	− 2
19	450	− 1	− 2
20	475	− 4	− 4
21	500	− 4	− 4

4 Conclusion

The identification of geometric error conditions on the CNC vertical milling machine tool DMTG XD 40-A was successfully conducted using the calibrated Renishaw ballbar QC 10 and Renishaw laser interferometer XL-80. The Kolmogorov–Smirnov method is used for normality testing of measurement result values on planes *YZ* and *ZX*. Its normality test results were greater than 0.05 (significance determined). The deviation value for each circular deviation item has normality distribution data. The measurement result values indicate the deviation value for each circular deviation item has a normality data distribution, and the value of the repeated measurement results at each position has the same population at each position in the same plane. The deviation parameters are obtained as the circular deviation, radial deviation, bi-directional circular deviation, and the mean bi-directional radial deviation in μm in the *YZ* plane are 9.822; 12.074; 31.944; and 11.972, while the *ZX* plane is 9.832; 16.899; 31.112; 17.192. The measurement results and analysis of the *XY* plane conclude that the most contributing deviation type to circular error is a scaling mismatch error of 15%.

Based on the analysis of the measurement results as well as the identification of the values and status of the machine geometric errors of the test object, it can be seen that the causal factor or disturbance that causes a geometric error on the XY plane machine are the result of damages and errors that occur in the machine's linear guideway. The values of the compensation parameter for geometric error from the identification and analysis based on the measurement results using a laser interferometer have been successfully defined and can be used as one of the recommendations for improvement to be given to the machine control of the machine object. The results of a deeper analysis based on the results of identification and analysis in the XY plane of the machine conclude that maintenance and repair activities need to be carried out on linear guideway on machine tools so that damage and errors that occur can be resolved. It is necessary to check the condition of the linear guideway on the machine first regarding the types of damage and errors in the linear guideway that occur. In-depth analysis activities with various methods need to be carried out so that they can find out the causal factors that affect geometric deviations in machine tools as a whole and not only in a certain plane.

References

1. Bryan, J.: International status of thermal error research. Ann. CIRP **39**, 645–656 (1990)
2. Tian, W., Gao, W., Zhang, D., Huang, T.A.: General approach for error modeling of machine tools. Int. J. Mach. Tools Manuf **79**, 17–23 (2014)
3. Ferreira, P.M., Liu, C.R.: An analytical quadratic model for the geometric error of a machine tool. J. Manuf. Sys. **5**(11), 51–63 (1986)
4. Okafor, A.C., Ertekin, Y.M.: Vertical machining center accuracy characterization using laser interferometer Part 1. Linear positional error. J. Mater. Process. Technol. **105**(3), 394–406 (2000)
5. Harja, H.B., Candra, A.: Assessment of positional error CNC machine tools using laser interferometer. In: Proceeding of the 4th International Conference on Applied Science and Technology on Engineering Science-ICAST-ES (2023). https://doi.org/10.5220/0010965300003260
6. Harja, H.B., Nurbaniah, A., Saksono, N., Noviandi, A.: Straightness geometric error assessment for CNC milling machine. Key Eng. Mater. **21**, 75 (2023)
7. Renishaw: XL 80 laser system training course manual, part 2. Renishaw Plc. (2000)
8. International Organization for Standrardization ISO 230-4: Circular tests for Numerically Controlled Machine Tools. International Organization for Standardization (2005)

9. Woźniak, M.M.: Reproducibility of machine tools' circularity test according to ISO 230-4 with respect to testing position. Arch. Mech. Technol. Mater. **37**, 22–25 (2017)
10. Renishaw: XL-80 laser system training course manual, part 1. Renishaw Plc. (2000)
11. Xinmeng, W., Jun, Y., Cedar, M., et al.: Analysis and prediction of thermal errors in feed system of precision coordinate boring machine. J. Xian Jiaotong Univ. **49**(10), 22–28 (2015)

Lignin Fractionation from Oil Palm Biomass Using Deep Eutectic Solvent as a Carbon Fibre Precursor

Siti Khadijah Amran, Afiqah Liana Sazali, Mohd Razealy Anuar, Khairul Faizal Pa'ee, and Tau-Len Kelly Yong

Abstract Deep eutectic solvents (DES) have emerged as a notable generation of environmentally benign solvents with properties similar to those of ionic liquids and organic solvents due to their minimal cost, high extraction efficacy, and technologically relevant capabilities. Pretreatment with DES is a viable option for removing biomass recalcitrance and increasing cellulose and lignin valorisation. DES can solubilise lignin from lignocellulose biomass and impart advantageous properties to the extracted lignin, such as precursors for carbon fibre production. However, the mechanisms of lignin-carbohydrate complex cleavage and lignin fractionation in lignocellulosic biomass are poorly understood. Furthermore, there has been limited investigation into the mechanism by which DES cleaves interlinkages in the lignin-carbohydrate complex structure. This is critical because the fundamental properties of isolated lignin determine its suitability as a carbon fibre precursor. This review aims to provide a new perspective on lignin extraction from lignocellulosic biomass using DES, with an emphasis on its potential as a carbon fibre precursor. The obstacles to tailoring lignin properties for various valorisation processes are also reviewed.

S. K. Amran · A. L. Sazali · M. R. Anuar · K. F. Pa'ee · T.-L. K. Yong (✉)
Universiti Kuala Lumpur Malaysian Institute of Chemical and Bioengineering Technology, Alor Gajah, 78000 Melaka, Malaysia
e-mail: kytlen@unikl.edu.my

S. K. Amran
e-mail: khadijah.amran02@s.unikl.edu.my

A. L. Sazali
e-mail: afiqah.sazali09@s.unikl.edu.my

M. R. Anuar
e-mail: mohdrazealy@unikl.edu.my

K. F. Pa'ee
e-mail: khairulfaizal@unikl.edu.my

T.-L. K. Yong
Centre for Women Advancement and Leadership, Universiti Kuala Lumpur, 50250 Kuala Lumpur, Wilayah Persekutuan, Malaysia

A. Ismail et al. (eds.), *Engineering Frontiers*, Advanced Structured Materials 208,
https://doi.org/10.1007/978-3-031-56844-2_24

Keywords Precursor · Carbon fibre · Oil palm biomass · Lignin · Deep eutectic solvent

1 Introduction

Carbon fibre has grown in demand in recent decades owing to its unique and desired mechanical properties. Carbon fibres are ten times stronger than steel but five times lighter. They exhibit minimal thermal expansion and density, are resistant to heat, and are biologically inert [1]. Carbon fibre also has a high strength-to-weight ratio; thus, it can be utilised to produce lightweight composite materials. It is used in various industries, including civil engineering, marine, aviation, aerospace, automotive, sports, and leisure, because of its unique properties [2].

However, the exponential increase seen in applications for carbon fibre can only be sustained if the material can be mass-produced cheaply without diminishing its fundamental properties. Polyacrylonitrile (PAN) and pitch are the primary precursors for carbon-fibre manufacturing. PAN-derived carbon fibres have high tensile strength but are prohibitively costly, contributing substantially to manufacturing costs. Carbon fibres derived from mesophase pitch have the greatest tensile moduli of any manufactured fibre and are less costly than PAN fibres. In contrast, these carbon fibres are brittle, have a short strain-to-failure time, and possess a lower tensile strength than carbon fibres made from PAN [3].

Concerns about material yield, solvent toxicity, and high manufacturing costs have driven the search for a technically and commercially viable carbon fibre precursor. Lightweight materials are in great demand, especially in the automotive and sports sectors, where competitive cost is more crucial than high-end strength properties. Therefore, studies have focused on biomass-derived carbon fibres due to their sustainability and commercial viability (Fig. 1).

Globally, lignocellulosic biomass has tremendous potential as a source of organic carbon for manufacturing biofuels and industrial chemicals. Therefore, effective conversion is crucial for reducing fossil-fuel dependency [4]. The primary emphasis is separating organic wastes into their constituents (cellulose, lignin, and hemicellulose) to maximise their potential for valorisation. Most biomass in Malaysia comes from oil palm plantations, yet this resource is underutilised. Oil palm biomass consists of mesocarp fibre, fronds, palm kernel shell, palm oil mill effluent, empty fruit bunch, and trunk (Fig. 2). Consequently, intensive studies have been conducted on lignin extraction from oil palm biomass [5].

Prior studies have shown that lignin can be a substitute precursor for carbon fibre [6, 7]. Carbon fibre from lignin costs USD 1.1 kg, but when PAN is used as a precursor, the cost increases to USD 6.0–10.0 kg [6]. Successful valorisation relies on the effective extraction of lignin with desirable properties. Conventional lignin extraction methods such as steam explosion and acid/base treatment are ineffective because they produce lignin with unsuitable properties for further upgrading, especially as a carbon fibre precursor. Attempts to valorise lignin have resulted in innovative pretreatment

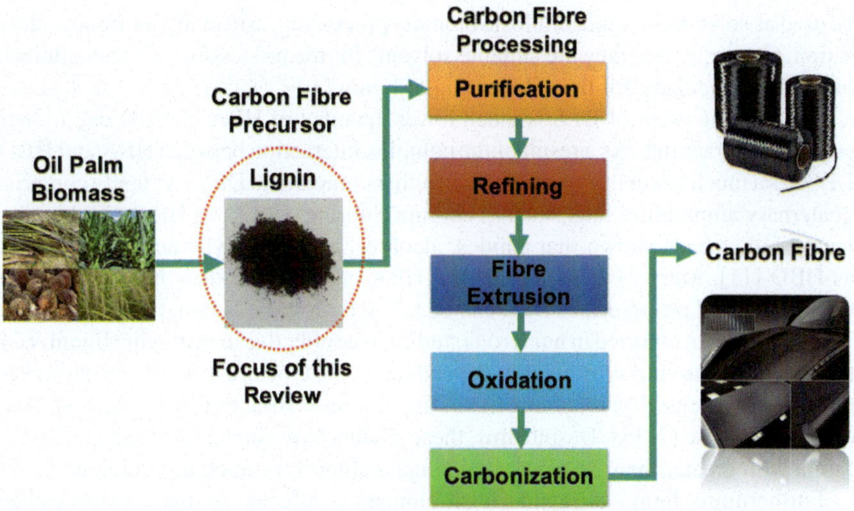

Fig. 1 Lignin from oil palm biomass as a carbon fibre precursor

Fig. 2 Oil palm biomass

methods that use a variety of solvents, including DES, organic solvents produced from biomass, and ionic liquids. Recently, interest in DESs has been sparked by their remarkable ability to dissolve lignin from biomass [7–9].

DESs are liquids with precise molar ratios of hydrogen bond donor (HBD) and hydrogen bond acceptor (HBA). DESs are non-toxic, biodegradable, low-vapour-pressure solvents with high thermal stability and low toxicity. As a result, they can be used as solvents in lignocellulosic biomass processing, especially for lignin valorisation. Consequently, they are suitable solvents for the processing of lignocellulosic biomass, particularly for the valorisation of lignin [10].

The melting point of DES is much lower than that of HBA or HBD due to their complex interactions. As a result of the complex interaction between HBA and HBD, DES has a much lower melting point than either component. DESs typically comprise quaternary ammonium salts, such as choline chloride (ChCl), as HBA. On the other hand, studies have shown that amides, alcohols, and carboxylic acids can be used as HBD [11]. Moreover, by altering the HBA or HBD and their molar ratios, the thermophysical properties of DES can be tailored for various uses [12].

DES has been reported in numerous studies to be effective in extracting lignin, with the efficiency varying substantially depending on the type of DES utilised. However, these studies focused on removing lignin from biomass structures to recover cellulose or hemicellulose [7, 13]. Despite this, these studies have shown that DES can isolate lignin with greater than 90% purity, although cellulose is much less soluble.

Furthermore, lignin extraction from biomass is affected by the choice of HBA and HBD and their molar ratios [14]. Since lignin was not the primary focus of these investigations, its structure, functional groups, and fundamental properties were not established. Tan et al. [15] observed that lignin yield was enhanced during the extraction of lignin from oil palm biomass using ChCl-based DES with HBDs of different functional groups, such as polyols (glycerol) and alpha-carboxylic functional groups (lactic acid) Tan et al. [15] investigated the feasibility of extracting lignin from oil palm biomass utilising ChCl-based DES with HBDs of various functional groups, such as polyols (glycerol) and alpha-carboxylic functional groups (lactic acid) and reported that the lignin yield was increased.

To date, there have been no studies on the influence of the DES molar ratio on the fundamental properties of lignin-derived from oil palm biomass, namely, its feasibility as a precursor for carbon fibre. Furthermore, the precise interactions between DES and lignin molecules during extraction, which are essential for understanding the reaction mechanism, are poorly defined.

2 Lignin Extraction from Oil Palm Biomass Using DES

The complexity of lignin structures presents significant obstacles to lignin extraction, despite the growing interest in DESs. However, a comprehensive and critical review of lignocellulosic biomass processing using DES has recently been conducted. The high selectivity of DES for lignin relative to cellulose is advantageous for high-value applications of isolated lignin [16, 17]. Strong hydrogen bonding in DES eutectic mixtures increases the likelihood of lignin ether bond cleavage, thereby enhancing its solubility in DES. Lignin is more soluble in acidic DES than basic due to stronger hydrogen bonds. ChCl is a frequently used HBA because of its strong hydrogen bonding and chloride ions (Cl–). The hydroxyl group and Cl– form strong hydrogen bonds that aid the dissolution of lignin [18].

DES can selectively depolymerise lignin fractions from lignocellulosic biomass by disrupting the ether linkages between the phenylpropane units [7]. Lignin extraction from biomass is highly dependent on various factors, including the DES molar ratio and reaction temperature. DES can be reused during lignin solvent extraction. This is feasible without significantly reducing the yield of extracted lignin. However, Malaeke et al. [13] recommended using freshly made DES because moisture can interfere with the dissolution of lignocellulosic constituents.

Tan et al. [19] showed that ChCl-based DES with harsher pH conditions extracted lignin more efficiently than DES with a near-neutral pH, with a high lignin yield of 88% reported when lactic acid was used as the HBD. However, lignin extraction with glycerol resulted in a 22% lower yield, which was thought to be due to the near-neutral pH of the eutectic system. In addition, Cl– from ChCl (HBA) forms hydrogen bonds with lignocellulosic hydroxyl groups, enabling the separation of lignin and hemicellulose from the lignin-carbohydrate complex [20].

Furthermore, the FTIR spectra of lignin indicated the presence of guiacyl and syringyl lignin units, depending on the lignin source and extraction method. The strengths of the peaks varied owing to the diverse HBD constituents. The lignin yield increased from 33% (1:1) to 61% (1:15) as the lactic acid molar ratio increased. In addition, the antioxidant activity of the extracted lignin was corroborated by its high phenolic hydroxyl content, which was comparable to that of technical and commercial lignin. The lignin yield increased from 33% (1:1) to 61% (1:15) as the lactic acid molar ratio increased.

Teh et al. [21] found that ChCl-based DES using lactic acid and glycerol as HBD successfully recovers lignin from EFB with a lignin yield of 74.61%. Several studies have demonstrated that a high HBD fraction and prolonged reaction time facilitate lignin extraction [17–19]. Furthermore, the study found that both DESs have the potential to break down biomass recalcitrance and reveal the surface area of the biomass for effective lignin extraction. However, ChCl with lactic acid was more effective than glycerol in breaking aromatic-dominated lignin linkages and cellulose crystallinity linked by β-(1,4)-glycosidic bonds. This is because glycerol molecules, despite having many hydrogen bond donor sites (–OH) in their hydrocarbon chains, are less able to form hydrogen bonds and are more difficult to dissolve [15, 20].

3 Mechanism of Lignin Extraction Using DES

The most promising precursor to carbon fibre is lignin, which can be derived from industrial crops. Nonetheless, lignin extraction from lignocellulosic biomass is extremely difficult. Due to poor extraction methods, commercial lignin utilisation faces several challenges. The properties and lignin composition of lignocellulosic biomass will likely differ depending on plant species and maturity, climate, growth environment, and most importantly, the extraction method.

Lignin is a three-dimensional amorphous polymer consisting of guaiacyl, syringyl, and p-hydroxyphenyl bonded by ether and carbon–carbon bonds. The complex arrangement of these units in the fibres and shells of oil palm biomass results in lignin with a more complex structure than that derived from wood [22]. Recent research has shown that lignin and hemicelluloses covalently bond to form lignin-carbohydrate complexes. There are eight linkage types: glycosidic/phenyl glycosidic, benzyl ester, hemiacetal/acetal bonds, benzyl ether, and ferulate/diferulate esters at the 4-OH and 4-O locations in lignin.

It has been proposed that lignin can be isolated by focusing on the lignin-carbohydrate complex, which is primarily covalently bonded to hemicellulose. Conventional methods for dissolving lignin-carbohydrate complexes include physicochemical, chemical, physical, and biological methods. The ability of DES to specifically rupture the ether linkages between lignin phenylpropane units facilitates lignin extraction from biomass [7].

DES extraction of lignin from biomass involves multiple steps, including breaking aryl ether and carbon–carbon linkages, hydroxylation, demethoxylation, and condensation/crosslinking of lignin fragments [23]. Thus, the interaction of depolymerisation and recondensation during lignin dissolution influences the final extracted lignin structure. Crosslinking causes the primary lignin structures to condense, resulting in a denser structure despite the loss of short side chains and small molecules from the original structure [24].

There has been considerable interest in lignin extraction from lignocellulosic biomass using ChCl-based DES, but few studies have examined this mechanism. The interaction between Cl– and polysaccharides in lignocellulose is a plausible mechanism. Cl– from ChCl (HBA) may establish strong hydrogen bonds with the-OH in polysaccharides, thereby facilitating the dissociation of the lignin-polysaccharide complex [25].

Schutyser et al. [26] proposed a proton-induced lignin bond dissociation mechanism. First, benzylic carbon protonation induces cleavage. Following dehydration and the formation of an active carbon cation, ether bonds are ruptured by one of two probable mechanisms: formaldehyde loss from the carbon and deprotonation/protonation of the carbon. Condensed lignin may also be formed through the cleavage of β-O-4 aryl ether intermediates, which may react with one another. Carbon–carbon bonds in lignin may also be cleaved, depending on the lignin source and DES type [24]. However, the precise mechanisms responsible for cleavage in DES remain unclear.

4 Lignin Functionalisation as Carbon Fibre Precursor

Lignin, the second most prevalent natural polymer in plants after cellulose, is a significant by-product of the lignocellulosic biorefinery and paper industry. Pulping mills generate approximately 50 million tonnes of lignin waste yearly. If the target of one billion tonnes of lignocellulosic biofuel is met, the lignin waste stream could exceed 300 million tons. Despite decades of study on its value-added products, lignin is most commonly used economically to recover heat from waste. Conversion of

Fig. 3 Cost of carbon fibre production

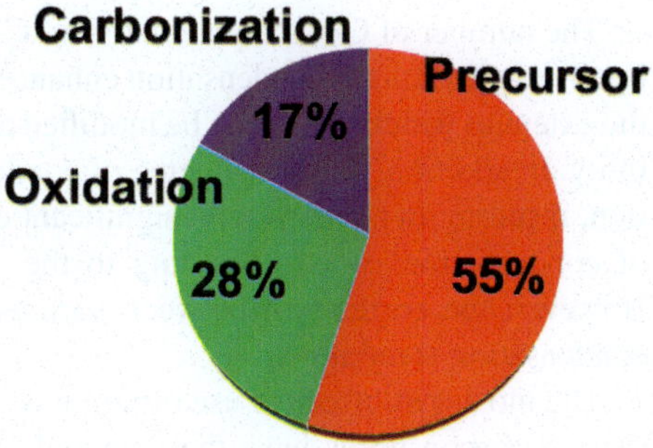

lignin into high-value commodities has long been challenging for the paper industry and lignocellulosic biorefineries [27]. Nevertheless, naturally occurring lignin was used as a composite reinforcement material.

Lignin is the least expensive precursor from which carbon fibre can be produced [6]. Approximately 55% of the total cost of manufacturing carbon fibre is attributable to the cost of the precursor. The cost of carbon fibre derived from lignin as a precursor is the lowest of any precursor [6] (Fig. 3). As a result, finding a cheaper precursor, such as lignin, is essential for lowering the cost of carbon fibre. Despite its low cost as a carbon fibre precursor, lignin has advantages and disadvantages. Biomass variability exacerbates the technological challenges in converting lignin into carbon fibre. Therefore, different types of lignin require different manufacturing processes. The intricate lignin structure and impurities may degrade the mechanical and functional properties of carbon fibres [28].

Lignin-derived carbon fibres may have lower production costs and are more environmentally friendly than those derived from traditional petroleum-based precursors. According to Luo et al. [29], lignin-derived carbon fibres with poor mechanical properties limit their use primarily because of their inability to be oriented. Furthermore, lignin with rich aromatic structures has a complex structure with varying functional groups, molecular weights, and distinct monolignol proportions, all linked by diverse interunit chemical bonds and higher carbon content [30].

5 Fundamental Properties of Lignin Isolated Using DES

The chemical pulping of wood generates large amounts of industrial or technical lignin as a by-product. Approximately 70 million metric tonnes of lignin is generated during cellulose extraction. However, only 2% is efficiently used. The remainder is discarded or burned for low-value applications or energy production. Therefore, understanding the fundamental properties of lignin after isolation from primary cell wall components is critical for enhancing lignin usage [31]. Lignin can be more beneficial if it has chemically labile moieties for functionalisation, low degree of condensation, with minimal impurities.

The number of C–C bonds relative to C–H and C–O bonds reflects the degree of lignin condensation. Condensation enhances the strength of the C–C bonds, limiting the extent to which lignin can be modified and, by extension, its prospective applicability. In addition, a higher degree of condensation inhibits enzymatic saccharification, resulting in residues with significant carbohydrate impurities [32]. The degree of condensation varies according to the chemistry and severity of pretreatment. However, the correlation between various pretreatment methods and inter-lignin condensation is unknown.

The most cost-effective use of lignin is in the production of carbon fibre. Lignin-derived carbon fibres have the potential to reduce production costs by more than 35% when using a renewable material with a high sustainability rating. Despite being less competitive than its current competitors, lignin-derived carbon fibre has evolved into a material with significant technological value, economic appeal, and environmental sustainability. It could be a viable substitute for several applications, particularly those in the low-cost but high-demand sectors [1]. Due to condensation, conventional extraction methods yield lignin with poor purity, high molecular weight, and limited functional group composition, limiting its potential as a precursor [33].

Therefore, the fundamental properties of lignin should be investigated to establish its potential as a carbon precursor. The Oak Ridge National Laboratory (ORNL) has developed a set of criteria to assess the viability of lignin as a carbon fibre precursor [34]. The specifications included more than 99% purity, less than 5% volatile content at 250 °C, less than 0.1% ash content, and the absence of particles larger than 1 μm. Experimental and empirical establishment of these conditions resulted in the formation of fusible-extrudable lignin. Consequently, they were designated as operational parameters [35].

According to the ORNL, lignin purity must exceed 99% as a carbon fibre precursor. However, the high structural heterogeneity and complexity of lignin hampered its extraction and use, primarily due to aryl ether linkages and carbon–carbon bonds. It has been shown that the ability of DES to break β-O-4 aryl ethers and specific lignin-carbohydrate covalent bonds considerably increases lignin solubilisation compared to conventional pretreatment methods [36].

Alvarez-Vasco et al. [7] discovered that a ChCl-based DES composed of multiple HBDs (glycerol, acetic acid, levulinic acid, and lactic acid) could selectively extract lignin from wood with a high purity of 95% and distinct structural properties. This study also suggests that DES preferentially disrupts ether bonds in wood lignin, facilitating lignin extraction. In addition, Li et al. [37] extracted lignin with a 94.46% purity using ChCl and lactic acid in a 1:10 molar ratio at 120 °C for 12 h.

As a carbon fibre precursor, lignin must be devoid of solid particles larger than 1 μm. Particulate matter is the amount of solid particles entrapped in lignin. The most common causes of structural defects in carbon fibre are natural particles, such as sand, clay, and cellulose. The modulus and tensile strength of carbon fibre are both decreased by a high particulate matter content, which leads to a higher breaking point [34]. The high particulate matter content of lignin necessitates extensive research on lignin structures.

Ash is a by-product of high-temperature combustion. As a carbon fibre precursor, lignin should have a low inorganic content. The ash content must be lower than 0.1 wt% for the carbon fibre to be free of significant flaws and defects. The lignin source and extraction method significantly affected the ash content of extracted lignin.

The volatile matter content of lignin must be determined because evaporation during high-temperature carbonisation may affect the carbon fibre structures. Low-molecular-weight fragments may induce structural defects during lignin carbonisation due to excessive heating and evaporation rates [38].

It is also important to evaluate the carbon content, glass transition temperature (T_g), and average molecular weight (M_w) of lignin, in addition to the fundamental properties provided by ORNL, when determining its viability as a carbon fibre precursor.

Compared to petroleum-based PAN, the lignin carbon content (60–65%) indicates a high yield following carbon fibre manufacturing [39]. A high carbon content suggests a low amount of non-carbon components, which is relevant given that lignin loses mass during carbonisation. A precursor with a carbon content of at least 92 wt.% after carbonisation is required for high-quality carbon fibre [38]. The carbon content of lignin recovered from oil palm biomass using DES has not yet been investigated. There have been no studies on the carbon content of lignin recovered from oil palm biomass using DES. However, only a few studies have reported the carbon content of lignin derived from various sources. For example, lignin extracted from willow using ChCl and lactic acid has a high carbon content of more than 60% [40].

T_g is a critical element that affects lignin mobility and thermal flow [41]. The ability of lignin to crosslink during oxidative thermostabilisation is dependent on its ability to maintain T_g above the process temperature, enabling it to be infused. The spinning rates, ramping characteristics, and processing temperatures must be precisely regulated to manufacture carbon fibres with excellent strength. Consequently, the melting temperature of lignin must be low enough to prevent polymerisation during extrusion, but the T_g must be high enough to allow for a reasonable rate of fibre stabilisation [34].

Carbon fibre precursors derived from lignin should have a high average M_w for structural homogeneity and superior mechanical capability. The selection of lignin with a low M_w often facilitates fibre manufacturing but produces carbon fibres with only average mechanical performance. Most lignins have a low average M_w and melting temperature because of their amorphous chemical structure [42]. Consequently, they often disintegrate when heated to 200–250 °C.

According to Li et al. [43], the poor mechanical performance results from the inherent heterogeneity of lignin, which is induced by its low average M_w. Although eliminating lignin with a low average M_w enhances the modulus of elasticity of carbon fibres, the molecular homogeneity of lignin affects its engineering properties. Therefore, lignin homogeneity has a greater effect on the spinnability, morphology, and mechanical properties of lignin-derived carbon fibres than on the average M_w.

Although the physicochemical and mechanical properties of carbon fibres derived from lignin do not meet industrial requirements, these properties indicate that lignin may be employed as a reinforcing component in composite transportation materials.

However, contamination must be considered if commercially manufactured lignin is used as the precursor [38].

6 Conclusion

A long-term strategy is required to develop a less expensive and comparable quality carbon fibre precursor to the current precursor. Lignin is a viable carbon fibre precursor due to its widespread availability and cheap cost. Although DES is a superior extraction solvent, it is necessary to optimise the solid loading, temperature, and reaction time to extract high-quality lignin from biomass. It is critical to evaluate the viability of carbon fibre produced from extracted lignin by comparing its properties with those specified by ORNL. This review concludes that lignin has excellent potential as a precursor to carbon fibre owing to its low cost, minimal environmental impact, and outstanding properties.

Acknowledgements The Ministry of Education, Malaysia, under the Fundamental Research Grant Scheme (FRGS), provided funding for this research (FRGS/1/2020/STG05/UNIKL/02/1).

References

1. Souto, F., Calado, V., Pereira, N.: Lignin-based carbon fiber: a current overview. Mater. Res. Express. **5**, 072001 (2018). https://doi.org/10.1088/2053-1591/aaba00
2. Trogen, M., Le, N.D., Sawada, D., Guizani, C., Lourençon, T.V., Pitkänen, L., Sixta, H., Shah, R., O'Neill, H., Balakshin, M., Byre, N., Hummel, M.: Cellulose-lignin composite fibres as precursors for carbon fibres. Part 1–manufacturing and properties of precursor fibres. Carbohydr. Polym. **252**, 117133 (2021). https://doi.org/10.1016/j.carbpol.2020.117133
3. Aldosari, S.M., Khan, M., Rahatekar, S.: Manufacturing carbon fibres from pitch and polyethylene blend precursors: a review. J. Mater. Res. Technol. **9**, 7786–7806 (2020). https://doi.org/10.1016/j.jmrt.2020.05.037
4. Tyagi, R., Das, A.K., Mandal, A.: Formation of superhydrophobic surface with enhanced hardness and wear resistance by electrical discharge coating process. Tribol. Int. **157**, 106897 (2021). https://doi.org/10.1016/j.triboint.2021.106897
5. Zulkefli, S., Abdulmalek, E., Rahman, M.B.A.: Pretreatment of oil palm trunk in deep eutectic solvent and optimization of enzymatic hydrolysis of pretreated oil palm trunk. Renew. Energy **107**, 36–41 (2017). https://doi.org/10.1016/j.renene.2017.01.037
6. Fang, W., Yang, S., Wang, X.L., Yuan, T.Q., Sun, R.C.: Manufacture and application of lignin-based carbon fibers (LCFS) and lignin-based carbon nanofibers (LCNFS). Green Chem. **19**, 1794–1827 (2017). https://doi.org/10.1039/C6GC03206K
7. Alvarez-Vasco, C., Ma, R., Quintero, M., Guo, M., Geleynse, S., Ramasamy, K.K., Wolcott, M., Zhang, X.: Unique low-molecular-weight lignin with high purity extracted from wood by deep eutectic solvents (DES): a source of lignin for valorization. Green Chem. **18**, 5133–5141 (2016). https://doi.org/10.1039/C6GC01007E
8. Kim, K.H., Dutta, T., Sun, J., Simmons, B., Singh, S.: Biomass pretreatment using deep eutectic solvents from lignin derived phenols. Green Chem. **20**, 809–815 (2018). https://doi.org/10.1039/C7GC03029K

9. Paiva, A., Craveiro, R., Aroso, I., Martins, M., Reis, R.L., Duarte, A.R.C.: Natural deep eutectic solvents–solvents for the 21st century. ACS Sustain. Chem. Eng. **2**, 1063–1071 (2014). https://doi.org/10.1021/sc500096j
10. Hong, S., Shen, X.J., Xue, Z., Sun, Z., Yuan, T.Q.: Structure–function relationships of deep eutectic solvents for lignin extraction and chemical transformation. Green Chem. **22**, 7219–7232 (2020). https://doi.org/10.1039/D0GC02439B
11. Yucui, H., Congfei, Y., Weize, W.: Deep eutectic solvents: green solvents for separation applications. Acta Phys. Chim. Sin. **34**, 873–885 (2018). https://doi.org/10.3866/PKU.WHXB201802062
12. Peyrovedin, H., Haghbakhsh, R., Duarte, A.R.C., Raeissi, S.A.: Global model for the estimation of speeds of sound in deep eutectic solvents. Molecules **25**, 1626 (2020). https://doi.org/10.3390/molecules25071626
13. Malaeke, H., Housaindokht, M.R., Monhemi, H., Izadyar, M.: Deep eutectic solvent as an efficient molecular liquid for lignin solubilization and wood delignification. J. Mol. Liq. **263**, 193–199 (2018). https://doi.org/10.1016/j.molliq.2018.05.001
14. Jablonský, M., Škulcová, A., Šima, J.: Use of deep eutectic solvents in polymer chemistry: a review. Molecules **24**, 3978 (2019). https://doi.org/10.3390/molecules24213978
15. Tan, Y.T., Ngoh, G.C., Chua, A.S.M.: Effect of functional groups in acid constituent of deep eutectic solvent for extraction of reactive lignin. Bioresour. Technol. **281**, 359–366 (2019). https://doi.org/10.1016/j.biortech.2019.02.010
16. Smith, E.L., Abbott, A.P., Ryder, K.S.: Deep eutectic solvents (DESs) and their applications. Chem. Rev. **114**, 11060–11082 (2014). https://doi.org/10.1021/cr300162p
17. Tang, X., Zuo, M., Li, Z., Liu, H., Xiong, C., Zeng, X., Sun, Y., Hu, L., Liu, S., Lei, T., Lin, L.: Green processing of lignocellulosic biomass and its derivatives in deep eutectic solvents. Chemsuschem **10**, 2696–2706 (2017). https://doi.org/10.1002/cssc.201700457
18. Smink, D., Juan, A., Schuur, B., Kersten, S.R.: Understanding the role of choline chloride in deep eutectic solvents used for biomass delignification. Ind. Eng. Chem. Res. **58**, 16348–16357 (2019). https://doi.org/10.1021/acs.iecr.9b03588
19. Tan, Y.T., Ngoh, G.C., Chua, A.S.M.: Evaluation of fractionation and delignification efficiencies of deep eutectic solvents on oil palm empty fruit bunch. Ind. Crops Prod. **123**, 271–277 (2018). https://doi.org/10.1016/j.indcrop.2018.06.091
20. Liu, Y., Chen, W., Xia, Q., Guo, B., Wang, Q., Liu, S., Liu, Y., Li, J., Yu, H.: Efficient cleavage of lignin–carbohydrate complexes and ultrafast extraction of lignin oligomers from wood biomass by microwave-assisted treatment with deep eutectic solvent. Chemsuschem **10**, 1692–1700 (2017). https://doi.org/10.1002/cssc.201601795
21. Teh, S.S., Loh, S.K., Mah, S.H.: Development of choline-based deep eutectic solvents for efficient concentrating of hemicelluloses in oil palm empty fruit bunches. Korean J. Chem. Eng. **36**, 1619–1625 (2019). https://doi.org/10.1007/s11814-019-0348-1
22. Rashid, T., Gnanasundaram, N., Appusamy, A., Kait, C.F., Thanabalan, M.: Enhanced lignin extraction from different species of oil palm biomass: kinetics and optimization of extraction conditions. Ind. Crops Prod. **116**, 122–136 (2018). https://doi.org/10.1016/j.indcrop.2018.02.056
23. Liu, W., Yao, Y., Fu, O., Jiang, S., Fang, Y., Wei, Y., Lu, X.: Lignin-derived carbon nanosheets for high-capacitance supercapacitors. RSC Adv. **7**, 48537–48543 (2017). https://doi.org/10.1039/C7RA08531A
24. Hong, S., Sun, X., Lian, H., Pojman, J.A., Mota-Morales, J.D.: Zinc chloride/acetamide deep eutectic solvent-mediated fractionation of lignin produces high-and low-molecular-weight fillers for phenol-formaldehyde resins. J. Appl. Polym. Sci. **137**, 48385 (2020). https://doi.org/10.1002/app.48385
25. Xia, Q., Liu, Y., Meng, J., Cheng, W., Chen, W., Liu, S., Liu, Y., Li, J., Yu, H.: Multiple hydrogen bond coordination in three-constituent deep eutectic solvents enhances lignin fractionation from biomass. Green Chem. **20**, 2711–2721 (2018). https://doi.org/10.1039/C8GC00900G
26. Schutyser, W., Renders, A.T., Van den Bosch, S., Koelewijn, S.F., Beckham, G.T., Sels, B.F.: Chemicals from lignin: an interplay of lignocellulose fractionation, depolymerisation, and upgrading. Chem. Soc. Rev. **47**, 852–908 (2018). https://doi.org/10.1039/C7CS00566K

27. Li, Q., Li, M., Lin, H.S., Hu, C., Truong, P., Zhang, T., Sue, H., Pu, Y., Ragauskas, A.J., Yuan, J.S.: Non-solvent fractionation of lignin enhances carbon fiber performance. Chemsuschem **12**, 3249–3256 (2019). https://doi.org/10.1002/cssc.201901052
28. Kadla, J.F., Kubo, S., Venditti, R.A., Gilbert, R.D., Compere, A.L., Griffith, W.: Lignin-based carbon fibers for composite fiber applications. Carbon N. Y. **40**, 2913–2920 (2002). https://doi.org/10.1016/S0008-6223(02)00248-8
29. Luo, Y., Qu, W., Cochran, E., Bai, X.: Enabling high-quality carbon fiber through transforming lignin into an orientable and melt-spinnable polymer. J. Clean. Prod. **307**, 127252 (2021). https://doi.org/10.1016/j.jclepro.2021.127252
30. Zhang, R., Du, Q., Wang, L., Zheng, Z., Guo, L., Zhang, X., Yang, X., Yu, H.: Unlocking the response of lignin structure for improved carbon fiber production and mechanical strength. Green Chem. **21**, 4981–4987 (2019). https://doi.org/10.1039/C9GC01632E
31. Fodil Cherif, M., Trache, D., Brosse, N., Benaliouche, F., Tarchoun, A.F.: Comparison of the physicochemical properties and thermal stability of organosolv and kraft lignins from hardwood and softwood biomass for their potential valorization. Waste Biomass Valorization. **11**, 6541–6553 (2020). https://doi.org/10.1007/s12649-020-00955-0
32. Yu, Z., Gwak, K.S., Treasure, T., Jameel, H., Chang, H.M., Park, S.: Effect of lignin chemistry on the enzymatic hydrolysis of woody biomass. ChemSusChemicals **7**, 1942–1950 (2014). https://doi.org/10.1002/cssc.201400042
33. Su, Y., Huang, C., Lai, C., Yong, Q.: Green solvent pretreatment for enhanced production of sugars and antioxidative lignin from poplar. Bioresour. Technol. **321**, 124471 (2021). https://doi.org/10.1016/j.biortech.2020.124471
34. Baker, D.A., Rials, T.G.: Recent advances in low-cost carbon fiber manufacture from lignin. J. Appl. Polym. Sci. **130**, 713–728 (2013). https://doi.org/10.1002/app.39273
35. Eberle, C., Webb, D.C., Albers, T., Chen, C.: Commercialization of new carbon fiber materials based on sustainable resources for energy applications. Oak Ridge National Laboratory. **10**, 1–24 (2013)
36. Tan, Y.T., Chua, A.S.M., Ngoh, G.C.: Deep eutectic solvent for lignocellulosic biomass fractionation and the subsequent conversion to bio-based products–a review. Bioresour. Technol. **297**, 122522 (2020). https://doi.org/10.1016/j.biortech.2019.122522
37. Li, T., Lyu, G., Liu, Y., Lou, R., Lucia, L.A., Yang, G., Chen, J., Saeed, H.A.: Deep eutectic solvents (DESs) for the isolation of willow lignin (salix matsudana cv. zhuliu). Int. J. Mol. Sci. **18**, 2266 (2017). https://doi.org/10.3390/ijms18112266
38. Luo, J., Genco, J., Cole, B.J., Fort, R.C.: Lignin recovered from the near-neutral hemicellulose extraction process as a precursor for carbon fiber. BioResources **6**, 4566–4593 (2011)
39. Bengtsson, A., Bengtsson, J., Olsson, C., Sedin, M., Jedvert, K., Theliander, H., Sjöholm, E.: Improved yield of carbon fibres from cellulose and kraft lignin. Holzforschung **72**, 1007–1016 (2018). https://doi.org/10.1515/hf-2018-0028
40. Liu, Y., Zheng, J., Xiao, J., He, X., Zhang, K., Yuan, S., Peng, Z., Chen, Z., Lin, X.: Enhanced enzymatic hydrolysis and lignin extraction of wheat straw by triethylbenzyl ammonium chloride/lactic acid-based deep eutectic solvent pretreatment. ACS Omega **4**, 19829–19839 (2019). https://doi.org/10.1021/acsomega.9b02709
41. Ruiz-Rosas, R., Bedia, J., Lallave, M., Loscertales, I.G., Barrero, A., Rodríguez-Mirasol, J., Cordero, T.: The production of submicron diameter carbon fibers by the electrospinning of lignin. Carbon N. Y. **48**, 696–705 (2010). https://doi.org/10.1016/j.carbon.2009.10.014
42. Jin, Y., Lin, J., Cheng, Y., Lu, C.: Lignin-based high-performance fibers by textile spinning techniques. Materials. **14**, 3378 (2021). https://doi.org/10.3390/ma14123378
43. Li, Q., Serem, W.K., Dai, W., Yue, Y., Naik, M.T., Xie, S., Karki, P., Liu, L., Sue, H.J., Liang, H., Zhou, F.: Molecular weight and uniformity define the mechanical performance of lignin-based carbon fiber. J. Mater. Chem. A. **5**, 12740–12746 (2017). https://doi.org/10.1039/C7TA01187C

The Decomposition of Calcium Carbonate in Cockle Shell: Study on Particle Size and Temperature

Nadia Razali, Nor Adilah Ahmad, Nurriswin Jumadi, Khairul Faizal Pa'ee, and Nurhanim Zulaikha Kamarulzaman

Abstract Cockle shells are perceived as having no important use and are commonly regarded as a waste. However, these calcium carbonate-rich materials are gaining favour as renewable resources and biomaterials precursors. The calcium carbonate ($CaCO_3$) in the cockle shells can be decomposed into calcium oxide (CaO) through calcination. This study examined the calcination conditions of the cockle shell on two parameters; particle size and temperature. The cockle shell samples were divided into granular (CSG) and powder (CSP), calcined at 600–900 °C. Physical (bulk density, pH, proximate analysis, mass loss) and chemical (FTIR, XRD, CaO purity) properties of calcined cockle shell for every temperature for both sizes were observed. An increase in temperature indicates there are significantly different colour transitions. FTIR results reported that the CaO spectrum was observed for samples and CSP_{900}. The XRD pattern for CSP_{900} shown was similar to industrial CaO at 2θ of 32.33°, 37.45°, 53.97°, 64.24°, 67.47°, 79.79° and 88.66°. The purity of calcined CSP_{900} and CSG_{900} showed the highest percentage, with 94.4 and 95.5%, respectively. Based on observation and data analysis, the cockle shell calcination process at high temperatures is not affected by the particle size of the cockle shell.

N. Razali (✉) · N. A. Ahmad · N. Jumadi · K. F. Pa'ee · N. Z. Kamarulzaman
Universiti Kuala Lumpur Malaysian Institute of Chemical and Bioengineering Technology, 78000 Alor Gajah, Melaka, Malaysia
e-mail: nadiarazali@unikl.edu.my

N. A. Ahmad
e-mail: adilah.ahmad@s.unikl.edu.my

N. Jumadi
e-mail: nurriswin@kkjel.edu.my

K. F. Pa'ee
e-mail: khairulfaizal@unikl.edu.my

N. Z. Kamarulzaman
e-mail: nurhanim.kamarulzaman@s.unikl.edu.my

N. Jumadi
Kolej Komuniti Jelebu, Jalan Seperi, 71600 Kuala Klawang, Negeri Sembilan, Malaysia

N. A. Ahmad
School of Engineering and Architecture, University of Zaragoza, 50009 Zaragoza, Spain

A. Ismail et al. (eds.), *Engineering Frontiers*, Advanced Structured Materials 208,
https://doi.org/10.1007/978-3-031-56844-2_25

Keywords Cockle shell · Calcination · Decomposition · Temperature · Size

1 Introduction

The blood cockle (*Anadara granosa*) is an important aquaculture species in Malaysia that has contributed to the national economy since 1948 [1]. According to a report from the Malaysian Department of Fisheries (DOF), blood cockles covered approximately 93% in 2015 and 70% in 2017 of total shellfish culture production in Malaysia. The strong demands for cockles in the food business are their supply availability, delicious taste, nutritional content, and inexpensive cost.

Since cockles are the cheapest source of protein compared to other shellfish, cockles are an affordable choice for consumers. Excessive demand for cockles has led to an increase in cockle shell (CS) waste. As it is considered a non-economic value, it was dumped in landfills without treatment, resulting in environmental problems such as unpleasant smell, leachate and pest attacks.

Calcium carbonate is a major mineral that monopolizes CS content. Approximately 96–99% of the total CS is calcium carbonate ($CaCO_3$); the rest are from other oxides such as silica oxide, magnesium oxide, and sulfur trioxide [2–4]. The same chemical composition but a different crystal structure called polymorphous was proven from SEM images of a dried cockle shell consisting of aragonite (needle-like) and calcite (cubic-like), and this could be used as a metal absorbent (manganese and copper) in industrial wastewater [5].

The $CaCO_3$ can be converted to CaO through a combustion process at high temperatures. The process of decomposition of $CaCO_3$ is called calcination. Many calcination studies have been conducted to determine the appropriate combustion temperature for a source of $CaCO_3$. Before calcination, the CS will be cleaned and dried to remove unwanted waste that may influence the results. In a study by Ghafar et al. [6], CaO powder was obtained by calcined granule CS at a temperature of 900 °C in the furnace for 4 h before being crushed and sieved.

On the other hand, Ghazali et al. [7] prepared CaO from CS by grinding and sieving CS prior to calcining process at 900 °C for five hours. It has been proven that CaO resulting from calcination temperature exceeding 800 °C for 10 h showed CaO characteristics through XRD, FTIR, and XRF tests [8]. In the other publication, FTIR results show that the calcination time does not affect the production of CaO [9]. Through the SEM–EDX diagram, a too-high calcination temperature does not produce a large quantity of CaO.

This study examines the cockle shell's calcination conditions based on particle size and temperature. Both granule and powder CS were calcined by various temperatures ranging from 600 to 900 °C. This study focused on the catalyst characterizations, functional group analysis, the catalyst's phase identification, and neutralization with an acid test.

2 Methodology

2.1 Preparation of Raw Material

The CS samples were collected from a seafood restaurant near Banting, Selangor, and were instantly washed under tap water as a pre-treatment procedure. This pre-treatment was conducted 12 h after CS was discarded to avoid mass decaying due to traces of interference materials and unpleasant odour. Then, the CS was dried under the sun for 3 h [10]. The CS was treated with boiled water for 15 min and scrubbed with a brush under tap water until all impurities and dirt were removed [11, 12]. The CS was immersed in weak acetic acid (vinegar) for two hours to remove excessive meat attached to the inner surface. Theoretically, the $CaCO_3$ elements would corrode (decrease in weight) when immersed in a solution of strong acetic acid [13]. Thus, the acetic acid solution was prepared with a ratio of acetic to water is 0.01:1.

Treated CS samples were prepared in powder (CSP) and granule (CSG) forms. The raw CS was grounded using an electric powder grinder (SY-25) of 2000 watts for approximately 20 s. Then, the CS was sieved with a stainless steel shaker unit test passing at 9.5 mm and retained at size 4.75 mm to obtain CSG with a size range of 0.5–1.0 cm. The smaller or bigger CS particles were grounded again to get CSP for another 20 s. Afterwards, both samples were placed in air-tight containers separately and ready to calcine. Figure 1 shows raw CS, CSG, and CSP.

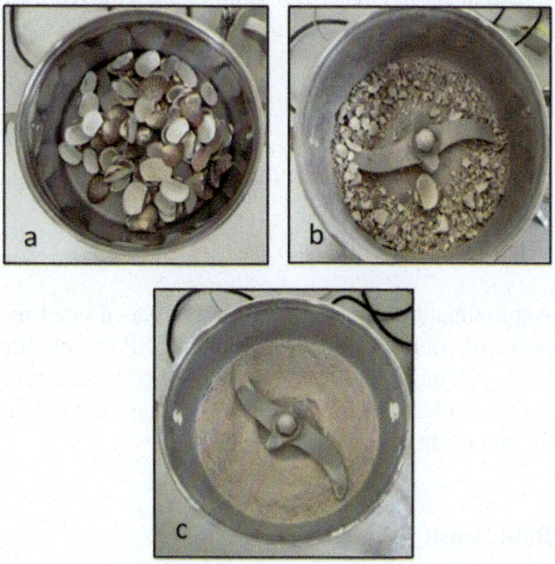

Fig. 1 **a** Raw CS, **b** CSG, and **c** CSP

Table 1 Calcination parameters

No	Sample code	Parameter	
		Temperature	Particle size
1	CSG_{600}	600 °C	Granule
2	CSG_{700}	700 °C	
3	CSG_{800}	800 °C	
4	CSG_{900}	900 °C	
5	CSP_{600}	600 °C	Powder
6	CSP_{700}	700 °C	
7	CSP_{800}	800 °C	
8	CSP_{900}	900 °C	

2.2 *Preparation of Catalyst*

A Shimadzu TGA-50 thermobalance was used to determine the decomposition temperature range of the sample by weight difference with increasing temperature. Approximately 13 g of CSP was used in this test. The samples were placed on the platinum pan with a heating setup rate of 10 °C/min for 30–1000 °C in a nitrogen atmosphere with a flow rate of 20 mL/min [14].

A 40 g of both CSG and CSP samples were placed in the different crucibles with covers and calcined at the desired temperature with a constant 4 h contact time, as shown in Table 1. Then, all samples are removed from the furnace and left to cool in a desiccator. The cooled sample is placed in an air-tight container to prevent it from reacting with oxygen and moisture from the surrounding.

2.3 *Catalyst Characterization*

pH

Approximately 10 g of each sample was diluted in 10 mL of distilled water in a different conical flask. The calcined CSP can be directly diluted in distilled water, while calcined CSG needs to be crushed before being dissolved. The solution was stirred using a magnetic stirrer for 60 min, and the pH reading was taken within 30 min using pH (Mettler Toledo).

Bulk Density

The bulk density of the samples was determined by the direct method. 10 mL of the sample was filled in a 25 ml measuring cylinder. Then, the density of the sample was calculated by Eq. (1):

$$\rho = \frac{\mathrm{m}}{\mathrm{V}}, \tag{1}$$

where ρ is the bulk density, m is the mass of the sample, and V is the volume of the sample.

Moisture Content

For the moisture content analysis, approximately 1 g of calcined CS sample was placed in the crucible with cover and exposed to a temperature of 105 °C for 5 h [15, 16]. The different weights of the sample were recorded using an analytical weight balance. The sample's moisture content (M_c) was calculated by Eq. (2):

$$M_c(\%) = \left[\frac{(A - B)}{A}\right] \times 100, \tag{2}$$

where A is the weight of the air-dried sample (g), B is the weight of the sample after oven-dried (g).

Functional Group Analysis

The Fourier transform infrared spectroscopy (FTIR) test determined the calcine samples' functional group. Before starting the test, all the apparatus and materials used were placed in a drying oven at 105 °C for at least 30 min to avoid moisture affecting the results. In this test, calcined CSP can be used directly while calcined CSG was grounded to powder before being mixed with potassium bromide (KBr) powder. The mixed ratio sample to KBr used in this study is 1:100.

Phase Identification of Catalyst

The X-ray powder diffraction (XRD) using a Shimadzu XRD LabX 6000 diffractometer was used to identify the crystalline substance of calcined CS. The sample was analyzed continuously from angle 10° to 90° by copper anode, which produced a Cu–kα wavelength of 1.54060 Å at a constant temperature of 25 °C, 30 mA current, and a 40 kV generator was used.

Neutralization with an Acid Test

In a 250 ml flask, 2.5 g of calcined CS sample was added into carbon dioxide-free water. After three minutes of boiling, the solution was cooled to room temperature (flask A). In another flask (flask B), 20 g of sugar was diluted in 20 ml of distilled water. Once the solution in flask A cooled, 10 ml of the sugar solution in flask B was added to flask A and stirred immediately for 30 min. After that, distilled water was added to flask A until the solution reached 250 ml. No 1 Whatman paper was used to filter the solution. The first 30 ml filtration was discarded. Then, 25 ml of the filtrated solution was pipetted into flask A. Five drops of phenolphthalein and 0.357 N sulphuric acid were added to the flask. Thus, the available CaO adopted by ASTM rapid sugar test for lime was calculated as in Eq. (3):

$$\begin{aligned} &\text{The available lime, as CaO,}\\ &\quad = \text{ml of } 0.357 \text{ N acid on the burette, } \times 4 \end{aligned} \tag{3}$$

3 Results and Discussion

3.1 Thermogravimetric Analysis

This test was carried out to determine a compound's weight change with increasing temperature at a constant rate. Based on Fig. 2, only two reaction steps occurred. A similar result was obtained by Wongsaenmai [17].

A small mass decrease was observed at the first step reaction. Approximately 0.036% of the weighted sample decreased at midpoint 196.43 °C. The weight reduction in the first step at a temperature range of 0–200 °C indicates eliminating moisture content [14, 18, 19]. The wide temperature range occurs in the second step, from 200

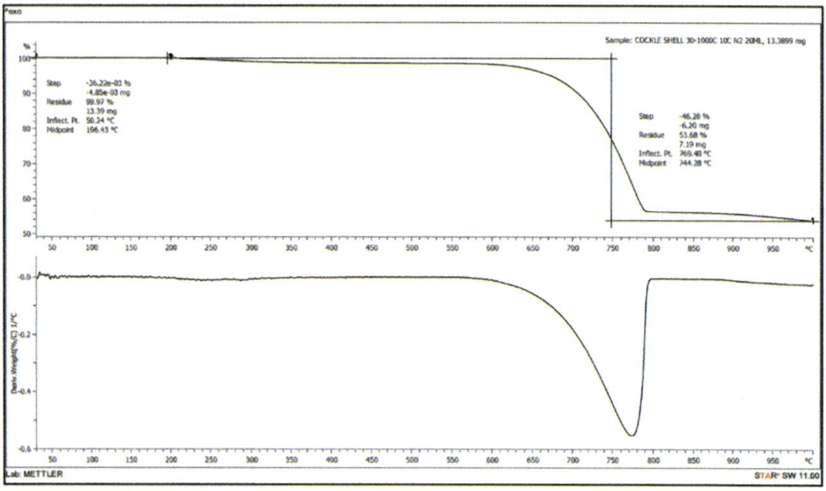

Fig. 2 Thermogravimetric analysis of CS

Fig. 3 pH value of calcined CS

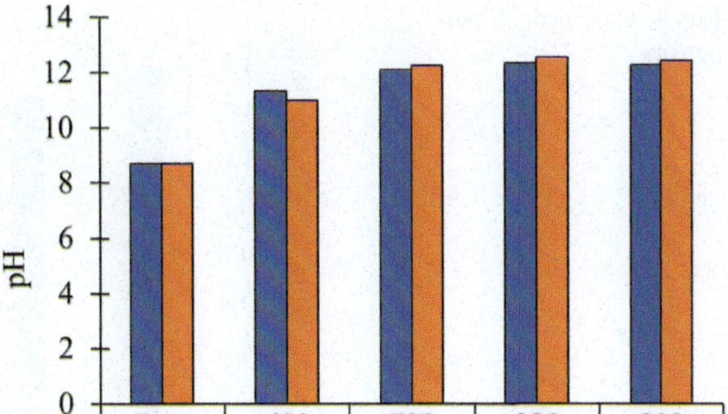

	Raw	600	700	800	900
CSP	8.7	11.3	12.1	12.4	12.3
CSG	8.7	11.0	12.3	12.5	12.4

to 1000 °C. Rapid weight change was observed starting from 600 °C. Approximately 50% of the initial weight decreased when the temperature reached 800 °C. The major weight change due to the decomposition of CS and CO_2 was released. A similar result was established by [4], where the decomposition of CS starts at a temperature of 668 °C. A small weight change was observed after passing the temperature of 780 °C. This indicates that almost all $CaCO_3$ have been decomposed. Based on Saharudin et al. [20], the constant line graph exceeding 850 °C indicated that the calcination process was done.

3.2 *pH*

A pH electrode meter (Mettler Toledo) was used to determine the alkalinity of the samples. Based on Fig. 3, the pH of calcined CS samples changed to strong alkali when the temperature increased. Both CSP and CSG samples with the same temperature have a similar pH. The samples with temperatures 700 °C and above were characterized as strong alkalinity corresponding to the presence of CaO. A similar result was established by Shao and Dong [21] the calcination at 700 °C produced sufficient CaO to turn the solution into high alkali.

3.3 *Bulk Density*

The bulk density of calcined CS with different temperatures is shown in Fig. 4. According to the graph, the sample with a higher calcination temperature produces a lower bulk density. There is a slightly decreased bulk density for CSP, while sample CSG remained constant at 600 and 700 °C. A significant amount of CaO presence in both sample CSP and CSG at high temperatures with a reduction of approximately

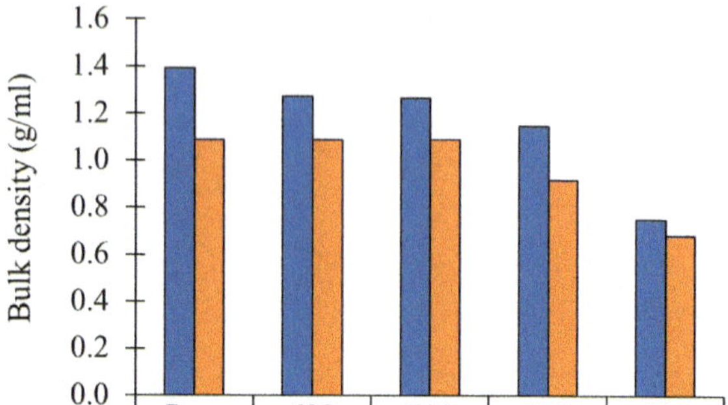

	Raw	600	700	800	900
CSP	1.39	1.27	1.27	1.15	0.75
CSG	1.09	1.09	1.09	0.92	0.68

Fig. 4 Calcined CS bulk density

17 and 15.6% for 800 °C and 46 and 38% for 900 °C, respectively. The obtained bulk density value shows that the size of calcined CS decreased when the temperature increased. This shows that the percentage of porosity and specific surface area decreases at high temperatures. It has been established that higher temperatures would affect average particle size and pore diameter formation [22].

3.4 Moisture Content

Figure 5 shows the calcine CS's percentage of moisture content at different temperatures. According to the graph, the higher the temperature, the higher moisture content percentages observed. TGA analyses showed that the moisture (O–H) bond was removed at 30–200 °C. In this case, the percentage moisture content of the calcined sample was related to the absorption of moisture in the atmosphere. As we know, one of the characteristics of CaO is moisture-sensitive. Thus, it can be concluded that the higher moisture absorption has a significant characteristic of CaO at temperatures 800 and 900 °C. Even though the sample was placed in a desiccator to avoid moisture and kept in an air-tight container, the calcined CS might absorb moisture during storage or testing procedures.

3.5 FTIR

The FTIR graph can be divided into four regions. The first region (single bond) from wavelength 4000–2500 cm^{-1} represents hydrogen bonding (O–H, N–H, C–H). The second and third regions from wavelength 2500–2000 cm^{-1} and 2000–1500 cm^{-1}, respectively, show the functional group present in the sample. The fourth region reveals the existence of biomaterials [23].

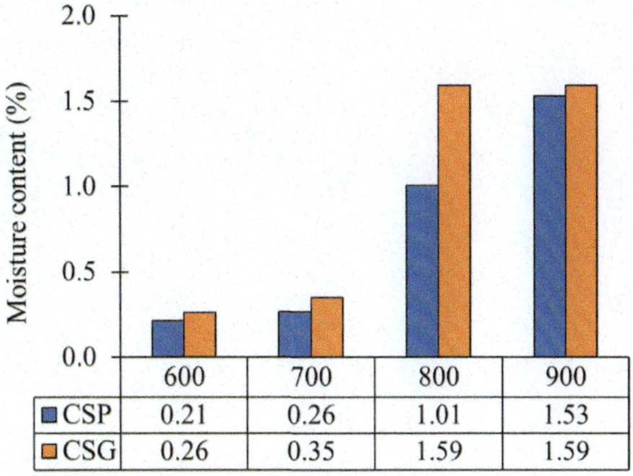

Fig. 5 Moisture content calcined CS

Based on pH, bulk density and moisture content results, the significant sample was chosen for the functional group identification test. Figure 6 shows the calcined FTIR test result for CSP and CSG at 800 and 900 °C. Based on the graphs, there are similar sharp bands shown in Figs. 6a, b, and d. A broad peak in the first region with a wavelength of 3401–3641 cm^{-1} indicates that the sample still contains moisture (O–H–O). The sharp band range of 702–875 cm^{-1} is associated with the presence C=O bond in carbonate anion [24]. According to Zhu et al. [25], the absorption band at 1459, 1417, 875, and 714 cm^{-1} correspond to the stretching vibration of the O–C–O bond. Thus, the sharp peak in the range 1400–1423 cm^{-1} indicates the presence of $CaCO_3$. So, by referring to the graphs, the samples with a broad peak at the first region and carbonate functional group can be considered as not fully decomposed yet.

The sharp narrow peak of 3401 cm^{-1} at the first region, along with the 1068 cm^{-1} band observed in Fig. 6c (CSP_{900}), corresponds to the presence of O–H vibration. A sharp transmittance band of C–O vibration occurs at 1423 cm^{-1}. Then, another significant peak of the presence of O=C=O bonds at 2361 cm^{-1} [26].

So, based on the FTIR graphs, it can be concluded that the significant FTIR result for calcines CS is sample CSP_{900}. The size of the particle contributes to the calcination process of the CS. The granular CS was not completely decomposed at 900 °C for 4 h of contact time. By referring to TGA, the decomposition of CS (step 2) is complete at a temperature of 1000 °C. According to Saharudin et al. [20], 5 h of contact time calcination shows better crystallinity than 3 h of contact time.

3.6 XRD

A significant sample was taken for the XRD test based on the FTIR result. The CSP_{900} and industrial CaO were analyzed and plotted in the same graph in this study. In Fig. 7, both samples have a distinct peak of CaO compound observed at 2θ of 32.3°, 37.5°,

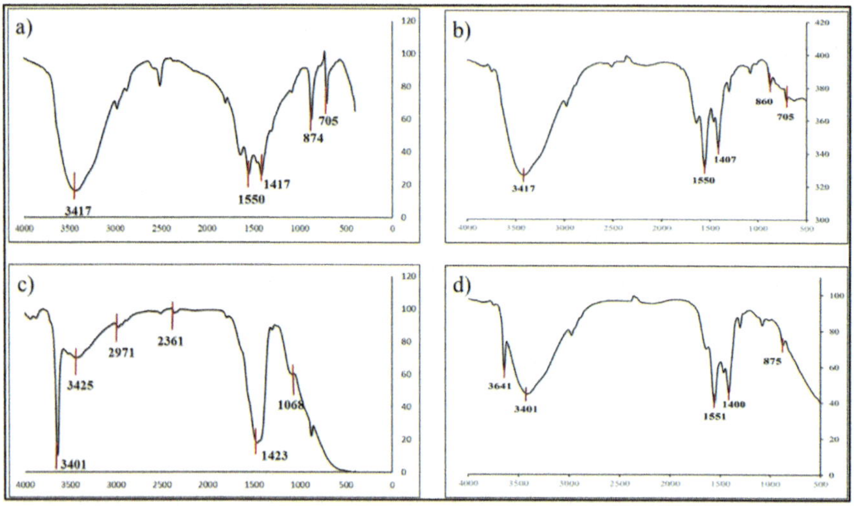

Fig. 6 FTIR spectra calcined CS **a** CSP_{800}, **b** CSG_{800}, **c** CSP_{900} and **d** CSG_{900}

54.0°, 64.2°, 79.8°, and 88.6°. For the CSP_{900} sample, calcium hydroxide ($Ca(OH)_2$) peak diffraction was observed. Generally, the reaction of CaO with water (H_2O) will produce $Ca(OH)_2$. In this case, most likely, sample CSP_{900} reacted with H_2O in the atmosphere during the procedures. Due to its moisture-sensitive characteristics, moisture absorption in the atmosphere easily occurs and changes the characteristics of CSP_{900}. Similar publications stated that $Ca(OH)_2$ could be found when the CaO absorbs moisture [27, 28]. Based on the moisture content results, samples calcined at high temperatures absorbed higher moisture content [29].

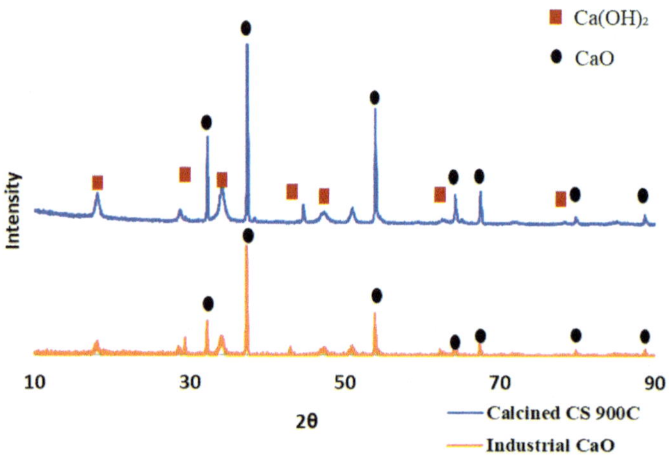

Fig. 7 XRD diffraction of CSP_{900} and industrial CaO

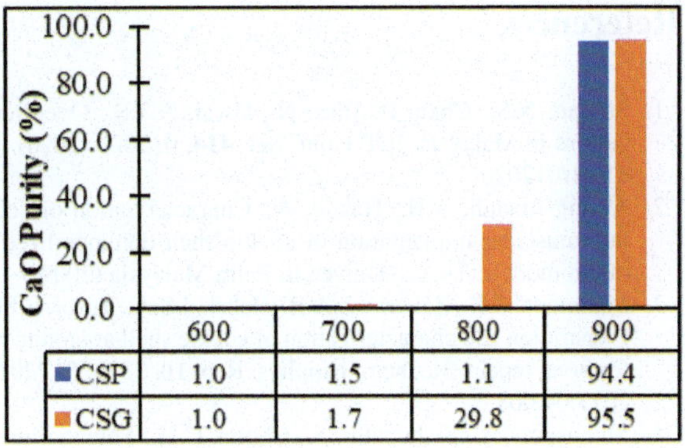

Fig. 8 Purity percentages for CSP and CSG at different calcination temperatures

3.7 Neutralization with an Acid Test

The result of determining the availability of CaO in the CSP and CSG sample by the rapid sugar testing is shown in Fig. 8 below. Based on the graph (see Fig. 7), samples CSP_{900} and CSG_{900} have the highest percentage of CaO, where the purity percentage is 94.4 and 95.5%, respectively. Sample CSG_{800} showed a higher CaO content than sample CSP_{800} even though the same temperature was used. So, it can be concluded here that surface area plays a role in calcination. As reported in Sect. 3.3 CSG samples have a low density (high porosity). So, during the calcination process, the heat can be distributed throughout the surface of the CS and cause the calcination process to occur faster.

4 Conclusion

Based on the research that has been conducted, the higher the calcination temperature, significant characteristics of CaO are obtained. Strong alkalinity, lower density, and higher moisture content percentage were obtained for the CSP_{900} sample. Furthermore, FTIR, XRD and purity testified that CSP_{900} had shown characteristics of CaO. On the other hand, even though CSG_{900} was calcined with the same condition as CSP_{900}, the characteristics of CaO are not clearly shown. It can be concluded that although CSG has a larger surface area and a larger porosity that supports heat distribution, its thick size makes the calcination process incomplete in decomposing the composition of $CaCO_3$ in CS. So, as a recommendation, the CSG calcination time should be increased to get results as CSP_{900}.

References

1. Saffian, N.S., Chen, P., Ilias, N., Hwai, A.T.S.: Overview and challenges of blood cockle culture in Malaysia. IOP Conf. Ser. **414**, 012020 (2020). https://doi.org/10.1088/1755-1315/414/1/012020
2. Awang-Junaidi, A.H., Hazmi, A.: Characterization of cockle shells and evaluation of post-intramuscular implantation of cockle shell demineralized bone matrix composite in rat and rabbit models. Thesis, Universiti Putra Malaysia (2009)
3. Mahmood, S.K., Zakaria, Z.A.B., Razak, I.S.A., Yusof, L.M., Jaji, A.Z., Isa, T., Hammadi, N.I.: Preparation and characterization of cockle shell aragonite nanocomposite porous 3D scaffolds for bone repair. Biochem. Biophys. Rep. **10**, 237–251 (2017). https://doi.org/10.1016/j.bbrep.2017.04.008
4. Munusamy, Y., Sethupathi, S., Choon, C.H.: Potential use of waste cockle shell as filler for thermoplastic composite. J. Mater. Cycles Waste Manag. **21**, 1063–1074 (2019). https://doi.org/10.1007/s10163-019-00867-9
5. Ridzwan, N.F.A.A., Aziz, N.A., Akbar, N.A., Ghani, Z.A., Hamzah, N., Zin, N.S.M.: Removal of MN and CU ions in synthetic wastewater using dried cockle shell. Environ. Ecol. Res. **10**, 31–43 (2022)
6. Ghafar, F., Sapawe, N., Jemain, E.D., Alikasturi, A.S., Masripan, N.A.: Study on the potential of waste cockle shell derived calcium oxide for biolubricant production. Mater. Today Proceed. **19**, 1346–1353 (2019). https://doi.org/10.1016/j.matpr.2019.11.148
7. Ghazali, S., Kem, K.L., Jusoh, R., Abdullah, S., Shariffuddin, J.H.: Evaluation of La-doped CaO derived from cockle shells for photodegradation of POME. Bull. Chem. React. Eng. Catal. **14**, 205 (2019). https://doi.org/10.9767/bcrec.14.1.3318.205-218
8. Linggawati, A.: Preparation and characterization of calcium oxide heterogeneous catalyst derived from *A. granosa* shell for biodiesel synthesis. KnE Eng. **1**, 494 (2016)
9. Dampang, S., Purwanti, E., Destyorini, F., Kurniawan, S.B., Abdullah, S.R.S., Imron, M.F.: Analysis of optimum temperature and calcination time in the production of CaO using seashells waste as $CaCO_3$ source. J. Ecol. Eng. **22**, 221–228 (2021)
10. Wang, C., Zhang, M., Yu, M., Zhu, R., Hu, H., Loncar, M.: Monolithic lithium niobate photonic circuits for Kerr frequency comb generation and modulation. Nat. Commun. **10**, 6 (2019). https://doi.org/10.1038/s41467-019-08969-6
11. Hussein, A.I., Ab-Ghani, Z., Mat, A.N.C., Ghani, N.A.A., Husein, A., Rahman, I.A.: Synthesis and characterization of spherical calcium carbonate nanoparticles derived from cockle shells. Appl. Sci. **10**, 7170 (2020). https://doi.org/10.3390/app10207170
12. Zain, N.B.M., Salleh, N.J.M., Hisamuddin, N.F., Hashim, S., Abdullah, N.H.: Adsorption of phosphorus using cockle shell waste. Ind. Domest. Waste Manag. **2**, 30–38 (2022)
13. Nithiyaa, N., Ho, A., Norhanis, M.R., Zulfigar, Y., Aileen Tan, S.H.: The effect of acidified seawater on shell characteristics of blood cockle, *Tegillarca granosa*. J. Surv. Fish. Sci. **7**, 133–142 (2021)
14. Razali, N., Jumadi, N., Jalani, A.Y., Kamarulzaman, N.Z., Pa'ee, K.F.: Thermal decomposition of calcium carbonate in chicken eggshells: study on temperature and contact time. Malay. J. Anal. Sci. 26, 347–359 (2022)
15. Razali, N., Musa, F.A., Jumadi, N., Jalani, A.J.: Production of calcium oxide from eggshell: study on calcination temperature, raw weight and contact time (2021)
16. Oulego, P., Laca, A., Calvo, S.M., Díaz, M.: Eggshell-supported catalysts for the advanced oxidation treatment of humic acid polluted wastewaters. Water **12**, 100 (2019). https://doi.org/10.3390/w12010100
17. Wongsaenmai, S.: Characterization of calcium oxide derived from cockle shells for carbon dioxide capture. SNRU J. Sci. Technol. **10**, 32–36 (2018)
18. Matejtschuk, P., Duru, C., Malik, K., Ezeajughi, E., Gray, E., Raut, S., Mawas, F.: Use of thermogravimetric analysis for moisture determination in difficult lyophilized biological samples. Am. J. Anal. Chem. **12**, 3023 (2016). https://doi.org/10.4236/ajac.2016.73023

19. Jewiarz, M., Wróbel, M., Mudryk, K., Szufa, S.: Impact of the drying temperature and grinding technique on biomass grindability. Energies **13**, 3392 (2020). https://doi.org/10.3390/en13133392
20. Saharudin, S.H., Shariffuddin, J.H., Nordin, N.I.A.A., Aqeela, I.: Effect of aging time in the synthesis of biogenic hydroxyapatite derived from cockle shell. Mater. Today Proceed. **19**, 1208–1215 (2019). https://doi.org/10.1016/j.matpr.2019.11.124
21. Shao, W.-C., Dong, Y.: A study on physicochemical properties and formaldehyde adsorption and degradation of purifying air quality by modified biocalcium. Int. J. Environ. Sci. Develop. **11**, 327–335 (2020)
22. Puspitasari, P., Rizkia, U.A., Sukarni, S., Permanasari, A.A., Taufiq, A., Putra, A.B.N.R.: Effects of various sintering conditions on the structural and magnetic properties of zinc ferrite (ZNFE2O4). Mater. Res. Ibero Am. J. Mater. **24**, 300 (2021). https://doi.org/10.1590/1980-5373-mr-2020-0300
23. Nandiyanto, A.B.D., Oktiani, R., Ragadhita, R.: How to read and interpret FTIR spectroscope of organic material. Indon. J. Sci. Technol. **4**, 97 (2019)
24. Hospodarova, V., Singovszka, E., Stevulova, N.: Characterization of cellulosic fibers by FTIR spectroscopy for their further implementation to building materials. Am. J. Anal. Chem. **09**, 303–310 (2018). https://doi.org/10.4236/ajac.2018.96023
25. Zhu, Q., Wang, J., Sun, J., Wang, Q.: Preparation and characterization of regenerated cellulose biocomposite film filled with calcium carbonate by in situ precipitation. BioResources **15**, 7893–7905 (2020)
26. Saleh, H.M., Salman, A.A., Faheim, A.A., Elsayed, A.: Influence of aggressive environmental impacts on clean, lightweight bricks made from cement kiln dust and grated polystyrene. Case Stud. Constr. Mater. **15**, e00759 (2021). https://doi.org/10.1016/j.cscm.2021.e00759
27. Tiandho, Y.: Quality improvement of water from post Tin mining based on a wasted cockle shell. IOP Conf. Ser. Mater. Sci. Eng. **496**, 012058 (2019). https://doi.org/10.1088/1757-899x/496/1/012058
28. Zul, N.A., Ganesan, S., Hamidon, T.S., Oh, W.-D., Hussin, M.H.: A review on the utilization of calcium oxide as a base catalyst in biodiesel production. J. Environ. Chem. Eng. **9**, 105741 (2021). https://doi.org/10.1016/j.jece.2021.105741
29. Tongwanichniyom, S., Kitjaruwankul, S., Phornphisutthimas, S.: Production of biomaterials from seafood waste for application as vegetable wash disinfectant. Heliyon. **8**, e09357 (2022). https://doi.org/10.1016/j.heliyon.2022.e09357

Sea Trial Evaluation of Kapok-Sawdust as New Green Oil Adsorbent in Portable Oil Spill Collectors

Alif Aizuddin Mohd Saper, Muhammad Abdul Munaim, Fatin Zawani Zainal Azaim, Norazlina Abdul Nasir, Putri Zirwatul Nadila Megat Zamanhuri, Syarmela Alaauldin, Norazizah Che Mat, Ismila Ishak, Md Salim Kamil, Nadiah Ghazali, Noraliya Hamizi, and Asmalina Mohamed Saat

Abstract In this project, a portable oil spill collector was developed using a combination of kapok and sawdust to be used as oil adsorbent filters. Both kapok and sawdust are water repellent and highly absorbent to oil owning to the hydrophobic and oleophilic character. Kapok was combined with sawdust according to 10:0, 7:3, 5:5, 3:7 and 0:10 ratio to evaluate their sorption performance in 30, 60 and 90 min. The portable oil spill collector was developed with a static filter system so that the adsorbent can absorb oil from water as the vessel moves. The optimum ratio for oil absorbent efficiency found for both laboratory and sea trial experimental recorded in kapok: sawdust samples with ratio of 7:3. As a conclusion, the samples with a ratio of 7:3 have a better oil absorption efficiency and stability during a longer period of immersion time, in various medium of environment and various filter positions.

A. A. M. Saper · M. A. Munaim · F. Z. Z. Azaim · N. A. Nasir · P. Z. N. M. Zamanhuri · S. Alaauldin · N. C. Mat · I. Ishak · M. S. Kamil · A. M. Saat (✉)
Universiti Kuala Lumpur, Malaysian Institute of Marine Engineering Technology, Lumut Perak, Malaysia
e-mail: asmalina@unikl.edu.my

M. A. Munaim
e-mail: mamunaim@unikl.edu.my

F. Z. Z. Azaim
e-mail: fatinawani@unikl.edu.my

N. A. Nasir
e-mail: norazlinaan@unikl.edu.my

P. Z. N. M. Zamanhuri
e-mail: puterizirwatul@unikl.edu.my

S. Alaauldin
e-mail: syarmela@unikl.edu.my

N. C. Mat
e-mail: norazizah@unikl.edu.my

A. Ismail et al. (eds.), *Engineering Frontiers*, Advanced Structured Materials 208,
https://doi.org/10.1007/978-3-031-56844-2_26

Keywords Oil adsorbent · Green filter · Kapok · Sawdust · Oil spill

1 Introduction

From raw materials to finished goods, the world's economy is highly dependent on maritime transportation. The shipping industry has been dubbed as the 'backbone of global trade and global economy' by the United Nations (UN) and ensures continuous and sustainable growth. An increase in trade would add more traffic to the already busy maritime world and the occurrences of collision between ships would also increase. Collison not only causes losses to the owner, but oil spills which give a devastating effect on the marine ecosystem and the environment. Clean-up is costly, however, it is hard to estimate as the amount depends on several factors such as the oil type, the amount of oil, location, sea conditions and the local response capacity. The cost also must include the machinery and equipment used which will be determined by the rescue team [1]. The methods that are developed for oil spill clean-up can be categorized into three main groups which are physical, chemical, and biological [2]. The physical method includes absorbent, boom, and skimmers, while for chemical method includes dispersion, in-situ burning and solidifiers. The biological method is categorized as the secondary method that uses microorganisms to eliminate oil in areas where there is limited access and when the primary method cannot be applied. The system in oil collector devices is purposely designed to achieve maximum oil separating and collecting efficiency. The effectiveness of the system depends on the material used. The most common material is a polymer as it can attract oil, however, scientists are currently searching for a more sustainable and greener alternative. The kapok (*ceiba petandra)* tree is originated from India but can also be found in the Southeast Asia region. The kapok fibers are used as fiberfill in a pillow, mattress and sometimes soft toys [3]. The fibers are underutilized they have some interesting characteristics such as low density, have good buoyancy, hallow lumen characteristic, and coated with surface wax [4, 5]. Sawdust is a waste product from the wood-based industry that has similar characteristics. It is a unique agricultural product because both materials have high hydrophobic and oleophilic characteristics. Low buoyancy and high surface area allow them to absorb oil more than their weight.

I. Ishak
e-mail: ismila@unikl.edu.my

M. S. Kamil
e-mail: ms.salim@unikl.edu.my

N. Ghazali · N. Hamizi
Mechanical Engineering Department, Faculty of Engineering, University of Malaya, 50603 Kuala Lumpur, Malaysia
e-mail: diyaghaz@um.edu.my

N. Hamizi
e-mail: aliyahamizi@um.edu.my

The combination of these characteristics allows the fiber to be an efficient oil–water separator and excellent absorbent material. Furthermore, kapok and sawdust are cheap, eco-friendly and biodegradable [6]. This study focuses on the evaluation of the combination kapok-sawdust efficiency as a new natural oil absorbent material in the portable oil spill collector. The efficiency was determined by several parameters such as the kapok-sawdust ratio, sorption time and sea trial.

2 Approach and Setup

2.1 Experimental Investigations

The maximum oil sorption efficiency is evaluated by mixing all the specimen ratios inside a beaker which contains 200 ml of oil (Fig. 1a). The maximum capacity of oil absorption and the reusability of the materials to absorb oil are evaluated at various ratios of composition kapok and sawdust at 10:0, 7:3, 5:5, 3:7 and 0:10. The specimens were packed in 7 cm × 7 cm meshes so that the surface areas are the same (Fig. 1b). 4 T engine oil is used due to its high density because the higher the oil density, the higher the adsorption capacity [7]. The weight of the specimens was taken after 60 min after draining the remaining oil for 2 min to get the maximum amount of efficiency. The oil sorption efficiency of the oil–water mixture is analyzed by mixing 500 ml of seawater from the Dindings River with 10 ml of oil. The measurements are taken at intervals of 30, 60 and 90 min, by leaving the water to drain for 2 min before taking the weights. The sorption efficiency is calculated below:

$$\text{Oil sorption efficiency, } \eta = \frac{\textit{weight after immersed} - \textit{weight before immersed}}{\textit{total weight of oil}} \tag{1}$$

2.2 Portable Oil Spill Collector Design

Figure 2a–c shows the design for the constructed floating vessel with two filters which are fitted at both sides of the vessel like a catamaran. A catamaran is impossible or very hard to sink due to the design of a closed cell which is waterproof, and this is crucial because the floating vessel bears the weight of the filters. The two filters are important in terms of the stability of the device. Stability is one of the most important characteristics as the vessel carries loads to ensure the vessel remains floating on water without any problem and the catamaran has a wide surface area that makes it more stable than other designs of floating vessels. The filter compartments are fitted in stages (Fig. 2c) to absorb the remaining oil that escapes from the first stage of the

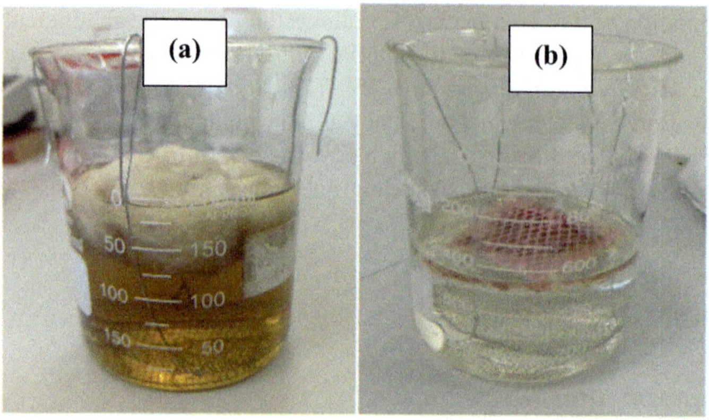

Fig. 1 **a** Absorbance specimen immersed in 200 ml of oil. **b** Absorbance specimen in an oil–water mixture

filter. At the front of the vessel, there is a curve opening to get a maximum amount of oil entering the filter. Figure 2d shows the real image of a fabricated portable oil spill collector. Figure 3a shows the placement of five absorbance materials in the filter compartment at the portable oil spill collector. Figure 3b shows a portable oil spill collector setup for a sea trial experiment on UniKL MIMET's beach. The two best ratios of adsorbents (10:0 and 70:30) from laboratory analysis were used in this experiment. Referring to the oil–water ratio, 1000 ml of oil is poured into the sea in an area of 15 m^2. The efficiency was determined by draining the water first for two minutes then the weight was measured at the time interval of 10, 20 and 30 min.

3 Results and Discussions

3.1 Maximum Oil Sorption Efficiency

Figure 4 shows the oil sorption efficiency for different kapok-sawdust ratios. Kapok has better absorbency compared to sawdust and upon the introduction of sawdust, a decline in sorption can be observed. The efficiency of pure kapok and pure sawdust is 21 and 1.97%. The higher kapok sorption efficiency is likely due to the long hallow structure, smooth surface layer and higher surface area which provides better capillary penetration [5, 8]. The sawdust structure consists of fine pores that should contribute to the oil retention properties. However, it is likely to note that the experiment was conducted for only 60 min and probably did not provide sufficient time to allow the adsorption and absorption process to occur [9].

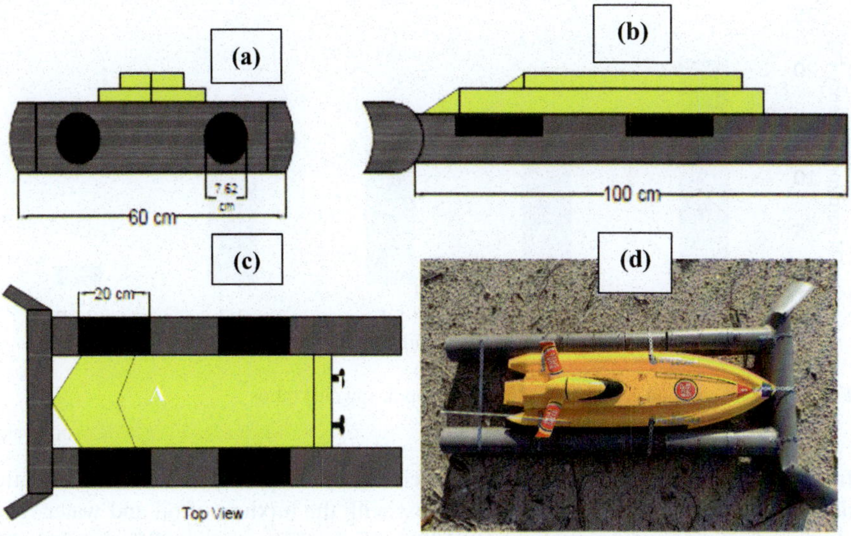

Fig. 2 Constructed design of portable oil spill collector. **a** front view **b** side view **c** top view **d** fabricated portable oil spill collector

Fig. 3 **a** Absorbance specimen fitted inside the filter compartment **b** the sea trial experiment

3.2 Oil Sorption Efficiency in Oil–Water Mixture

Figure 5 shows the oil sorption efficiency effect on (a) time and (b) the ratio of adsorbent. The results show that the efficiency drops as time increases to 60 min in samples with a ratio of 10:0, 7:3, 5:5 and 3:7. Meanwhile samples with a ratio of 0:10 sawdust show an increased efficiency as time increases to 60 min. The observation

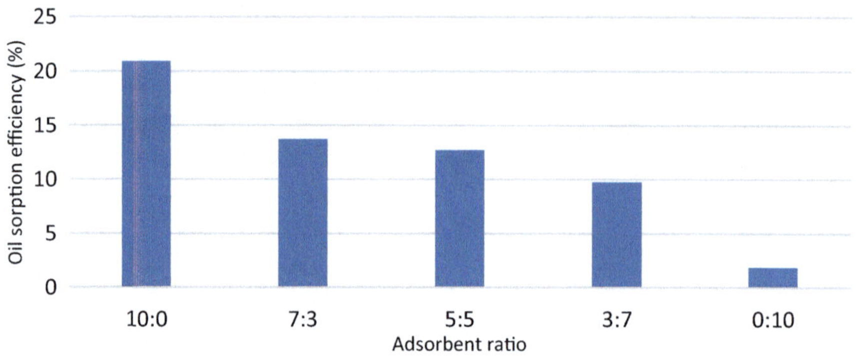

Fig. 4 The oil sorption efficiency for different kapok-sawdust ratios

from 60 to 90 min shows a decreased efficiency in sample 10:0, 3:7 and 0:10 ratio due to the interaction of kapok and sawdust with the mixture of oil and water. The change of efficiency for samples 0:10 (100% sawdust) is not much different from 30 to 60 min with only a small amount of oil absorbed. Samples with 7:3 and 5:5 ratios show an increase in efficiency as the time increased up to 90 min. This finding shows that the addition of sawdust to kapok leads to the stability of kapok in absorption for more than 60 min. The sawdust needs more time to absorb more oil and it seems to be efficient in less quantity because the efficiency of 7:3 is higher than 5:5 in 30, 60, and 90 min.

3.3 Oil Sorption Efficiency for Filter in Sea Trial

Figure 6 shows the oil efficiencies for samples with a ratio of 10:0 and 7:3 conducted in a sea trial experiment. Both samples show oil efficiencies that increased linearly within 30 min of the sea trial absorption experiment. The calculated oil sorption rate for both ratios is 6.3 ml/min and 7.7 ml/min for 10:0 and 7:3, respectively. Referring to the laboratory experiment earlier, the efficiency for 10:0 is higher than 7:3 in 30 min but after 90 min, 7:3 is more efficient. In the sea trial, the ratio 7:3 shows higher efficiency than 10:0. Samples with a ratio of 7:3 show better absorbency at a longer time and in a mixed medium environment.

Figure 7 shows the performance of the oil sorption efficiency for four filter compartments (A, B, C, D) conducted within 30 min. The efficiencies of each compartment are varying due to their position for both samples ratio of 10:0 and 7:3. The oil efficiency for samples 7:3 is higher than samples 10:0 in all compartments. Samples with a ratio of 7:3 show that compartments A and B absorb more oil than compartments C and D. This is due to the position of compartments A and B which is in front of both compartments C and D. Therefore, this ratio meets the recommendation of a good multistage filter in designing a portable oil spill collector

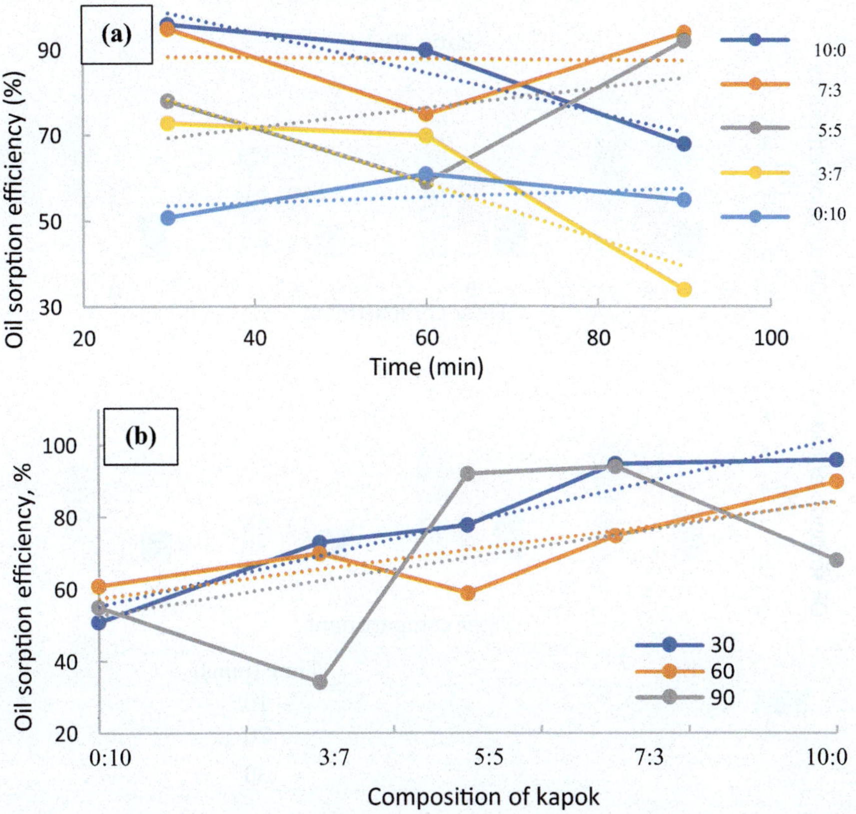

Fig. 5 The oil sorption efficiency effects on **a** the ratio of adsorbent and **b** time

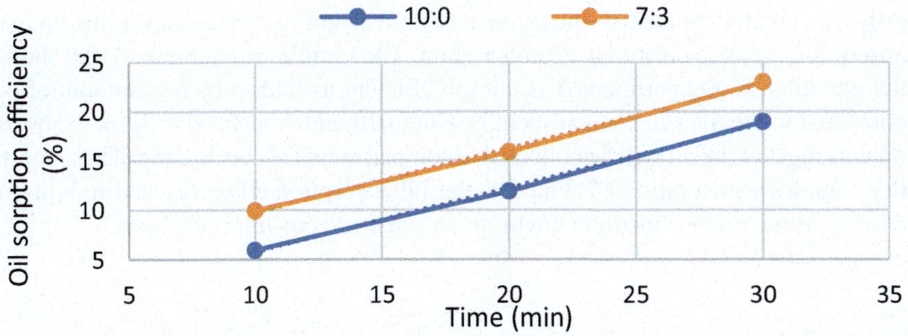

Fig. 6 The oil sorption efficient for a ratio of 10:0 and 7:3

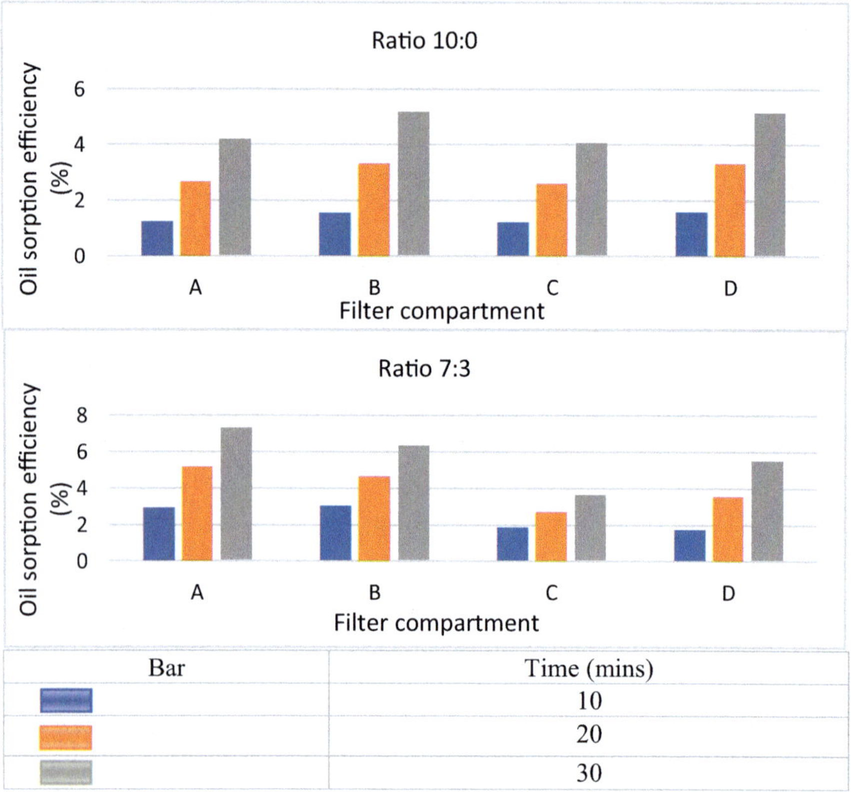

Fig. 7 Performance of oil sorption efficiency for four (A, B, C, D) filter compartments

[10]. The oil absorption also shows increased oil absorption efficiency as the time of absorption increases from 10, 20 and 30 min. The sample with a ratio of 10:0 shows that the filter in compartment A is not too efficient as it absorbs less amount of oil compared to the filter in compartment D. Compartment A reached its limit to absorb oil thus most of the oil will be collected at compartment D. This finding also supports that samples with a ratio of 7:3 have better oil absorption efficiency and stability in a longer time, various medium environments and various filter positions.

4 Conclusion

There are many kinds of methods that were introduced to clean up the oil spill at sea where each of them has its own advantages and disadvantages. Portable oil spill collectors have been developed to make the cleanup process become much easier, but the cost of installation is expensive due to complex systems such as skimmers. This project purposely introduces a simple operating system of filters and selection of adsorbent materials for costs saving. Based on the experiments, the performance of the adsorbent filter materials combination of kapok and sawdust has been determined. Reducing the amount of the kapok quantity leads to a better oil sorption efficiency. The optimum ratio for oil absorbency efficiency found for both laboratory and sea trial experimental recorded samples kapok is sawdust with a ratio of 7:3. As a conclusion, samples with a ratio of 7:3 have a better oil absorption efficiency and stability in longer period immersion time, various medium environments, and various filter positions. Both natural organic materials have proven that they are very capable of absorbing oil, so the use of this material is highly recommended for cleaning up oil spills plus they are very cheap and environmentally friendly.

References

1. Trang, C.T.T.: Oil Spills and Clean-Up Costs. Institute of Marine Environment and Resource, pp. 1–4 (2006)
2. A. Bayat, G. Vakili Nezhaad, S. Aghamiri: Oil Spill Cleanup from Sea Water by Sorbent Materials (2005)
3. Hori, K., Flavier, M.E., Kuga, S. et al.: Excellent oil absorbent kapok [*Ceiba pentandra* (L.) Gaertn.] fiber: fiber structure, chemical characteristics, and application. J. Wood Sci. **46**, 401–404 (2000)
4. Wang, J., Zheng, Y., Wang, A.: Superhydrophobic kapok fiber oil-absorbent: Preparation and high oil absorbency. Chem. Eng. J. **213**, 1–7 (2012)
5. Quek, C.S., Ngadi, N., Zaini, M.A.A.: The oil-absorbing properties of kapok fibre—a commentary. J. Taibah Univ. Sci. **14**, 1 (2020)
6. Zang, D., Liu, F., Zhang, M., Gao, Z., Wang, C.: Novel Superhydrophobic and Superolephilic Sawdust as a Selective Oil Spill Cleanup (2015)
7. Abdelwahab, O., Nasr, S.M., Thabet, W.M.: Palm Fibers and Modified Palm Fibers Adsorbents for different Oils (2016)
8. Lim, T.-T., Huang, X.: Evaluation of Hydrophobicity/Oleophilicity and its Performance in Oily Water Filtration: Comparison of Raw and Solvent-Treated Fibers (2007)
9. S.N.C.M. Hussein, N. Hidayati Othman, A. Dollah, A. Nazihah Che Abdul Rahim, N. Shuhadah Japperi, N. Syamimi Mohd Asymawi Ramakrishnan, Study of acid treated mixed sawdust as natural oil sorbent for oil spill. Mater. Today Proc. **19**, 1382–1389 (2019)
10. Safety.cat.com.: Caterpillar Filters Recommendations (2013)

Investigation on the Interaction of Controllable Pitch Propeller with Boss Cap Fins at Slow-Ahead, Half-Ahead, and Full-Ahead Position

Khairunisa' Ismail, Nur Eqa Fatihah Ab Patah, Maisarah Mohtar, Iwan Zamil Mustaffa Kamal, Engku Ahmad Azrulhisham Engku Mohamed, and Yaseen Adnan Ahmed

Abstract As one of the main components in propulsion systems, the propeller works by converting the engine revolution into a forward force. A lot of fuel and power might be needed to propel the propeller due to various forces acting underwater, hence there is a need to increase the propeller performance. One of the methods is to use propeller boss cap fins (PBCF). The effectiveness of PBCF usage has been studied for years and has been proven to improve the propeller efficiency at in-design condition. However, there are limited research studies and proves on the efficiency of PBCF at off-design conditions. In this study, an investigation is carried out on the interaction of the controllable pitch propeller (CPP) with PBCF at full ahead, half ahead, and slow ahead position. The hypothesis claimed that the application of PBCF to CPP is beneficial to the performance of the CPP in full ahead position, but deteriorating to the performance of CPP-PBCF at half ahead and slow ahead position. The result was achieved through computational fluid dynamic analysis using the ANSYS CFX software. Comparison is made between CPP-PBCF to the one without PBCF. The result shows that the use of PBCF at full ahead, half ahead, and slow ahead positions, created higher efficiency. Therefore, the usage of PBCF is recommended to all CPP.

Keywords Controllable pitch propeller · Propeller boss cap fins · ANSYS CFX

K. Ismail · N. E. F. A. Patah · M. Mohtar · I. Z. M. Kamal (✉)
Universiti Kuala Lumpur Malaysian Institute of Marine Engineering Technology, Jln Pantai Remis, 32200 Lumut Perak, Malaysia
e-mail: iwanzamil@unikl.edu.my

E. A. A. E. Mohamed
Universiti Kuala Lumpur Malaysian France Institute, Section 14, Jalan Damai, Seksyen 14, 43650 Bandar Baru Bangi, Selangor, Malaysia
e-mail: engku@unikl.edu.my

Y. A. Ahmed
Department of Naval Architecture, Ocean and Marine Engineering, University of Strathclyde, 100 Montrose St, Glasgow G4 0LZ, United Kingdom
e-mail: yaseen.ahmed@strath.ac.uk

A. Ismail et al. (eds.), *Engineering Frontiers*, Advanced Structured Materials 208,
https://doi.org/10.1007/978-3-031-56844-2_27

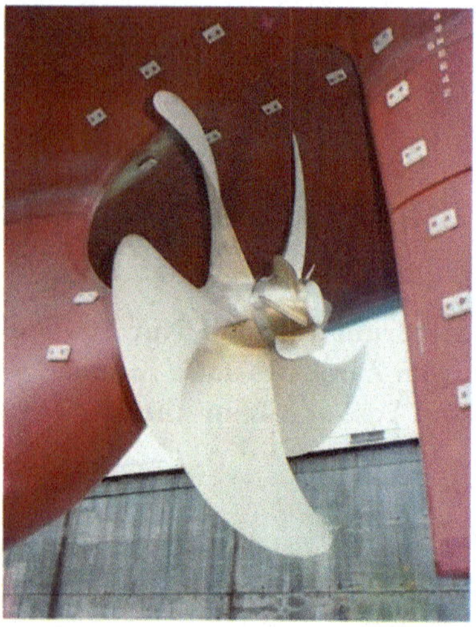

Fig. 1 Propeller with PBCF—MOL techno trade

1 Introduction

The need to increase propulsive energy especially in larger vessels has grown tremendously due to the environmental issue and economic impact. As a solution, various energy saving devices (ESD) have been innovated to accommodate the escalating problem, such as bulbous bow and propeller boss cap fins (PBCF).

PBCF, shown as in Fig. 1, which was first introduced in 1987, does not require any vessel modification during the installation on account of the original boss cap that can directly be replaced by the PBCF. There is no need of propeller modification or welding work. In fact, there is no specific maintenance necessary after installation.

PBCF-installed vessel tends to consume less fuel compared to non-PBCF vessel, by recapturing energy from the hub vortex created behind the propeller, illustrated in Fig. 2. The induced hub vortex will reduce and sometimes be completely broken up, hence lowering the cavitation risk upon the propeller and rudder.

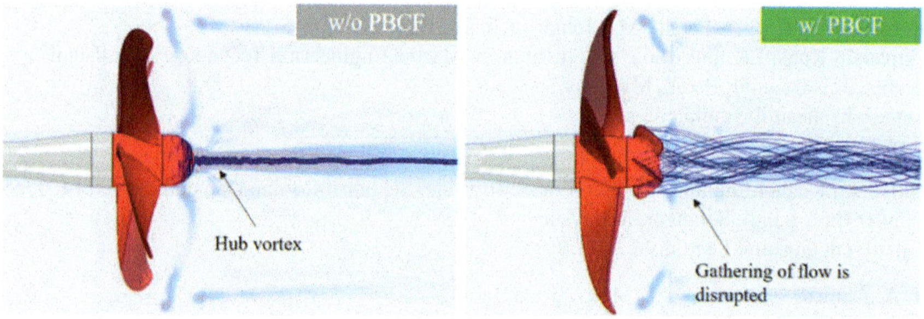

Fig. 2 The hub vortex visualized by CFD simulation (left: normal cap, right: with PBCF)

PBCF works by using the ground effect to increase the thrust and giving a powerful push to the fins in the same direction of the propeller rotation, eliminating the hub vortex and decreasing the resistance. According to Gearhart and McBride [1], the increased propulsive effectiveness will rise by 6%, where 2% out of the total 6% efficiency gain comes from the increased thrust while the remaining 4% is due to decreased torque.

PBCF is usually used for a fixed-pitch propeller (FPP) instead of CPP due to special design requirement for the C.P propeller boss cap fins. A conventional propeller with PBCF installation can easily be found at in-design condition; full ahead. At off-design condition, the pitch of CPP changes while the pitch of PBCF stays the same, causing the thrust and the torque generation to become a concern effecting the propeller performance. The interaction of the C.P propeller blade at in-design and off-design with a boss cap fin is shown in Fig. 3. When the pitch angle of the C.P propeller is changing, the PBCF fin will remain in its position without altering its pitch angle. The thrust and torque produced by the propeller may have changed even slightly, especially when the propeller is at zero-pitch condition. Although at zero-pitch condition, the rotating shaft will still allow the PBCF to generate a small amount of thrust for the ship to move forward instead of holding its stationary position. This mechanism is worrying when operating in CPP especially in a standby mode, therefore the amount of the thrust created by the PBCF needs to be quantified. The accuracy of propeller performance using a theory-based method such as blade element momentum theory and lifting line theory is unfortunately not as satisfying as CFD simulation.

For that reason, the performance of CPP-PBCF is analysed in this research project using the computational fluid dynamics (CFD) method, focusing only on the interaction at in-design condition; full-ahead, half-ahead, and slow-ahead. Although past studies have proven the increasing efficiency of using PBCF at full-ahead condition, it may cause a different outcome when the PBCF is installed at the CP propeller

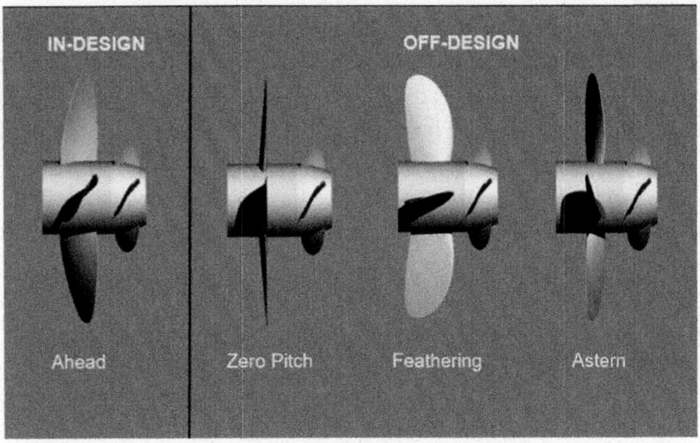

Fig. 3 CPP-PBCF interaction at in-design and off-design conditions

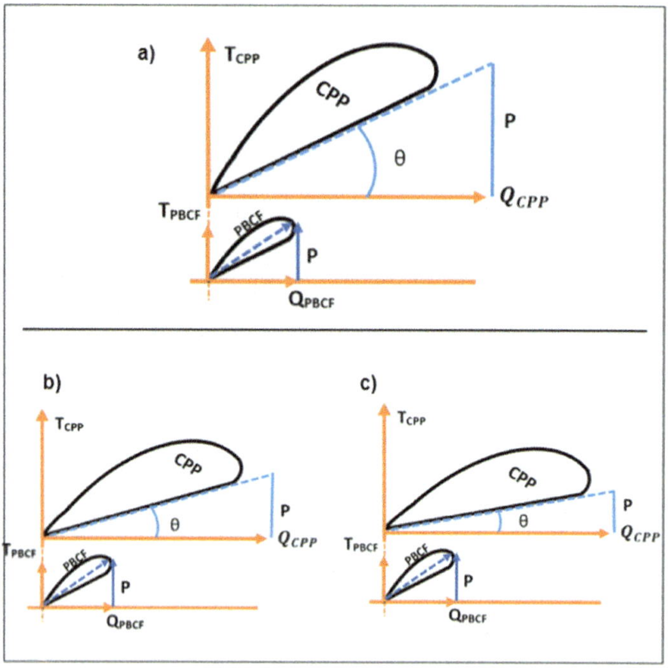

Fig. 4 Force diagram of the propeller blade and boss cap fin at in design condition **a** full ahead, **b** half ahead, **c** slow ahead

especially when it changes the blade pitch from full-ahead to half-ahead and slow-ahead position. This is due to different thrust and torque forces created by different blade pitch as shown in the force diagram in Fig. 4.

Problem Statement:

PBCF is usually equipped with fixed-pitch propellers. However, there are limited application of PBCF with controllable-pitch propellers as it may be beneficial or non-beneficial. F.P propeller-boss cap fins are proven to being cost-wise and benefit the environment, but it is unsure how they effect the C.P. propeller-boss cap fins.

Hypothesis:

- The application of PBCF to CPP is beneficial to the performance of the CPP in full ahead position.
- However, the application of PBCF to CPP will deteriorate the performance of the CPP in half ahead, and slow ahead positions.
- Therefore, this research will try to approve the hypothesis above.

Table 1 Propeller parameters

Parameters	Dimension
Type	B-series
Number of blade	4
Blade area ratio	0.35
Propeller diameter	250 mm
Pitch/diameter	Full ahead: 1.1
	Half ahead: 1.0
	Slow ahead: 0.8

Table 2 PBCF parameters

Parameters	Dimension
Number of blade	4
Blade area ratio	0.35
Propeller diameter	109 mm
Pitch/diameter	1.5

2 Methodology

2.1 Geometry Modelling

At the early stage, the propeller and PBCF is modelled using PropCAD Rhino and 3D based on the parameters shown in Tables 1 and 2.

After the propeller modelling is done in Rhino 3D, the SOLIDWORKS software is then used for the computational setup. The stationary domain was divided into three parts; inlet, middle, outlet. The inlet is defined as the axial up stream in the flow. The middle part contained the rotating domain shown in Fig. 5 which is located at the centre of the coordinate which is aligned to the fluid flow. And lastly, the outlet is defined as the downstream with a distance of six times of the propeller diameter.

2.2 Design of Experiment

This experiment is designed in detail to ensure the parameters and constant variables stay unchanged throughout all the simulations. The advanced coefficients that will be used are tabulated in Table 3.

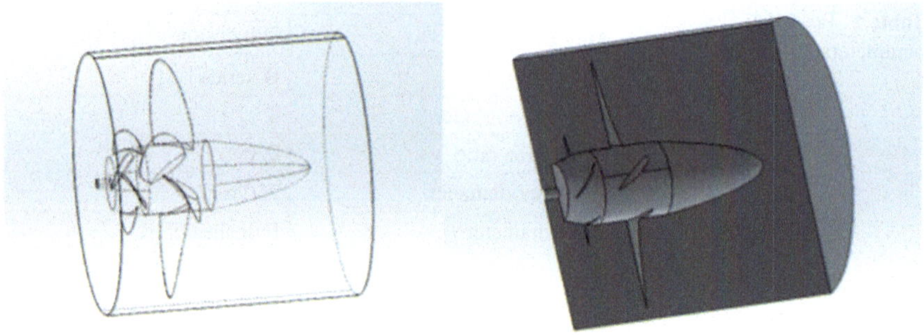

Fig. 5 Rotating domain

Table 3 Advance coefficient

	Full-ahead		Half-ahead		Slow-ahead	
	Without PBCF	With PBCF	Without PBCF	With PBCF	Without PBCF	With PBCF
Advance coefficient (J)	0.4, 0.6, 1.0, 1.2, 1.4					
Number of simulations	5	5	5	5	5	5
Total simulations	30					

2.3 Computational Fluid Dynamics

At this stage, the simulation process begins with mesh generation using ANSYS meshing. The multiple reference frames (MRF) approach is applied in the simulation of the propeller performance where interfaces are used to distinguish between a stationary and rotating zone. Figure 6 shows the computational domains for the open water test setup in the ANSYS-CFX solver.

Ten inflation layers are used to capture the velocity gradient information and to acquire an accurate solution to the fluid flow problems. The computational setup followed the setup suggested by Tarmizi [2] of the ANSYS CFX as shown in Table 4.

The last stage of the CFD methodology is data collection. As the simulation is completed, the thrust and torque values are taken, and calculated in formula to find the thrust coefficient K_T, torque coefficient K_Q, and open water efficiency η_O. A $K_T - K_Q - J$ chart is a typical chart of the open water characteristics. This chart is usually used to plot the hydrodynamic performances as function of the advance coefficient. The non-dimensional characteristics can be calculated by using the formulae:

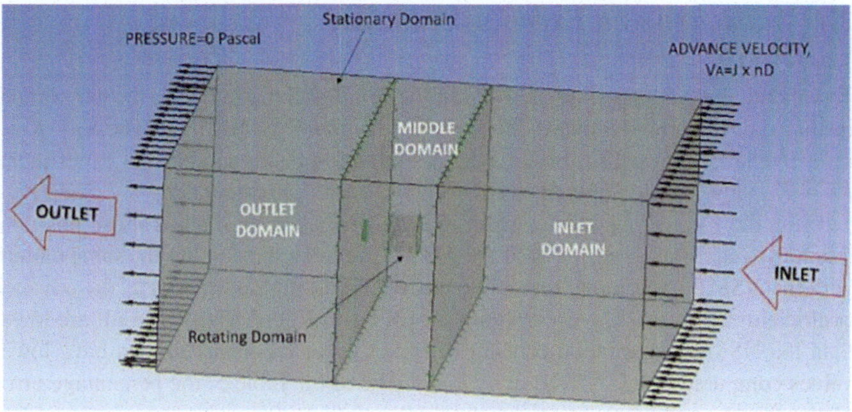

Fig. 6 Computational domains for open water test setup in ANSYS-CFX solver (suggested by Tarmizi [2])

Table 4 Turbulent setup

Description	Condition	Values
Fluid model	Heat transfer none	–
Turbulent	Shear stress transport (SST)	–
Mass and momentum (Inlet)	Normal speed (m/s)	1.5, 2.25, 3, 3.75, 4.5
	Turbulent	Low intensity = 1%
Mass and momentum (Middle)	No slip wall	–
Mass and momentum (Outlet)	Static pressure (Pa)	0

$$\text{Thrust Coefficient,}\quad \mathrm{K_T} = \frac{\mathrm{T}}{\rho n^2 \mathrm{D}^4} \tag{1}$$

$$\text{Torque Coefficient,}\quad \mathrm{K_Q} = \frac{\mathrm{Q}}{\rho n^2 \mathrm{D}^5} \tag{2}$$

$$\text{Advance Coefficient,}\quad \mathrm{J} = \frac{\mathrm{V_a}}{\mathrm{nD}} \tag{3}$$

$$\text{Open Water Efficiency,}\quad \eta_O = \frac{J}{2\rho} \times \frac{K_T}{K_Q} \tag{4}$$

where T is the thrust (in N), Q is the torque (in Nm), ρ is the density of water (in kg/m^3), n is the number of propeller shaft revolution, and D is the diameter of propeller (in m).

2.4 Validation with Experimental Data

To validate the ANSYS simulated result, the result is compared to to the experiment result from Netherland Ship Model Basin Experiment (NSMB 1962). The $K_T - K_Q - \eta_O - J$ chart is plotted for both ANSYS simulation result and NSMB experiment. $K_T - K_Q$ values are compared and percentage errors are calculated.

From the $K_T - K_Q - \eta_O - J$ chart in Fig. 7, which is for the full ahead position, the thrust and torque coefficient of the ANSYS simulated result has the same pattern with the NSMB experiment result but carried increasing differences as the advance coefficient moves up. The overall values of KT and KQ of ANSYS result are lower than the NSMB experiment, causing the open water efficiency to also have lower values compared to the NSMB experiment. Based on Table 5, the percentage error of the thrust coefficient is lowest at J = 0.4 with 8.90%. The error increased as the advance coefficient increases, cutting at J = 1.2 with 663.80% error. As for the torque coefficient, the lowest error is also at J = 0.4 with 6.86%, while the highest percentage error is 97.75% at J = 1.2.

Based on Fig. 8, all the K_T, K_Q and *ETA* O graphs which is for the half ahead position, the results from the ANSYS experiment followed the same pattern as the NSMB experiment, with ANSYS results reveal lower values. The percentage error for the thrust coefficient between ANSYS simulated result and the NSMB experiment as shown in Table 6 is 10.46% for J = 0.4, increased to 122.68% at J = 1.2%. Meanwhile, the torque coefficient has 7.57% error at J = 0.4 and 40.30% error at J = 1.0.

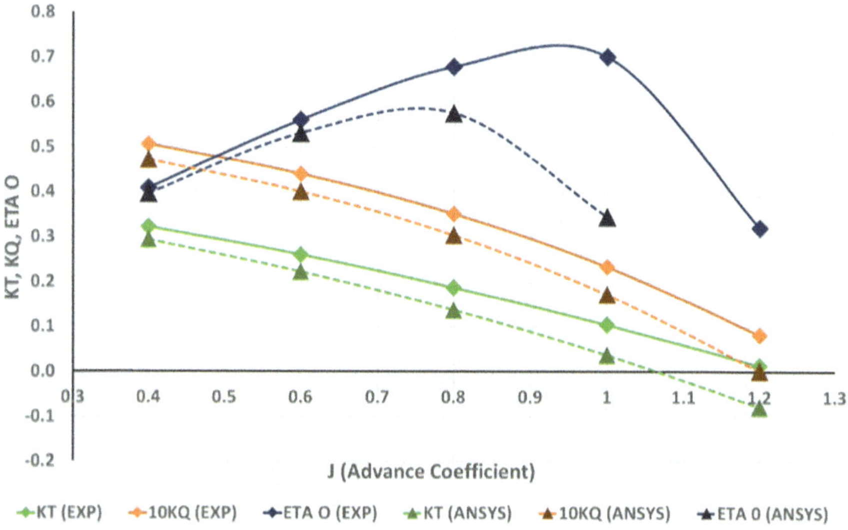

Fig. 7 $K_T - K_Q - \eta_O$ versus J of full ahead between ANSYS simulated result and NSMB experiment

Table 5 Percentage error for full ahead

J	Kt (EXP)	Kt (ANSYS)	Percent Error %	10Kq (EXP)	10Kq (ANSYS)	Percent error %
0.4	0.32229	0.29362	8.90	0.50650	0.47177	6.86
0.6	0.26034	0.22238	14.58	0.43990	0.40030	9.00
0.8	0.18704	0.13704	26.73	0.35100	0.30341	13.56
1.0	0.10424	0.03716	64.35	0.23380	0.17175	26.54
1.2	0.01383	– 0.07797	663.80	0.08240	0.00185	97.75

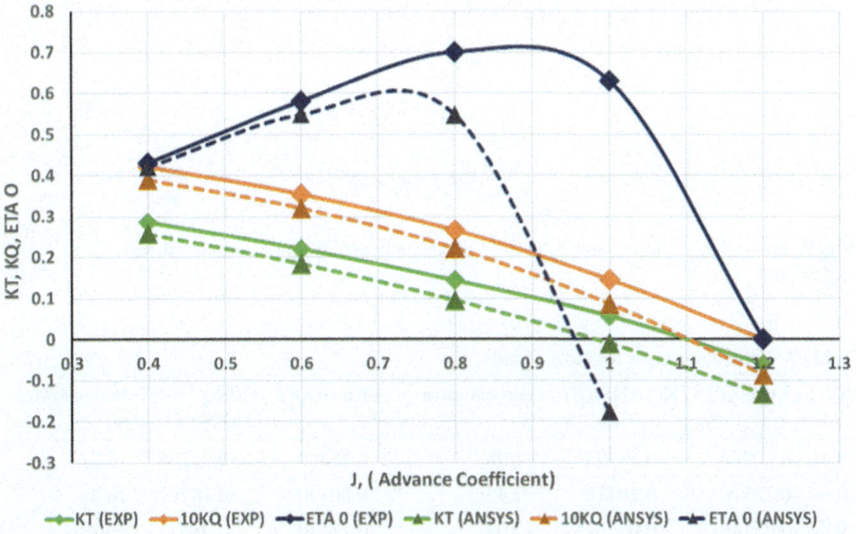

Fig. 8 $K_T - K_Q - \eta_O$ versus J of half ahead between ANSYS simulated result and NSMB experiment

Table 6 Percentage error for half ahead

J	Kt (EXP)	Kt (ANSYS)	Percent Error %	10Kq (EXP)	10Kq (ANSYS)	Percent error %
0.4	0.28660	0.25662	10.46	0.42100	0.38913	7.57
0.6	0.22060	0.18348	16.83	0.35400	0.32006	9.59
0.8	0.14350	0.09481	33.93	0.26500	0.22105	16.59
1.0	0.05700	– 0.00964	116.91	0.14500	0.08656	40.30
1.2	– 0.06000	– 0.13361	122.68	0.00000	– 0.08663	0.00

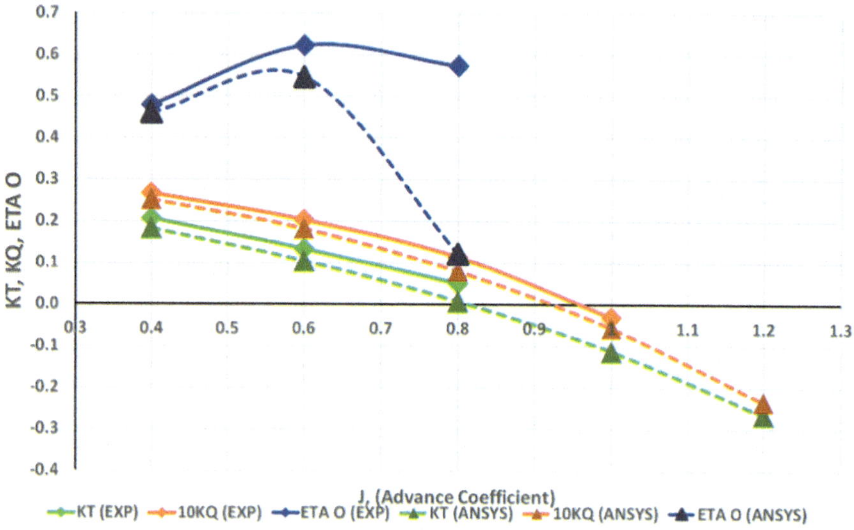

Fig. 9 $K_T - K_Q - \eta_O$ versus J of slow ahead between ANSYS simulated result and NSMB experiment

Table 7 Percentage error for slow ahead

J	Kt (EXP)	Kt (ANSYS)	Percent error %	10Kq (EXP)	10Kq (ANSYS)	Percent error %
0.4	0.20783	0.18293	11.98	0.26800	0.25268	5.71
0.6	0.13510	0.10478	22.45	0.20440	0.18297	10.48
0.8	0.05146	0.00767	85.09	0.11570	0.08078	30.19
1.0	0.00000	− 0.11150	0.00	− 0.03000	− 0.05619	87.31
1.2	0.00000	− 0.26563	0.00	0.00000	− 0.23370	0.00

The data comparison for the K_T, K_Q, and ETA O graphs for the slow ahead position of the ANSYS simulation data also has the same pattern with the NSMB experiment, with the ANSYS result having lower values as shown in Fig. 9. The percentage error from Table 7 for the slow ahead result at J = 0.4 is 11.98% for the thrust coefficient, and 5.71% for the torque coefficient. At J = 0.8, the thrust coefficient has 85.09%, and 87.31% error of torque coefficient at J = 1.0.

3 Results and Discussion

The result between the propeller without PBCF is compared to the propeller with PBCF. The differences in thrust and torque are noted as it will determine the propeller efficiency and the optimum condition.

Table 8 Data collection of full ahead without PBCF

Full ahead without PBCF							
J	Va (m/s)	T (N)	Q (N m)	K_T	K_Q	$10K_Q$	ETA 0
0.4	1.50	264.515	– 10.62510	0.29362	– 0.04718	0.47177	0.39622
0.6	2.25	200.338	– 9.01556	0.22238	– 0.04003	0.40030	0.53050
0.8	3.00	123.456	– 6.83349	0.13704	– 0.03034	0.30341	0.57507
1.0	3.75	33.474	– 3.86805	0.03716	– 0.01717	0.17175	0.34433
1.2	4.50	– 70.244	– 0.04171	– 0.07797	– 0.00019	0.00185	– 80.40536

Table 9 Data collection of full ahead with PBCF

Full ahead with PBCF							
J	Va (m/s)	T (N)	Q (N m)	K_T	K_Q	$10K_Q$	ETA 0
0.4	1.50	263.691	– 10.5086	0.29270	– 0.04666	0.46659	0.39937
0.6	2.25	200.636	– 8.8658	0.22271	– 0.03936	0.39365	0.54026
0.8	3.00	125.361	– 6.69324	0.13915	– 0.02972	0.29719	0.59618
1.0	3.75	37.705	– 3.75654	0.04185	– 0.01668	0.16679	0.39937
1.2	4.50	– 65.198	0.02635	– 0.07237	0.00012	0.00117	– 118.1496

3.1 *Full-Ahead Position*

Based on the Tables 8 and 9, the thrust of full-ahead with PBCF is lower at first, but slowly increased as the advance coefficient increases. The torque is lower at full ahead with PBCF compared to without PBCF. Although the differences are small and almost zero, the open water efficiency in still higher in PBCF than without PBCF.

The thrust comparison for full ahead position without and with PBCF is shown in Figs. 10 and 11, while the torque comparison is shown in Figs. 12 and 13. The open water efficiency is shown in Fig. 14.

3.2 *Half-Ahead Position*

Based on Tables 10 and 11, the thrust of half-ahead with PBCF is higher than without PBCF, and the torque is lower at half-ahead with PBCF compared to without PBCF. Therefore, the open water efficiency is higher with PBCF.

The thrust comparison for half ahead position without and with PBCF can be found in Figs. 15 and 16, while the torque comparison is shown in Figs. 17 and 18. The open water efficiency is shown in Fig. 19.

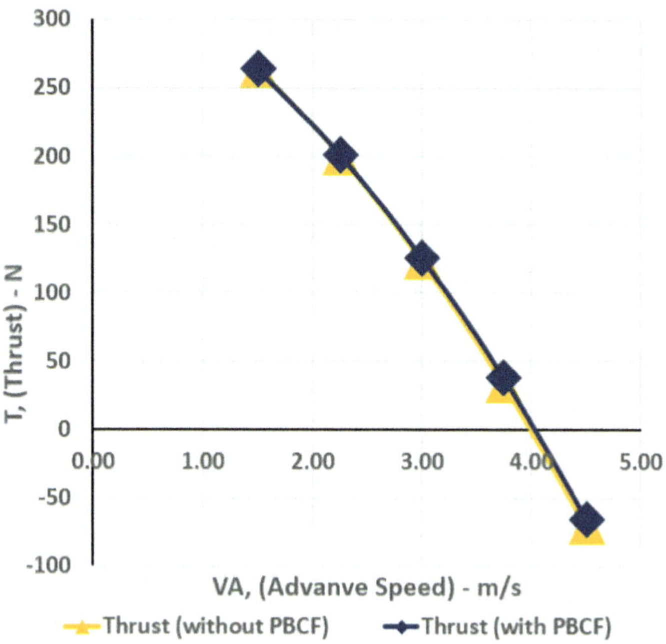

Fig. 10 Thrust versus advance speed for full ahead with and without PBCF

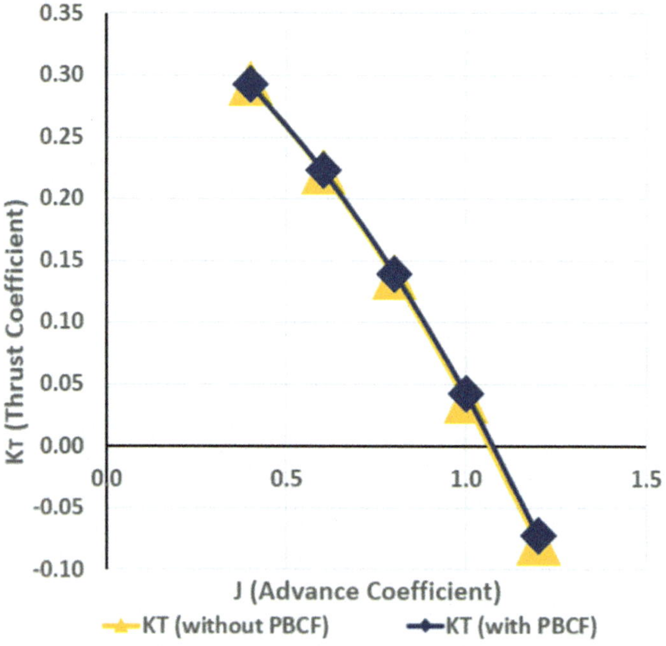

Fig. 11 Thrust coefficient versus advance coefficient for full ahead with and without PBCF

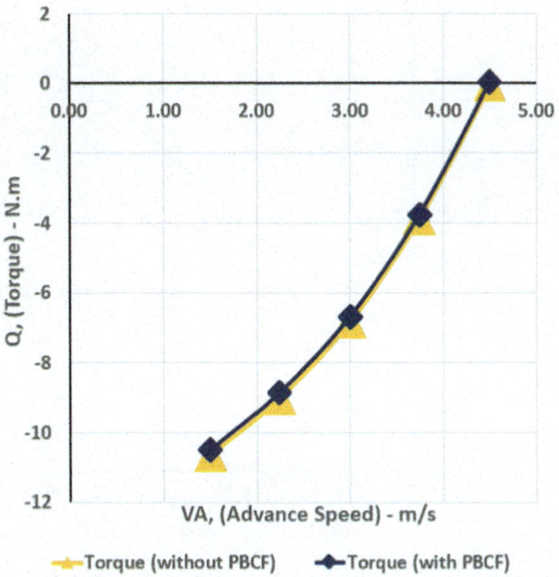

Fig. 12 Torque versus advance speed for full ahead with and without PBCF

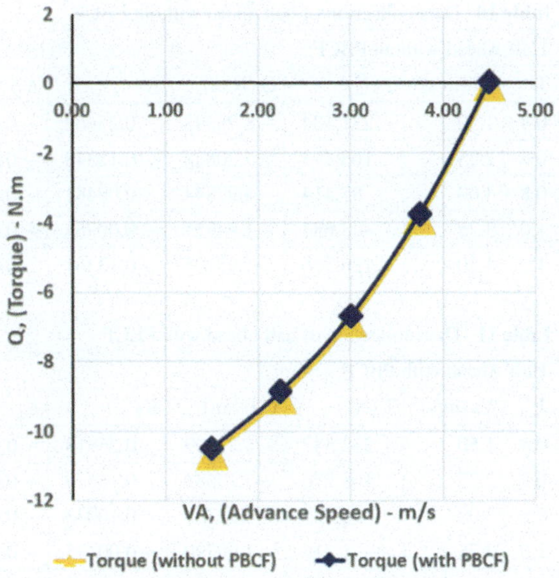

Fig. 13 Torque coefficient versus advance coefficient for full Ahead with and without PBCF

3.3 Slow-Ahead Position

Based on Tables 12 and 13, the thrust of slow-ahead with PBCF is higher than without PBCF. The torque is slightly higher at slow-ahead with PBCF compared to without

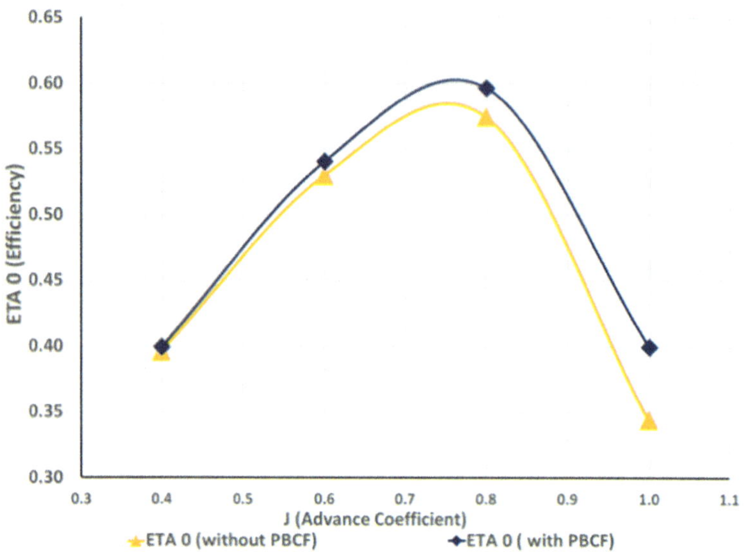

Fig. 14 Open water efficiency versus advance coefficient for full Ahead with and without PBCF

Table 10 Data collection of half ahead without PBCF

Half Ahead without PBCF							
J	Va (m/s)	T (N)	Q (N.m)	K_T	K_Q	$10K_Q$	ETA 0
0.4	1.50	231.184	− 8.76398	0.25662	− 0.03891	0.38913	0.41983
0.6	2.25	165.293	− 7.20828	0.18348	− 0.03201	0.32006	0.54744
0.8	3.00	85.414	− 4.97844	0.09481	− 0.02210	0.22105	0.54612
1.0	3.75	− 8.683	− 1.94951	− 0.00964	− 0.00866	0.08656	− 0.17722
1.2	4.50	− 120.366	1.95105	− 0.13361	0.00866	− 0.08663	− 2.94562

Table 11 Data collection of half ahead with PBCF

Half Ahead with PBCF							
J	Va (m/s)	T (N)	Q (N.m)	K_T	K_Q	$10K_Q$	ETA 0
0.4	1.50	232.547	− 8.71109	0.25813	− 0.03868	0.38678	0.42487
0.6	2.25	168.276	− 7.13688	0.18679	− 0.03169	0.31689	0.56289
0.8	3.00	90.385	− 4.91381	0.10033	− 0.02182	0.21818	0.58550
1.0	3.75	− 1.216	− 1.90790	− 0.00135	− 0.00847	0.08471	− 0.02536
1.2	4.50	− 112.051	1.96230	− 0.12438	0.00871	− 0.08713	− 2.72641

Fig. 15 Thrust versus advance speed for half ahead with and without PBCF

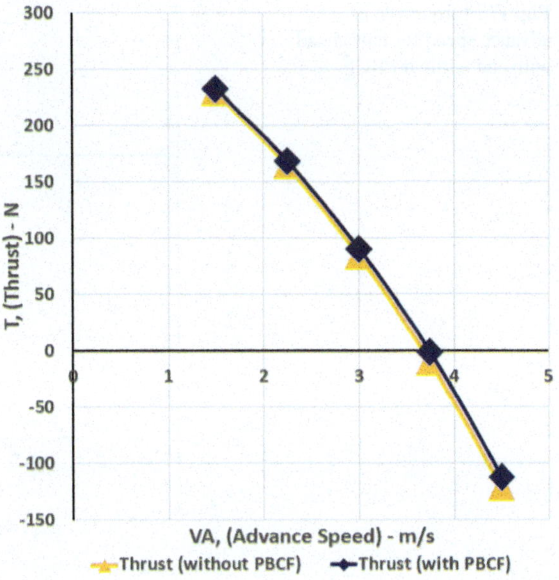

Fig. 16 Thrust coefficient versus advance coefficient for half ahead with and without PBCF

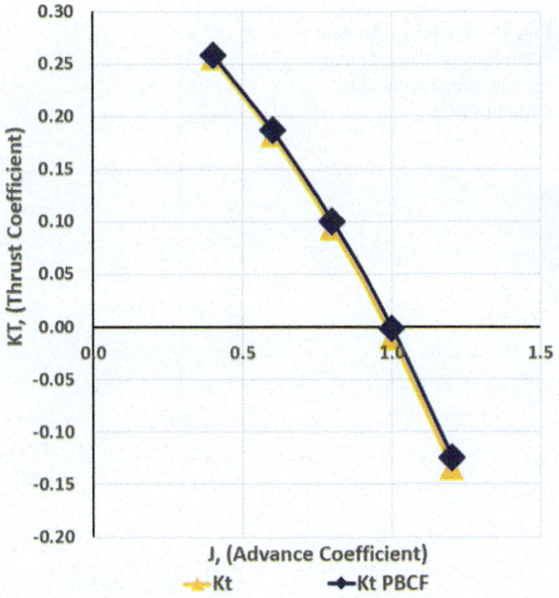

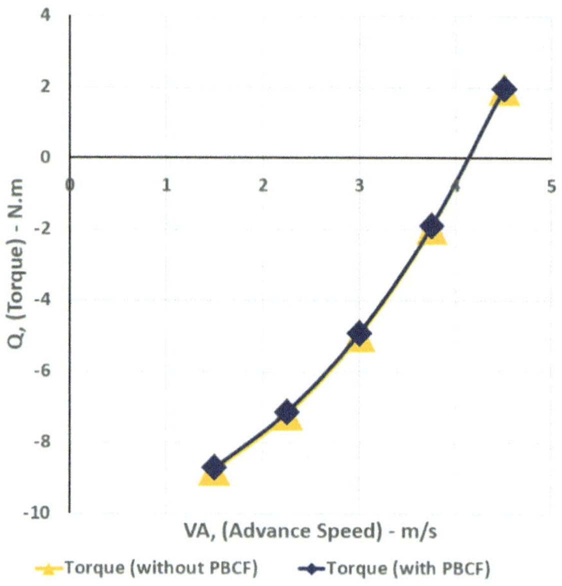

Fig. 17 Torque versus advance speed for half ahead with and without PBCF

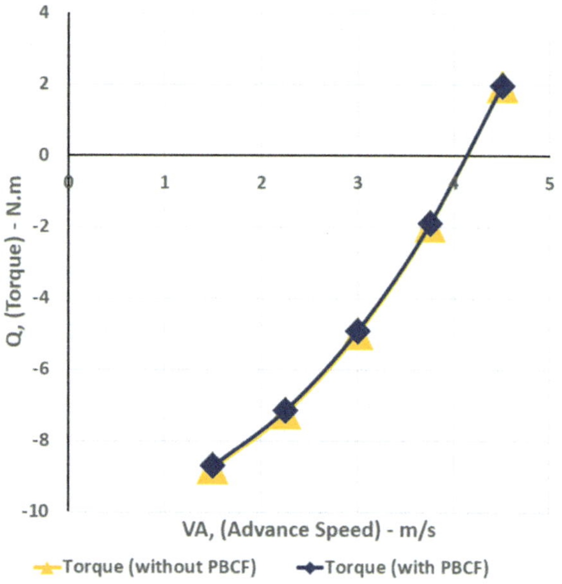

Fig. 18 Torque coefficient versus advance coefficient for half ahead with and without PBCF

PBCF. Nonetheless, the open water efficiency is still higher with PBCF compared to without PBCF.

The thrust comparison for the half ahead position without and with PBCF can be found in Figs. 20 and 21, while the torque comparison is shown in Figs. 22 and 23. The open water efficiency is shown in Fig. 24.

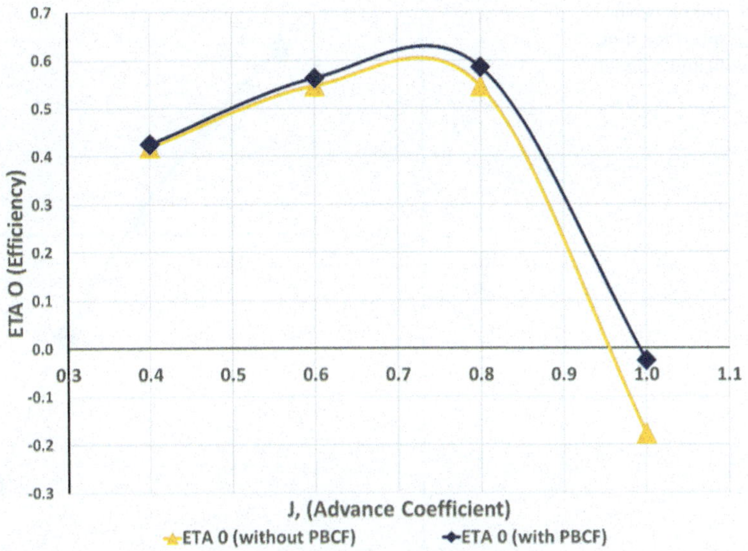

Fig. 19 Open water efficiency versus advance coefficient for half ahead with and without PBCF

Table 12 Data collection of slow ahead without PBCF

Slow ahead without PBCF							
J	Va (m/s)	T (N)	Q (N.m)	K_T	K_Q	$10K_Q$	ETA 0
0.4	1.50	164.802	− 5.69095	0.18294	− 0.02527	0.25268	0.46089
0.6	2.25	94.390	− 4.12092	0.10478	− 0.01830	0.18297	0.54682
0.8	3.00	6.912	− 1.81923	0.00767	− 0.00808	0.08078	0.12094
1.0	3.75	− 100.446	1.26555	− 0.1115	0.00562	− 0.05619	− 3.15801
1.2	4.50	− 239.302	5.26344	− 0.26563	0.02337	− 0.2337	− 2.17079

Table 13 Data collection of slow ahead with PBCF

Slow ahead with PBCF							
J	Va (m/s)	T (N)	Q (N.m)	K_T	K_Q	$10K_Q$	ETA 0
0.4	1.50	166.424	− 5.58948	0.18474	− 0.02482	0.24818	0.47388
0.6	2.25	98.826	− 4.03779	0.10970	− 0.01793	0.17928	0.58430
0.8	3.00	15.051	− 1.76231	0.01671	− 0.00782	0.07825	0.27185
1.0	3.75	− 88.443	1.29028	− 0.09817	0.00573	− 0.05729	− 2.72734
1.2	4.50	− 223.762	5.28031	− 0.24838	0.02345	− 0.23445	− 2.02334

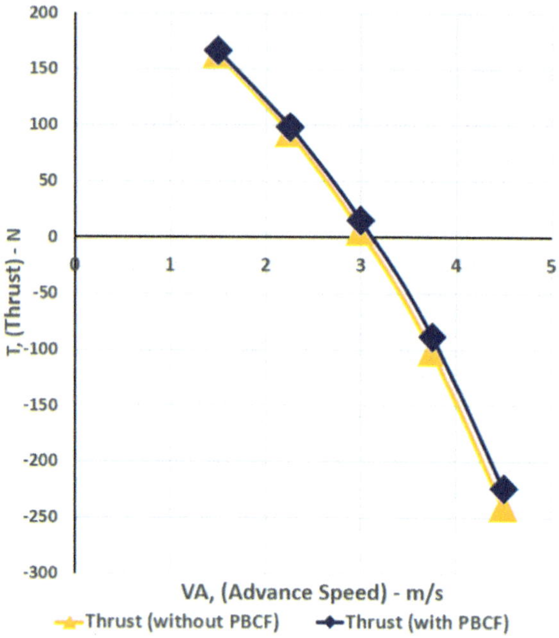

Fig. 20 Thrust versus advance speed for slow ahead with and without PBCF

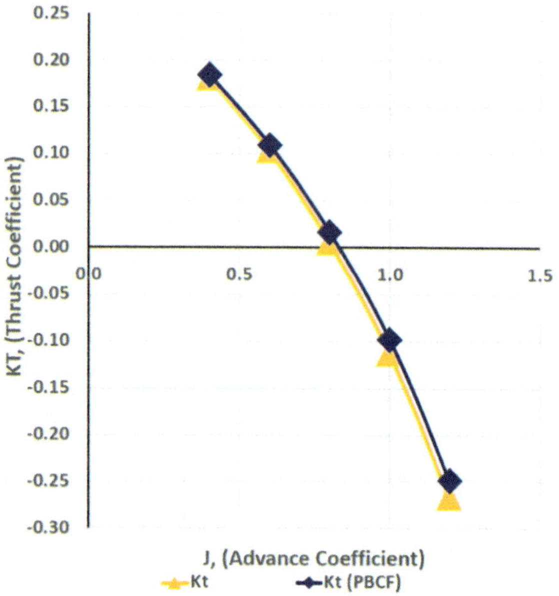

Fig. 21 Thrust coefficient versus advance coefficient for slow ahead with and without PBCF

Fig. 22 Torque versus advance speed for slow ahead with and without PBCF

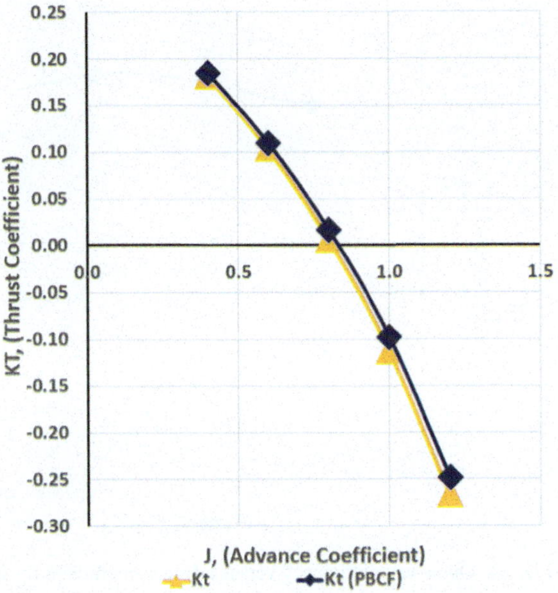

Fig. 23 Torque coefficient versus advance coefficient for slow ahead with and without PBCF

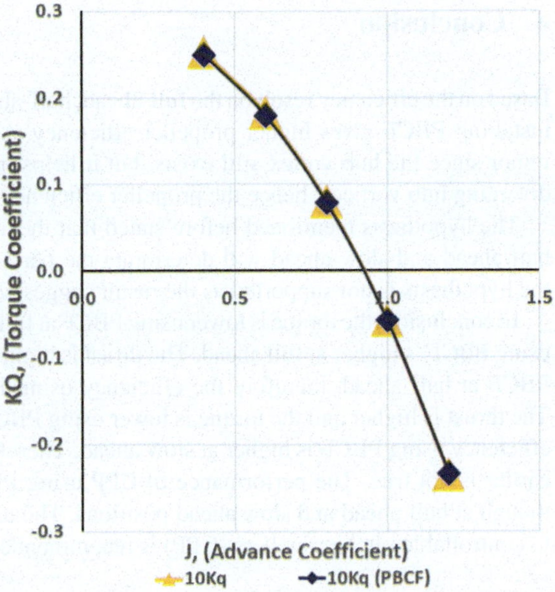

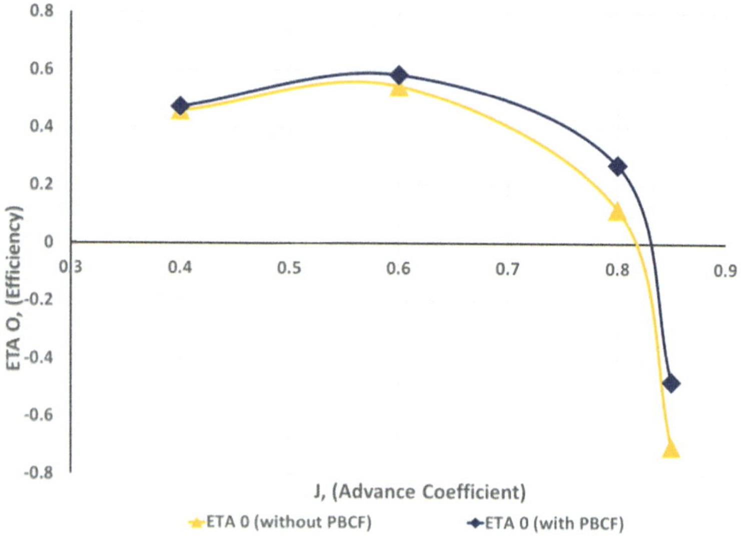

Fig. 24 Open water efficiency versus advance coefficient for slow ahead with and without PBCF

4 Conclusion

Based on the efficiency result on the full-ahead, half-ahead, and slow-ahead positions, installing PBCF gives higher propeller efficiency. Although the efficiency is only minor since the hub vortex still exists, but it helps in breaking up and creates less diverging hub vortices, hence the propeller efficiency is increased.

The hypothesis mentioned before stated that the application of PBCF to CPP in half-ahead and slow-ahead will deteriorate the performance of the CPP. However, the hypothesis is not supported as the result suggested the opposite.

In conclusion, the torque is lower using PBCF at full ahead, therefore the efficiency using PBCF is higher at full ahead. The thrust is higher and the torque is lower using PBCF at half ahead, therefore the efficiency using PBCF is higher at half ahead. The thrust is higher and the torque is lower using PBCF at slow ahead, therefore the efficiency using PBCF is higher at slow ahead. Therefore, the hypothesis mentioned earlier is not true. The performance of CPP using PBCF will not deteriorate even though at half ahead and slow ahead positions. The usage of propeller boss cap fins to controllable pitch propellers (CPP) is recommended.

References

1. Gearhart, W.S., McBride, M.W.: Performance assessment of propeller boss cap fin type device. Paper presented at the 22nd American towing tank conference, St. John's, Newfoundland, Canada, 8–11 Aug 1989
2. A.Tarmizi, N.N.A.: Computational fluid dynamics simulation of open water propeller test—postdam propeller test case. Dissertation, University of Kuala Lumpur (2020)

Design and Strength Analysis of Boat Trailer Sliders Using the Finite Element Method

Norfadhlina Khalid, Noor Hidayah Farhaneem Hussin, Asmalina Mat Saat, Puteri Zarina Megat Khalid, and Syed Radzi Rahamaddulla

Abstract This research project aims to enhance the usability and performance of boat trailers by incorporating existing mechanisms and employing advanced design analysis techniques. The primary objective is to analyze the design of the boat trailer using equation-based load analysis, considering its maximum load capacity of 1400 kg and length of 6 m. Additionally, a 3D modeling simulation in Autodesk Inventor with the finite element method utilized to evaluate the trailer's performance under various loading conditions. This simulation, which employs finite element analysis, generates a wealth of data, including stresses and strains, principal stresses, and safety factors. This data provides valuable insights into the boat trailer design's structural behavior and safety margins. The project incorporates different stages of the design process to determine the final design, ensuring that it meets the required criteria for strength, stability, and load-bearing capacity. By leveraging advanced simulation techniques and comprehensive data analysis, this research aims to optimize the boat trailer's performance and ensure its suitability for various boat transportation needs.

Keywords Boat trailers · Usability · Performance enhancement · Design analysis · Finite element analysis

N. Khalid (✉)
Maritime Engineering Technology Section, Universiti Kuala Lumpur, Malaysian Institute of Marine Engineering Technology, Jalan Pantai Remis, 32200 Lumut, Perak, Malaysia
e-mail: norfadhlina@unikl.edu.my

N. H. F. Hussin · A. M. Saat
Marine and Electrical Engineering Technology Section, Universiti Kuala Lumpur Malaysian Institute of Marine Engineering Technology, Jalan Pantai Remis, 32200 Lumut, Perak, Malaysia
e-mail: asmalina@unikl.edu.my

P. Z. M. Khalid
Department of English Language and Literature, Faculty of Languages and Communication, Universiti Pendidikan Sultan Idris, 35900 Tanjung Malim, Perak, Malaysia
e-mail: puteri.zarina@fbk.upsi.edu.my

S. R. Rahamaddulla
Faculty of Industrial Management, Universiti Malaysia Pahang Al Sultan Abdullah, Lebuhraya Persiaran Tun Khalil Yaakob, 26300 Kuantan, Pahang, Malaysia
e-mail: syedradzi@ump.edu.my

A. Ismail et al. (eds.), *Engineering Frontiers*, Advanced Structured Materials 208,
https://doi.org/10.1007/978-3-031-56844-2_28

1 Introduction

Boat trailers have long been essential equipment for boat owners and enthusiasts, facilitating the seamless transfer of boats between water and land [1]. However, the conventional designs of boat trailers often pose challenges in terms of loading and unloading, stability, and overall usability. To address these limitations, this research project focuses on developing an enhanced boat trailer design that incorporates innovative features, including a tilting mechanism, and employs advanced analysis techniques. The first objective of this study is to analyze the design of boat trailers using equation-based load analysis. By considering factors such as the maximum load capacity of 1400 kg and the length of 6 m, the trailer's structural integrity and load-bearing capabilities can be accurately assessed. This analysis serves as a foundation for identifying potential areas for improvement and determining the necessary modifications to enhance the trailer's performance.

In addition to load analysis, a 3D modeling simulation is created using Autodesk Inventor, enabling a detailed examination of the boat trailer's design. Finite element analysis evaluates the stresses, strains, and safety factors on all planes, including the crucial region where the tilting mechanism is located. This analysis provides valuable insights into the structural behavior and performance of the trailer, specifically concerning the tilting mechanism, under various loading conditions. By assessing the stresses at the fulcrum of the trailer, the research aims to ensure the durability and reliability of the tilting mechanism to accommodate the boat's weight during transportation safely. It can ensure that the design can withstand the anticipated forces, enhancing its overall strength and minimizing the risk of failure or deformation for future recommendations [2].

By incorporating existing mechanisms such as electric winches, keel holders, and sliding mechanisms, in combination with the tilting mechanism, the proposed design aims to streamline the loading and unloading process, improve stability during transportation, and enhance the overall usability. Through this research, the goal is to optimize the performance and functionality of boat trailers, providing boat owners and operators with a reliable and efficient solution for their transportation needs [3]. The subsequent sections of this article will delve into the methodology employed, including the stages of the design process and the detailed analysis techniques utilized. The results and findings obtained from the simulations are discussed, shedding light on the effectiveness of the proposed design enhancements [4], specifically focusing on the tilting mechanism.

1.1 Problem Statement

The problem addressed in this project is the lack of durability and structural strength in boat trailers, particularly in the frame structure. Previous designs commonly utilized square hollow bars, which are not very durable. The project proposes using I-beams as the primary material for the entire body frame to improve the trailers due to its superior load-carrying capacity and reduced material consumption. The project also emphasizes the importance of prioritizing durability by selecting robust materials and employing construction techniques that can withstand the challenging marine environment, including prolonged exposure to water and salt. In addition to the corrosive effects of saltwater, boat trailers are exposed to various environmental factors such as water, wind, heat, and abrasives, which can compromise their durability. The project suggests incorporating components designed to operate in such conditions, including UV-resistant coatings or protective finishes to prevent fading and degradation from sunlight. Moisture-resistant materials and sealing procedures should be used to protect against rust and corrosion. Materials that are resistant to hot and cold conditions and abrasive substances can help avoid scratches and surface damages. Figure 1 shows a previous project boat trailer with this primary mechanism.

Another aspect the project addresses is the lack of winches in boat trailers for loading and unloading. Manual loading methods are laborious and time-consuming, especially for larger or heavier boats. The inclusion of winches in trailers streamlines the process, increases efficiency, and contributes to injury prevention. Winches also enhance the boat safety, reduce the risk of damage, and potentially increase the boat's longevity. Electric winches offer additional advantages in terms of time and effort savings.

Furthermore, the project acknowledges the challenges in achieving a seamless glide-on and glide-off experience during the unloading process. The lack of rollers in trailers poses difficulties, particularly in rough conditions. The project proposes an improved retrieval system that combines keel support slides with bunks or rollers for lateral support. This ensures that the keel, which is the most solid component of the boat's hull, bears the heaviest weight and guarantees safe and stable transit. Balancing the load evenly over the keel and side supports is crucial to preventing damage and ensuring a smooth sliding experience.

Fig. 1 Previous project of the boat trailer (UniKL MIMET)

2 Methodology

The stage of design development was presented in the previous literature [5]. Three critical components of this stage include product benchmarking, quality function deployment (QFD), and concept selection matrix (CSM). These steps are crucial in ensuring that the design meets the desired objectives and fulfils the customers' expectations. Figure 2 exhibits the design process of this study.

2.1 Boat Trailer Specifications

The specifications outlined below can safely transport a boat, ensuring it meets the required criteria and specifications while ensuring proper weight distribution, stability, and performance under dynamic forces. Its dimensions, load capacity, suspension system, motor winch, and rollers work together to ensure safe and smooth sliders when loading and unloading the boat. Safety and security are prioritized,

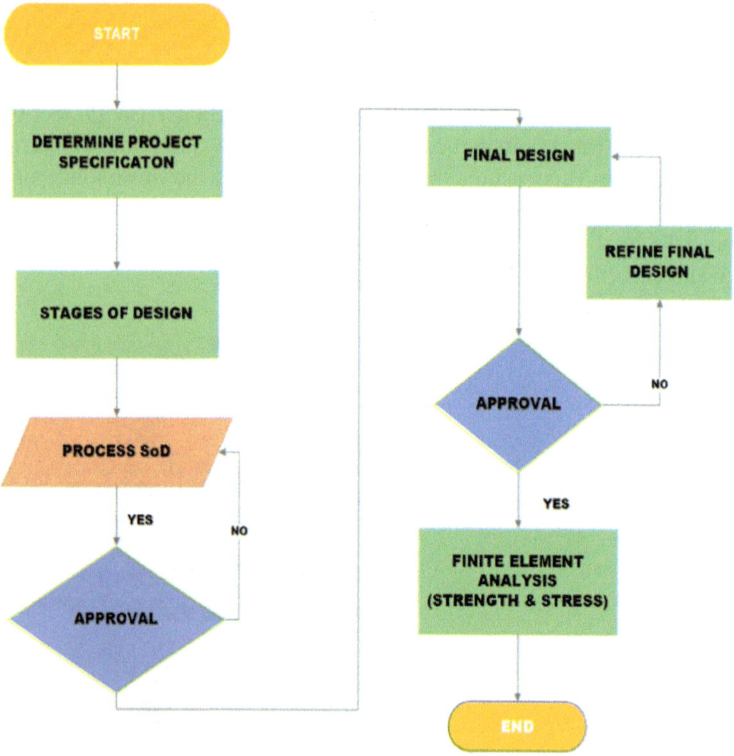

Fig. 2 Flowchart of the design process

Table 1 Boat trailer specification

Parameter	Description
Total length	6 m
Total width	1.87 m
Material	Aluminum 5063 I-channel
Number of axle	Single axle
Suspension	leaf spring
Trailer weight	1200 kg
GVWR	1400 kg
Roller type	Wobble roller and carpet bunk
Features	Electric winch
Winches capacity	1361 kg
Length boat Angle tilt	Approximately 5.6 m 15°

ensuring smooth boat movement and protecting against marine environment corrosion. These specifications aim to provide boat owners with a reliable and hassle-free transportation solution. Table 1 shows boat trailer specifications.

3 Results and Discussion

A load analysis formula will guarantee the boat trailer's durability and security. The dynamic forces and distribution of mass that occur during towing are accounted for in these equations. After considering all the information obtained, the final boat trailer design will be unveiled and reviewed. Finite element analysis (FEA) is used to verify the design and guarantee the system's robustness. This computational method may evaluate stress distribution, deformation, and safety consideration within the boat trailer.

3.1 Product Benchmarking

First, unique identifiers, such as name or model numbers, were assigned to each trailer to ensure convenient referencing and comparison throughout the analysis. Subsequently, the focus shifted to the evaluation of trailer axles. This involved determining whether trailers were equipped with single or tandem axles and comprehending their influence on weight capacity and stability. Material selection was a critical aspect of the analysis, considering durability and resistance to corrosion. The dimensions of the trailers, including length, width, and height, were carefully considered. The suspension systems of each trailer was evaluated to determine their effectiveness in absorbing shocks and vibrations during transportation. A smooth and stable ride protects both the boat and the trailer from damage. The winch, an essential component for boat loading and unloading, was evaluated in terms of its design, strength, and ease of use. A well-designed and robust winch system contributes to efficient and safe boat handling. The type of wheel support, whether traditional rollers or bunk-style supports, was examined to determine its effectiveness in securely holding the boat during transit. The winch stand, which supports the winch system was scrutinized for its stability and adjustability. In complying to safety regulations, the lighting systems of the trailers are considered. The braking system of each trailer was evaluated for its efficiency in providing reliable and responsive braking performance. Lastly, the type of frame used in the trailers were assessed to ensure structural integrity and durability. The frame serves as the foundation for the entire trailer structure and should be capable of withstanding the rigors of transportation.

3.2 Quality Functional Deployment

Boat trailer customer requirements can be summarized in three essential aspects that drive customer satisfaction: user-friendliness, safety and durability, and performance and functionality. When it comes to user-friendliness, customers value trailers that are easy to operate and set up quickly. They seek intuitive controls, clear assembly procedures, and concise instructions. By incorporating modular components, standardized parts, and efficient locking mechanisms, manufacturers can provide a seamless assembly experience. Additionally, effortless boat loading and unloading, user-friendly winches, and quick-release mechanisms enhance convenience and elevate the overall user experience. Safety and durability are paramount for boat owners who depend on trailers to securely transport their precious boats in various weather conditions and situations.

Customer demand trailers that can withstand shipping rigors without compromising boat security. To meet this requirement, robust and corrosion-resistant materials like aluminum, reinforced joints, and cross-members ensure a stable and durable trailer frame. Properly positioned bunks or rollers, adjustable tie-down systems, and secure boat placement guarantee safe transportation, and are minimizing excessive movement during transit. Performance and functionality are also crucial considerations for customers when selecting a boat trailer. Aluminum trailers, appreciated for their lightweight nature, improve fuel efficiency and ease of towing. A well-designed suspension system that minimizes impact and vibrations ensures a smooth ride for both the boat and towing vehicle, enhancing safety and overall enjoyment. Features like electric winches, sturdy winch stands, convenient tongue jacks, robust I-beam construction, and strategically positioned rollers and bunks contribute to optimal performance, functionality, and ease of use. Figure 3 shows the functional deployment of the visual quality.

	TECHNICAL DETAIL														
		FUNCTIONAL							PERFORMANCES						
Customer Requirement	customer important [1-5]	aluminium	suspension system	Electric winch	winch stand	swivel jack	I beam	roller and bunk	lightweight	smooth ride	pulling heavy item	Secure Attachment Points	lifting & lowering	high payloads	smooth operation
User Friendly	5	5	5	5	5	5	5	5	5	5	5	5	5	5	5
Durability	5	5	5	4	5	4	5	5	5	4	5	4	4	5	5
Easy to Assemble	4	5	5	5	4	5	4	5	5	5	5	4	4	5	5
Move/Slide safely	5	5	5	5	5	5	5	5	4	5	4	4	5	5	5
Life Span	5	5	5	5	5	5	5	5	5	5	5	5	5	5	5
Low Maintenance	4	5	5	4	4	5	5	5	5	5	4	4	5	4	5
Save time	5	5	5	4	4	5	5	5	4	5	4	5	5	4	5
Low cost	3	5	4	5	4	5	4	5	4	5	4	5	4	5	4
Max load [capacity]	5	5	4	5	5	4	5	5	5	4	4	5	4	5	5
corrosion resistant	5	5	4	5	5	5	4	5	5	4	5	5	4	5	4
Secure Battery Compartment	5	5	5	5	5	5	5	4	5	4	5	5	5	5	5
ABSOLUTE WEIGHT	275	255	242	266	239	245	243	275	242	260	233	212	233	246	247
%		93%	88%	97%	87%	89%	88%	100%	88%	95%	85%	77%	85%	89%	90%
Ranking Order		2	4	1	4	4	4	1	4	2	5	7	5	4	3

Fig. 3 Quality functional deployment

3.3 Concept Selection Matrix (CSM)

In phase 1, it is necessary to assign a score contingent upon the customer's requirements and their level of importance. It is not advisable to award full marks to any design at this stage, as it may not be the final choice and may require further implementation. Any score exceeding 9 marks in phase 1 will be carried over to phase 2. In order to progress to the subsequent phase, it is possible to amalgamate design elements from designs 3, 6, and 7 to enhance the overall design. In phase 2 of the project, half of the initially proposed designs are implemented, due to a thorough evaluation of the project's requirements. Designs, 2, 5, and 8 remain. The amalgamation of design 3, design 6 and design 7 have been undertaken to enhance the boat trailer's design with respect to its features. The third and conclusive phase entails the crucial determination of the design that best suits the customer's requirements and priorities. Phase 3 will result in selecting of two designs, one of which resulted in a novel design that incorporates five distinct elements within its structure. The design underwent enhancements pertaining to the boat trailer roller and carpet bunk, operational fluidity, mechanical components and novel functionalities.

3.4 Finite Element Analysis (FEA)

For the given weight loads of 300 kg, 500 kg, 700 kg, 1000 kg, 1200 kg, and 1400 kg, the shear stress values were calculated for different planes (XX, XY, XZ, YY, YZ, and ZZ). These values represent the shear stress experienced in different directions within the material. When analyzing the results, it was observed that the shear stress values vary based on the weight load. As the weight load increases, the shear stress tends to increase in most planes. This is expected, as higher loads exert greater forces on the trailer components, resulting in increased stress. By identifying areas with high shear stress, it can focus on reinforcing those specific areas or incorporating structural modifications to ensure the trailer can withstand the applied loads during tilting and maintain its structural stability. Figure 4 shows a visual finite element analysis for shear stress and the shear stress result shown in Table 2.

Table 3 shows the result of the strain analysis when performing the finite element analysis. The analysis of shear strain values provides valuable insights into the distribution of deformation and strain across different planes in the trailer. In the XX plane, shearing along the length of the trailer increases with higher weight loads. The XY plane shows significantly higher shearing at 500 kg, indicating more significant deformation along the width. The XZ plane exhibits consistent shearing regardless of the load. The YY plane experiences increased shearing along its width with higher weight loads. The YZ plane shows variations in vertical shearing with different weights. Finally, the ZZ plane consistently experiences more horizontal shearing as

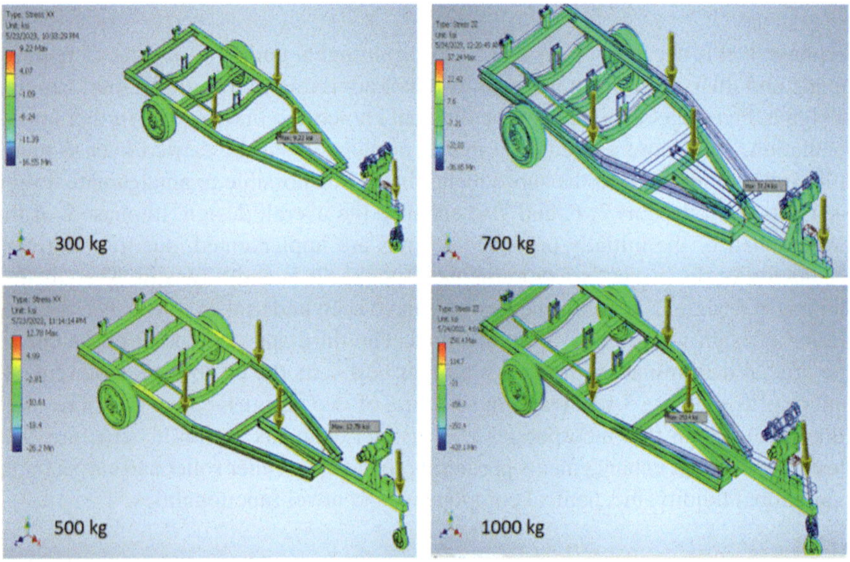

Fig. 4 Shear stress on all planes

Table 2 Shear stress result

Weight (kg)	Shear stress					
	xx	xy	xz	yy	yz	zz
300	63.5719	41.7837	27.973015	68.1226	36.21254	110.8716
500	88.1181	45.57595	87.91125	139.34795	91.4277	199.8171
700	146.174	66.39885	68.46735	159.06765	109.69945	256.494
1000	208.84955	94.80625	98.1848	275.59315	22.86	366.9519
1200	250.63325	113.83645	116.45655	272.4904	186.5787	439.901
1400	292.27905	132.6598	138.24475	318.48005	220.2263	513.60855

the load on the trailer increases. Understanding these shear strain patterns helps evaluate the trailer's structural integrity and performance under different load conditions. Figure 5 shows the visual strain for FEA.

The first principal stress represents the maximum normal stress acting on a material or structure in a specific direction. It is a critical parameter that helps engineers understand the state of stress and potential failure mechanisms within the analyzed component. Table 4 shows 1st principal stress results for loads of 300, 500, 700, 1000, 1200 and 1400 kg. Figure 6 shows the visual measurement of the first principal stress at various weights. The result provides crucial information about the maximum stress experienced by the trailer structure. At 300 kg, the stress is measured at 112.1127 units, serving as a critical reference for assessing structural performance. As the weight increases to 500 kg, the stress rises significantly to 293.3133 units, highlighting the need for solid and rigid components. At 700 kg, the stress slightly decreases to 259.6657 units but remains substantial. Further analysis shows that at 1000 kg, the stress increases to 371.01995 units, and at 1200 kg, it reaches 444.93435 units, emphasizing the importance of reinforcement and durability. Finally, at 1400 kg, the stress reaches 519.40035 units, indicating critical stress concentration areas. These findings underscore the necessity for meticulous design considerations to ensure the trailer can withstand the applied loads effectively while maintaining structural integrity. Third-principal stresses are the minimum stresses

Table 3 Strain result

Weight (kg)	Strain					
	xx	xy	xz	yy	yz	zz
300	0.00574216	0.00556427	0.0037233	0.00692948	0.00047576	0.01108027
500	0.00630893	0.0540568	0.00117077	0.01783737	0.01216968	0.01976797
700	0.01317635	0.0088325	0.00911519	0.01617567	0.01460361	0.02566319
1000	0.01882335	0.0129626	0.01306603	0.02311204	0.02097459	0.03666761
1200	0.02258802	0.01514832	0.01544480	0.02771101	0.02483579	0.04398321
1400	0.02634580	0.0176512	0.02529776	0.03236513	0.02931065	0.05134017

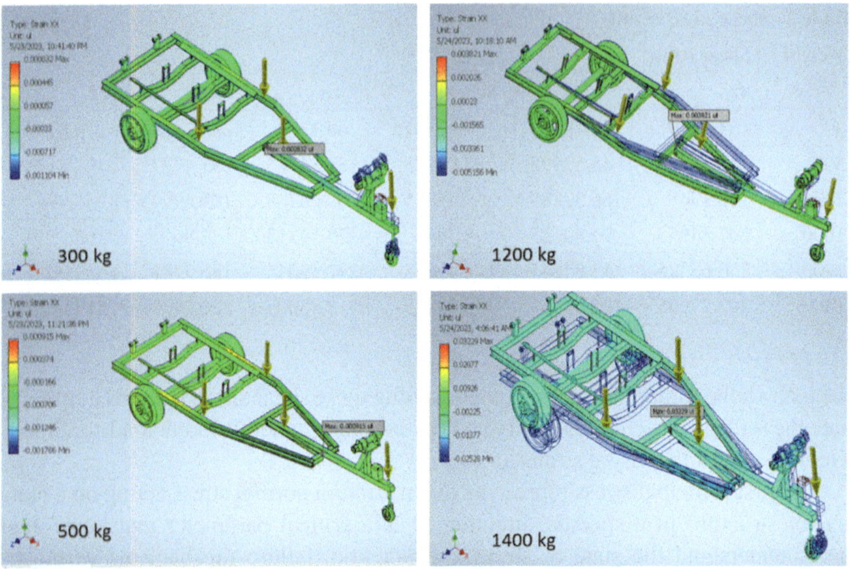

Fig. 5 Strain on all planes

acting in orthogonal directions within a material. The third principal stress represents the stress magnitude along the third principal axis, perpendicular to the first two principal axels. Table 5 shows the result of the 3rd principal stresses.

The measurement of the third principal stress at different weight loads provides valuable insights into the stress distribution perpendicular to the primary loading directions. At 300 kg, the stress value is 9.72195 units, indicating potential areas of concern that may not be immediately apparent. As the weight increases to 500 kg, the stress rises to 34.0613 units, highlighting an increase in stress magnitude. At 700 kg, the stress slightly decreases to 28.89005 units but still represents a significant stress component. The stress continues to increase at 1000 kg (41.2321 units), 1200 kg (49.02345 units), and 1400 kg (57.918 units), emphasizing the need to consider all stress components for ensuring structural integrity. Understanding these stress distributions helps to identify potential weaknesses and informs the design and analysis

Table 4 Result of 1st principal stress

Weight (kg)	1st principal stress
300	112.1127
500	293.3133
700	259.6657
1000	371.01995
1200	444.93435
1400	519.40035

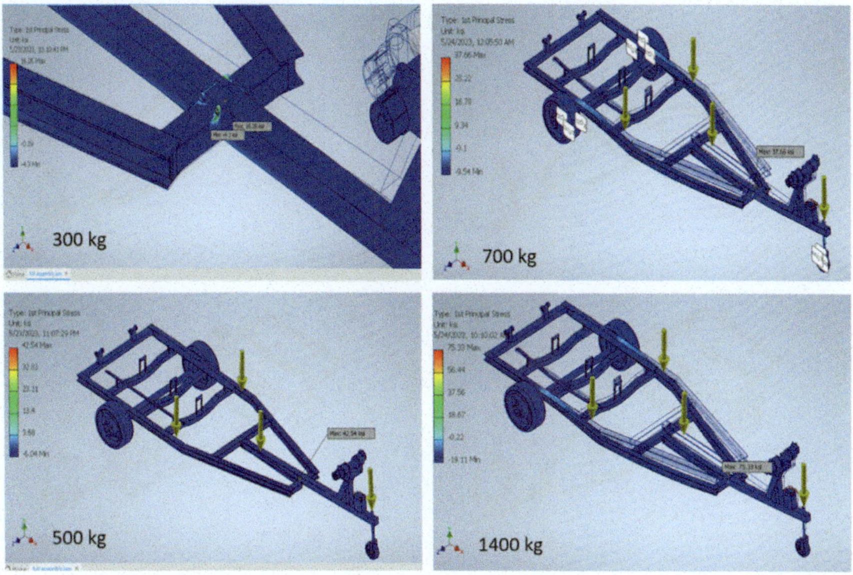

Fig. 6 1st principal stress

Table 5 Result of 3rd principal stress

Weight (kg)	3rd principal stress
300	9.72195
500	34.0613
700	28.89005
1000	41.2321
1200	49.02345
1400	57.918

process to ensure the trailer can withstand the applied loads effectively. Figure 7 shows the visual of the third-principal stress. The safety factor is a measure used to assess component or structure's structural integrity and safety. It indicates the factor by which the actual stress in the material is below its yield or failure point. Table 6 shows the results of the safety factor.

The safety factor for the trailer has been carefully determined and found to have a consistent maximum value of 15 across various weight configurations. This means that the trailer has been designed with a sufficient safety margin to handle loads up to its maximum rated weight. Critical components like the frame and fulcrum are engineered with ample strength and load-bearing capacity to withstand the anticipated forces during towing. Complex simulations and calculations are used to evaluate stress and strain distribution within the trailer, comparing them to material strength

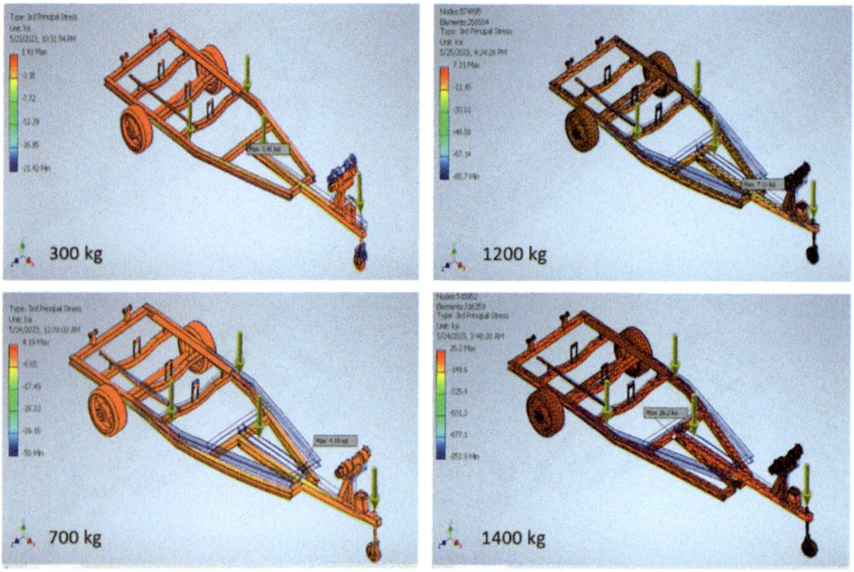

Fig. 7 3rd principal stress

Table 6 Result of safety factor

Weight (kg)	Safety factor
300	15
500	15
700	15
1000	15
1200	15
1400	15

properties to determine the safety factor. With a safety factor of 15, there is a significant margin of safety, providing confidence in the trailer's ability to handle loads securely and effectively. Figure 8 shows the visual safety factor by finite element analysis.

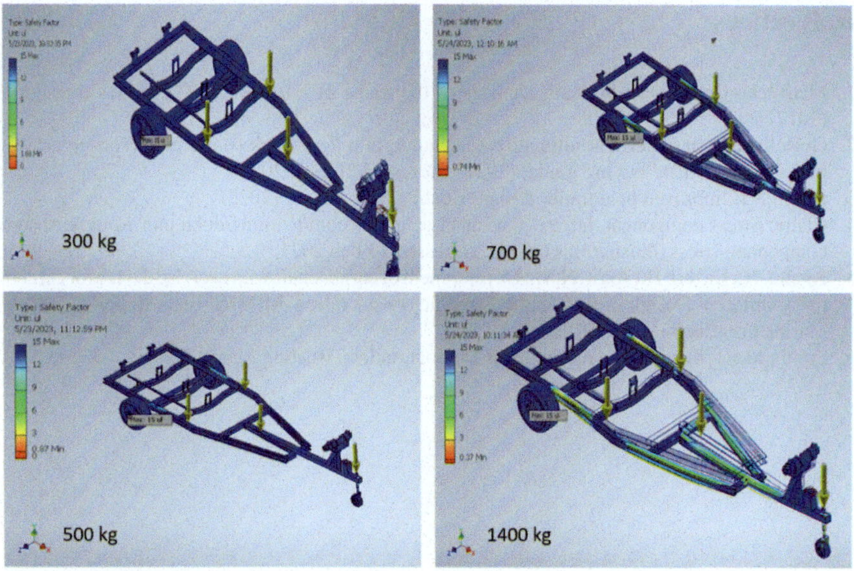

Fig. 8 Safety factor

4 Conclusion

The fulcrum area of the trailer plays a critical role as it bears the responsibility of holding the boat securely during transportation. This specific area is of utmost importance due to the weight-bearing nature of the boat and the need for a robust and reliable support system. Through finite element analysis, it has been determined that the fulcrum experiences the highest stress levels compared to other areas. Considering the potential risks associated with exceeding the maximum weight capacity at the fulcrum is essential. Placing a boat with a weight beyond the trailer's designated capacity puts excessive strain on the supporting components, such as the nut. This heightened stress increases the risk of nut breaking or experiencing failure, affecting the overall stability and safety of the trailer and boat during transport. Tilt mechanisms provide a convenient solution for simplifying the process of loading and unloading boats. By tilting the boat upwards, accessing the trailer ramp and lifting the boat onto the trailer becomes significantly easier. This saves valuable time and effort and helps minimize the potential for injuries. Tilt mechanisms greatly enhance the overall user experience and contribute to a safer and more efficient boat loading and unloading process [6].

Acknowledgements We are immensely grateful to University Kuala Lumpur, Malaysian Institute of Marine Engineering Technology (UniKL MIMET) for providing us with the opportunity to undertake and successfully complete this research.

References

1. Medvecka-Benova, S.: Strength analysis of the frame of a trailer. SJSUT. Series transport 9 (2017)
2. GoRollick: Boat trailer: The ultimate buying guide. http://gorollick.com/articles/consumer/boat-trailers-the-ultimate-buying-guide/ (2021). Accessed 21 June 2022
3. Doe, J.S.J: Improved boat trailer design. Boating Mag. 42–47 (2022)
4. Claire: Stress component. http://www.intact-solutions.com/learning/doku.php?id=wiki:snspro:component_stress_illustrations (2014). Accessed 14 Feb 2023
5. Khalid, N., Hazeri, H., Azrir, N.M.: Development of corner chair with exercise device for cerebral palsy children: Design process and ergonomic consideration. MIMET Tech. Bull. **4**(1), 35–57. Marine Frontier@UniKL (2013)
6. Smith, M.J.J: The best tilt mechanisms for boat trailers. Boating Mag. (2022)

Ship Pipeline Corrosion Surface Evaluation Using Image Processing in MATLAB

Puteri Zirwatul Nadila Megat Zamanhuri, Muhd Syafiq Ridhwan Suhaimi, Asmalina Mohamed Saat, Norhanifah Abdul Rahman, and Mat Zaini Katon

Abstract Image Processing Toolbox is one of the apps in MATLAB which allow a user to automate standard image processing processes. The user can segment image data in real-time, benchmark image registration methods, and batch process large datasets. In shipyard industry, the image processing technique is a potential method to do corrosion inspection instead of visual inspection to achieve reliable method and give quick results. This paper investigates a corrosion method specifically surface area. The corrosion image with different areas is improved by using image enhancement and noise removal technique. All images are then evaluated using image analysis whereby extracting useful information from images, such as shape recognition, object tracking, coloring, or object characteristics measurement. For quality results, MSE (Mean Square Error) and PSNR (Peak Signal Noise Ratio) are applied. The results demonstrate that image processing effectively calculates the corroded area.

Keywords Image enhancement · Image analysis · Image quality · Image processing toolbox

P. Z. N. M. Zamanhuri (✉) · M. S. R. Suhaimi
Marine Engineering Technology Section, Universiti Kuala Lumpur Malaysian Institute of Marine Engineering Technology, Lumut, Perak, Malaysia
e-mail: puterizirwatul@unikl.edu.my

M. S. R. Suhaimi
e-mail: syafiq.suhaimi03@s.unikl.edu.my

A. M. Saat
Marine and Electrical Engineering Technology Section, Universiti Kuala Lumpur Malaysian Institute of Marine Engineering Technology, Lumut, Perak, Malaysia
e-mail: asmalina@unikl.edu.my

N. A. Rahman · M. Z. Katon
School of Mechanical Engineering, College of Engineering, Universiti Teknologi MARA, Cawangan Terengganu, Kampus Bukit Besi, 23200 Bukit Besi, Dungun, Terengganu, Malaysia
e-mail: norhanifah@uitm.edu.my

M. Z. Katon
e-mail: matzaini@uitm.edu.my

A. Ismail et al. (eds.), *Engineering Frontiers*, Advanced Structured Materials 208,
https://doi.org/10.1007/978-3-031-56844-2_29

1 Introduction

Every pipe and associated structure that is exposed to corrosive elements will eventually experience corrosion on both the inside and outside [1]. Various forms of corrosion can occur in pipelines. A number of corrosion types, some more common than others, affect pipelines. These include, but are not limited to, corrosion that has been impacted by microbes, pitting corrosion, uniform corrosion, galvanic corrosion, and fissure corrosion [2].

Previous study has shown Nondestructive Testing (NDT) method [3] to investigate the existence of corrosion for metal. This could involve visual examinations carried out locally or remotely using digital images. Other than visual inspection, the inspector can approach another procedure to detect the defects on the pipeline surfaces. Using a digital camera, the inspection can be carried out on the entire image acquisition process. Defects can then be analyzed, and the image quality improved using image processing techniques [4].

This study discusses the method to enhance the image acquired from the digital camera. The focus was on finding the best method for digital image enhancement using MATLAB as it makes the defect clearer and more visible. Another part of the study is to create the graphical user interface (GUI) that follows the method of image processing used to enhance the digital image.

In the visual inspection, not all deficiencies are visible, hence the inspector may miss some important points to look at. From the results, it is shown that the MATLAB image processing toolbox was able to study the enhancement of the image. In addition, some of the defects can be traceable and much more visible [5]. Furthermore, with the additional GUI features, the enhancement process was easier.

2 Problem Statement

Visual examination is one of the approaches that may be used to inspect for corrosion detection. Visual inspection is the most popular method used to find visual flaws by experienced and qualified inspectors. The flaw area needs to be treated and marked by the inspector to find the results.

Therefore, the defects can easily be inspected by an expert inspector. However, some of the corrosion defects cannot be clearly seen by the naked eye. Hence, the inspection needs to use another method such as non-destructive testing (NDT) which requires expensive equipment for the testing. For example, ultrasonic and radiographic tests are the methods used for the inspection that require expensive equipment. Furthermore, the maintenance cost for small and medium-sized enterprises across the country is too high to afford the inspection.

3 Literature Review

The breakdown of a metal because of chemical reactions between it and its surroundings is known as corrosion. The kind of metal and the surrounding environment, especially the gases that come into contact with the metal, affect both the shape and rate of deterioration.

One study stated that corrosion is the degradation of a metal because of chemical reactions between it and its surrounding atmosphere. The shape and rate of deterioration are determined by both the metal type and the environmental conditions, especially the metal-contact gasses [6]. Therefore, corrosion attacks need to be treated immediately so that the pipeline can have greater efficiency.

The field of image processing is expanding and has a wide range of uses. Rust detection is one area where image processing is used [7]. Rust detection based on processing images has been the topic of numerous investigations. In addition, image processing is an important technique to improve the images to have a better appearance and pictorial information for human interpretation [8].

4 Research Methodology

There are multiple approaches to gathering the information and producing the results. Trial and error (experimentation), as depicted in Fig. 1, is a practical approach since the observation results demonstrated a clear image.

The processing of images is an essential part of this research. There are many image processing methods and techniques used to improve images, but this study focused only on three methods which are image enhancement, image analysis and image quality [9, 10].

For the image enhancement method, three methods can be used: contrast adjustment, noise removal and sharpening. As for the image analysis, it is the segmentation method. Finally, image quality consists of two methods which are MSE and PSNR [11].

The interactive tool created in MATLAB is the final component of this study. The creation of this interface makes it easier to improve images by simplifying all image processing techniques.

5 Results and Discussion

The data for the MATLAB and physical measurements of specimens A, B, C, D, and E are shown in Tables 1 and 2, respectively. The specimen was physically measured using the mapping method, which employed a 1 cm by 1 cm grid. Although there are differences between physical measurement and MATLAB measurement, both results

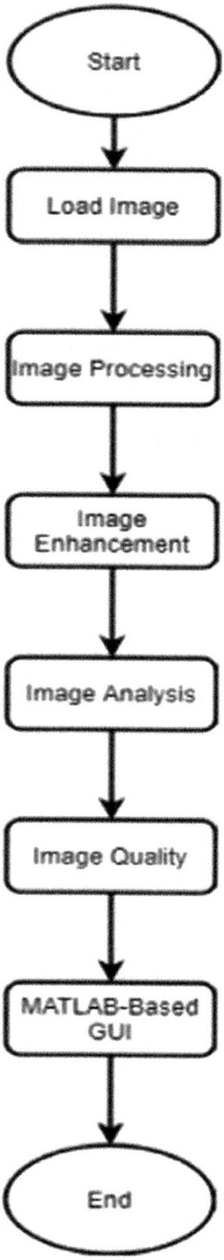

Fig. 1 Research methodology

Table 1 Physical measurement

Specimen	Total area in cm^2	Total corroded area in cm^2	Percentage corroded area (%)
A	792	228	28.79
B	408	216	52.94
C	130	41	31.54
D	600	181	30.17
E	300	197	65.67

Table 2 MATLAB measurement

Specimen	Total corroded area in pixels	Percentage area corroded (%)
A	1.1447×10^6	45.87
B	1.0448×10^6	54.31
C	878,915	26.67
D	1.5072×10^6	68.32
E	4.3604×10^6	65.62

Table 3 Result for MSE and PSNR

Specimen	Noise removal					
	Median filtering		Gaussian filtering		Wiener filtering	
	MSE	PSNR	MSE	PSNR	MSE	PSNR
A	16.92	35.84	19.65	35.19	18.31	35.5
B	13.11	36.95	13.65	36.77	11.7	37.45
C	61.44	30.24	68.75	29.75	52.97	30.9
D	17.68	35.68	20.2	35.07	16.41	35.97
E	24.98	34.15	29	33.5	23.52	34.41

show an increase of percentage area. The higher the percentage area at the results, the larger the area. Table 3 shows the MSE and PSNR. The interpretation of image quality can be concluded as the results are better when the MSE value is smaller. The higher the PSNR value, the better the image quality. The comparison of the corrosion defects is shown in Fig. 2. It is observed that the MATLAB measurement is better than the physical measurement in terms of their value of percentage and area.

6 Conclusion and Recommendation

In this study, the types of image processing methods and the percentage of corroded area for piping were able to identify by using MATLAB.

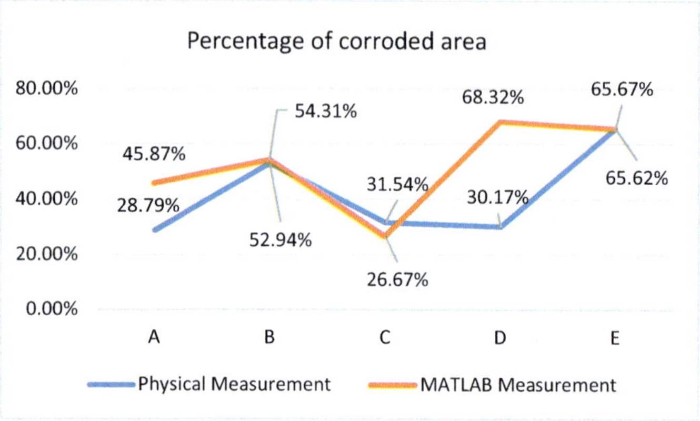

Fig. 2 The percentage of corroded area

Fig. 3 Image of specimen A; left (original image) and right (processed image)

Fig. 4 Image of specimen B; left (original image) and right (processed image)

The image was processed using the image enhancement and image segmentation tools in the Image Processing Toolbox. Image quality can be used to analyze any signal error of percentage and corroded area using the noise removal method.

In conclusion, this study successfully developed a toolbox for image processing by using MATLAB graphical user interface. It is recommended that the Image Processing Toolbox be implemented in the Marine Industry (Figs. 3, 4, 5, 6 and 7).

Fig. 5 Image of specimen C; left (original image) and right (processed image)

Fig. 6 Image of specimen D; left (original image) and right (processed image)

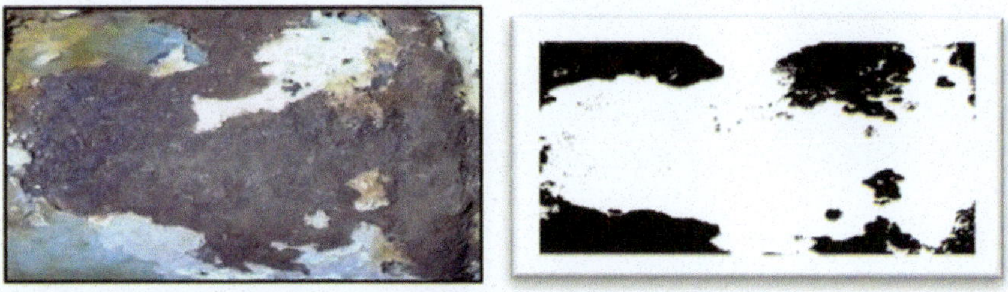

Fig. 7 Image of specimen E; left (original image) and right (processed image)

Acknowledgements The authors would like to express their gratitude and special thanks to UniKL—Malaysian Institute of Marine Engineering Technology (UniKL-MIMET) member and others member at Universiti Teknologi MARA for contribute idea in this research paper.

References

1. Bondada, V. Pratihar, D., Kumar, C.: Detection and quantitative assessment of corrosion on pipelines through image analysis. In: International Conference on Robotics and Smart Manufacturing, 2018. vol. 133, pp. 804–811. Elsevier (2018)

2. Hoang, N., Tran, V.: Image processing- based detection of pipe corrosion using texture analysis and metaheuristic—optimized machine learning approach. Comput. Intell. Neurosci. July 2019. pp. 1–13. (2019)
3. Wu, R., Zhang, H., Yang, R., Chen, W., Chen, G.: Nondestructive testing for corrosion evaluation of metal under coating. J. Sens. 2021 (2021)
4. Manurung, H.P.M., Puteri Zirwatul Nadila, M.Z., Suhaila, A.H., Sunhaji, K.A., Ghalib, T., Esa, H., Zulfahmy, A.: Development of stand-alone application tool for analyzing and reporting weld imperfection captured by m-focused digital radiography using MATLAB-based GUI. J. Appl. Sci. **12**(7), 612–626 (2012)
5. Halim, S.A., M.Z.P.Z. Nadila, Ibrahim, A., Manurung, Y. H. P.: Performance of noise removal methods with image quality parameter on μ-focused digital radiographic. Faculty Comput. Math. Sci. 313–317 (2012)
6. Sharma, V., Thind, T.: Techniques for detection of rusting of metals using image processing: A survey. Int. J. Emerg. Sci. Eng. (IJESE), Feb 2013, pp. 60–62. (2013)
7. Diaz, J.A.I., Ligeralde, M.I., Jose, J.A.C., Bandala, A.A.: Rust detection using image processing via Matlab. TENCON 2017–2017 IEEE region 10 conference, pp. 1327–1331. IEEE (2017)
8. S. Sukhjinder, R.K. Bansal, B. Savina. Comparative study and implementation of image processing techniques using MATLAB. Int. J. Adv. Res. Comput. Sci. Soft. Eng. 244–250 (2012)
9. Stella, A., Bhushan T.: Implementation of order statistic filters on digital image and OCT image: a comparative study. Int. J. Mod. Eng. Res. (IJMER) 3143–3145 (2012)
10. Dhariwal, S.: Comparative analysis of various image enhancement techniques. Int. J. Electron. Commun. Technol. Sept 2011 **2** (2011)
11. Mohamad, S., Abd Halim, S.: Image enhancement process on digital radiographic image with weld discontinuities. J. Mech. Eng. (JMechE) **SI 5**(4), 275–292 (2018)

Engine Room Indoor Air Quality Assessment Onboard a Medium-Speed Vessel

Zakiman Zali, Nor Ashimy Mohd Noor, and Ahmad Azrizal Mohd Ariffin

Abstract Human comfort is important not only in a building but also in the engine room of a ship. This issue has been discussed for a long time to ensure that seafarers onboard ships can work with good performance and comfort. In this assessment, the indoor air quality (IAQ) in the engine room area of 55.2 square meters and 2.54 m high on a medium-sized vessel was investigated. The present IAQ status of this study attempts to provide information in the engine room during the engine running ongoing task. This assessment included visual observations, chemical contaminant parameters (carbon dioxide), and physical condition parameters (air velocity) were measured using a thermal comfort multi-station, TCM. This device was able to record eight indoor parameters, i.e. the air velocity, CO_2 level, air pressure, relative humidity, sound level, air temperature, luminance, and globe temperature. The monitoring of CO_2 levels in indoor air as well as the air movements in the engine room are used as one part of determining the effectiveness of a ventilation strategy and providing information on indoor air quality in compliance with IAQ standards and guidelines. The recorded data were compared with corresponding parameters in the Code of Practice on Indoor Air Quality set by the Department of Occupational Safety and Health, Ministry of Human Resources, Malaysia, Malaysia DOSH 2010, Guide for Crew Habitability on Ships from American Bureau of Shipping, ABS 2016, and American Society of Heating, Refrigerating, and Standard Ventilation and Acceptable Indoor Air Quality from Air-Conditioning Engineers, ASHRAE 2016. The assessment is conducted to evaluate the exposure of the engine crews to the carbon dioxide contaminant and to recommend further appropriate control measures to prevent or reduce risks. Based on the assessment conducted, it shows that the average 3-h CO_2 concentration did not exceed the ceiling point recommended by the Department of Occupational Safety and Health, Ministry of Human Resources, Malaysia, 2010. However,

Z. Zali (✉) · N. A. Mohd Noor · A. A. Mohd Ariffin
Centre of Technology in Marine Engineering, Politeknik Ungku Omar, Perak, Malaysia
e-mail: zakizali@puo.edu.my

N. A. Mohd Noor
e-mail: ashimy@puo.edu.my

A. A. Mohd Ariffin
e-mail: azrizal@puo.edu.my

A. Ismail et al. (eds.), *Engineering Frontiers*, Advanced Structured Materials 208,
https://doi.org/10.1007/978-3-031-56844-2_30

due to the exhaust system leakage factor and the inefficiency of the ventilation in the engine room caused the CO_2 gas to accumulate.

Keywords Indoor air quality · IAQ · CO_2 · Engine room ventilation

1 Introduction

An important factor in determining the performance of seafarers working on a ship is also influenced by the indoor air quality and thermal comfort. This matter has been discussed for a long time and is a topic that is constantly discussed in the shipping industry. The working atmosphere in the engine room of a ship by the engine room crew is almost the same as the working atmosphere of a staff ashore. It is often stated that the human element accounts for at least 80% of all catastrophic marine casualties. The importance of addressing the human element in maritime safety has been recognized by the International Maritime Organization, IMO [1]. The engine department onboard is responsible for the generation of electrical power, the propulsion of the ship, and the repair and maintenance work on various mechanical equipment. The main engine for the ship's propulsion is located in the engine room, ER. The monitoring and operation of all types of machinery in the engine room are performed from the engine control room, ECR in the engine room. The engine room is defined as an enclosed space under the revised IMO, Resolution A.1050 (27) that the engine room is not designed for continuous worker occupancy. The importance of indoor air quality is because it directly affects the health of workers, comfort, and work performance. The engine crew may be exposed to dangerous pollutants while on board due to the engine room's small and cramped quarters, which can lead to pollution buildup and occasionally higher concentrations than outdoors. The IAQ monitoring process with the aim of regulating a health workplace environment is very important as recommended by DOSH Malaysia, ABS, and ASHRAE. This study attempts to provide information about the present IAQ status in the engine room while the engine runs during the vessel's ongoing task and is conducted from a selected sampling site using the guidelines prescribed by the DOSH Malaysia 2010, ABS 2016, and ASHRAE 2016.

2 Literature Review

IAQ is significant on ships since crew members spend comparatively more time there than in other types of transportation. Carbon monoxide and carbon dioxide are the most acutely hazardous indoor air pollutants. Higher concentrations of carbon dioxide is toxic, 1% (10,000 ppm) will make some people feel drowsy. Concentrations of 7–10% cause dizziness, headache, visual and hearing dysfunction, and unconsciousness within a few minutes to an hour [2]. Carbon dioxide is very important to the human respiratory system. The oxygen is transported to the body tissues while carbon dioxide is taken out of the body when the breathing process occurs. Carbon dioxide is also produced by the fermentation of liquids, the combustion of coal or hydrocarbons, and the breathing of animals. The internal combustion engine is an engine in which the combustion of a fuel, normally a fossil fuel occurs with an oxidizer, usually air in a combustion chamber that is an integral part of the working fluid flow circuit. The main derivatives of the process are CO_2, water, and some soot also called particulate matter, PM. During combustion it is fundamental to ensure safe and complete combustion by maintaining appropriate airflow. The demand and economic development in the shipping industry also has a high impact covering the engineering and ship design industry. Because of this, the shipowner has focused on the ability of a ship to operate with comfort of the ship's crew in performing their duties on board [3, 4]. The crew who work in the engine room and carry out engine and machine operation are individuals who are exposed to CO_2 pollutants. In addition to that, if the air flow is limited in a room, the rate of concentration of pollution easily increases compared to the outside. Crew members can be exposed to hazardous pollutants onboard with limited and enclosed spaces [5–8].

One of the first checks, often overlooked, in a comfort study is making sure that sufficient air is moving in space. Air velocity is a good indicator that air is being appropriately distributed or balanced throughout the building and reaching all the spaces intended. The rate of air exchange between the indoor and outdoor spaces will affect the volume of air flow in a space [9]. It has a significant role in thermal comfort because stagnant air frequently makes one feel stuffy and encourages the accumulation of chemical pollutants. Monitoring the concentration of CO_2 levels in indoor spaces is one of the effective methods of determining ventilation strategies through compliance with IAQ standards and guidelines. If the ventilation rate in a space decreases, then the CO_2 concentration will increase. The measurement and interpretation of indoor CO_2 concentration can provide information on building indoor quality and ventilation [10]. Excessive carbon dioxide levels can indicate insufficient fresh air supply or occupancy levels beyond the designed capacity of the room. In previous reports regarding IAQ in the engine room [4, 6, 7], discussed how engine rooms had significantly higher levels of carbon dioxide, CO_2, carbon monoxide, CO, and particulate matter, PM than what the corresponding standards called for. The combined impact of inadequate ventilation and exhaust emission system leaks may have caused this.

In Malaysia, indoor air quality has been recognized by the Department of Occupational Safety and Health as a part of the requirements. The department has established a code of practice on indoor quality to guarantee that all employees are shielded from indoor air pollutants [10].

3 Methodology

The methodology of this study is through data collection in the field and measurements taken onboard the ship. The vessel is powered by two units of 4-stroke MTU 20 V 538 TB91 marine diesel engines capable of giving an output power of 4501 HP/3357 kW. Each main engine was supplied with two independent turbo blowers supplying intake air from the engine room indoor air for the combustion. The engine room is equipped with two ventilation fans, each capable of supplying fresh air running alternately as a supply and exhaust fan respectively. The room has two entrances and one skylight opening or natural ventilator, the entrances are closed during the engine running except the skylight as shown in Fig. 1.

In order to determine the current indoor air quality status in the engine room, an investigation was conducted into the engine room's 55.2 m^2 area and 2.54-m height on a medium-sized vessel. Besides visual observations and inspections during an assessment, chemical contaminant parameter (carbon dioxide) and physical condition parameter (air velocity) were measured simultaneously using the thermal

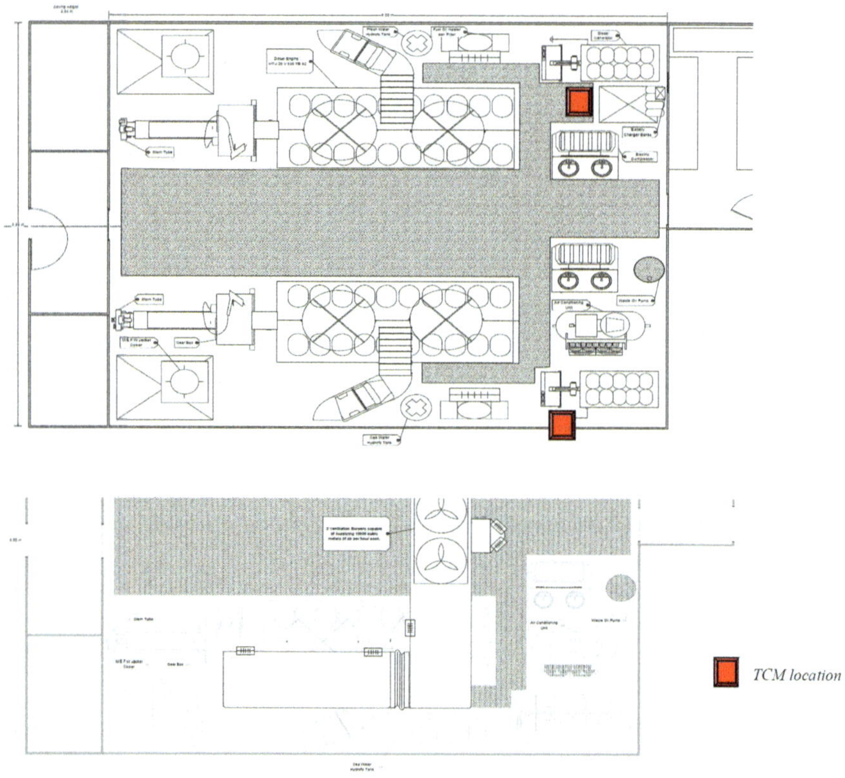

Fig. 1 Engine room (ER) and ventilation layout plan

Table 1 Guidelines for the air quality parameters

No	Parameter	DOSH Malaysia (2010)	ABS (2016)	ASHRAE (2016)
1	Carbon dioxide, CO_2 (ppm)	1000	–	1000
2	Air velocity (m/s)	0.15–0.5	0.5	0.8

comfort multi-station, TCM, ADAM-4017 16-bit, 8-channel analog input module that provides programmable input ranges capable of sampling eight environment parameters such as sound level, air temperature, relative humidity, air velocity, air pressure, CO_2 level, luminance, and globe temperature. Extensive measurements were performed before starting the diesel engines in order to identify the possible indoor sources that might affect the initial mass and particle number concentrations in the room. When the engine room's doors and skylight were kept closed, no appreciable changes in indoor concentrations were observed.

The sampling equipment, TCM was positioned in front of the port side main engine as shown in Fig. 1, and programmed to take sample readings at 30-s intervals. Mounted on a 1.5-m tripod the experiment was conducted for 3 h while the engine ran during the vessel's ongoing task. Only three hours of time allocation for this assessment was given by the ship's management due to the end voyage process.

This study also uses indoor air quality assessment guidelines from DOSH, Malaysia, 2010 with the Code of Practice on Indoor Air Quality. In addition, the guidelines set by the ABS 2016, Guide for Crew Habitability on Ships referred to indoor climate requirements and ASHRAE 2016 with Ventilation and Acceptable Indoor Air Quality. Table 1 displays a comparison of the data recorded by TCM according to their respective standards.

4 Results and Discussion

In addition to the inspection report and visual observations, the IAQ of the chosen sampling site is evaluated in this section by comparing the measurement findings to the permitted limits of the relevant parameters established by Refs. [10–12].

From the visual observation and inspection, as shown in Fig. 2, the leakages of exhaust gases through the exhaust bellow the trunk to the ship side pipe fitting from the port side main engine and the exhaust pipe outlets from diesel generators were identified. These leakages were unattended and became more serious during continuous engine running. While there was visual evidence of pipe fitting leaking on the exhaust gas system, there was a strong exhaust gas odor surrounding the engine room during the engine room walk through.

During this study and assessment, both the engine speeds, port and starboard engines were recorded relative to time. Speed of both the port and starboard engine were recorded throughout the experiment and data was collected as shown in Table 2. The relationship between CO_2 in Fig. 3, air velocity in Fig. 4, and ship speed in

Fig. 2 Exhaust fittings leakages

Table 2 during maneuvering out from the port at engine speed increases. At idling conditions in port, most of the CO_2 concentration from the leakages accumulated inside the engine room due to long periods of operation of the ship when performing routine missions before the experiment takes place. Since the air velocity is lower than what is required by the guideline and CO_2 is 1.5 times heavier than air, the air movement could not be drawn out of the CO_2 concentration from the room. From Fig. 3, during idling, the accumulated CO_2 generated is higher than the limit set by DOSH and ASHRAE. When the vessel started increasing the speed while maneuvering out from the port, most of the diesel engines onboard used the large volume of engine room indoor air as the intake air for the combustion processes causing a decrease in air velocity or air movement inside the engine room which was lower than what is required by the guideline. Due to the limitation of air velocity, some of the CO_2 concentrations in the room were extracted with the indoor air into the engine's combustion chambers thus, reducing the amount of CO_2 concentration in the engine room.

When the vessel speed is at the full away condition with constant maximum speed, the required indoor air for combustion processes is almost steadily and uniformly

Table 2 Ship speed

Time (hr: min)	Engine speed (rpm)			Description
	Port	Starboard	Mean	
8.15 am	150	150	150	Idling speed
8.30	150	150	150	Idling speed
8.45	500	500	500	Start manoeuvring from the port
9.45	800	800	800	Final manoeuvring from port
10.00	900	900	900	Full away at open sea
11.00	900	900	900	Full away at open sea
11.15	600	600	600	Decrease speed berth maneuvering
11.45 am	0	0	0	Engines stop

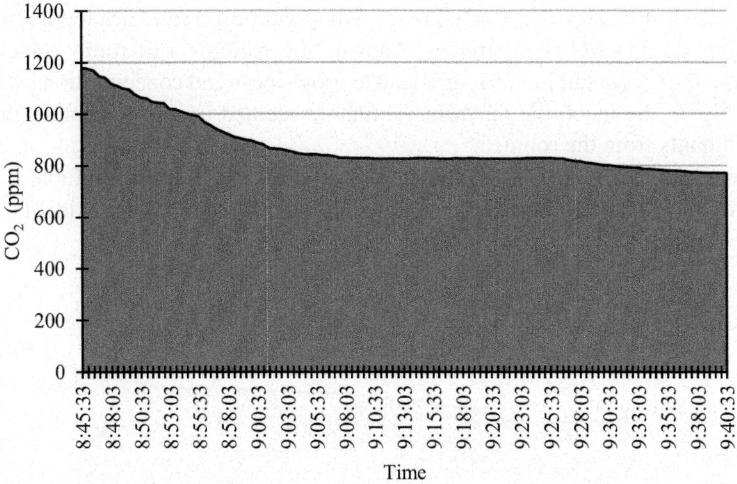

Fig. 3 CO_2 versus time compared to standard value 1000 ppm during start maneuvering out from port

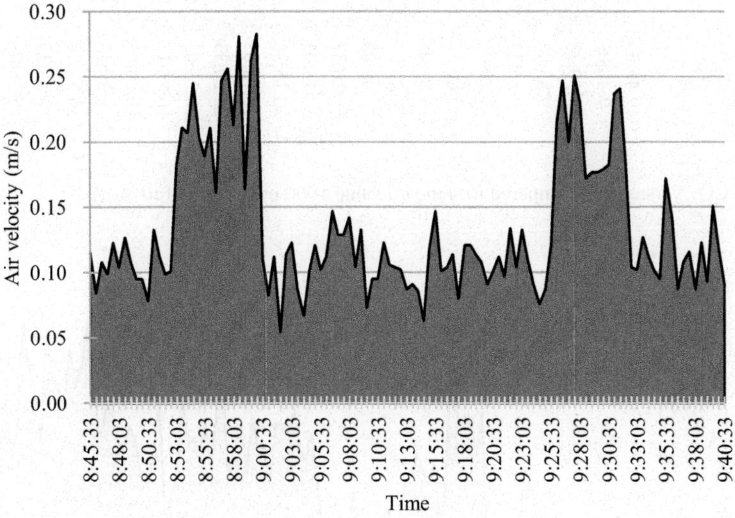

Fig. 4 Air velocity versus time compared to standard value 0.5–0.8 m/s during start maneuvering out from port

distributed. Due to the leakages in the exhaust piping systems, the generated CO_2 inside the room started to accumulate, refer to Fig. 5. The high-density characteristic of CO_2 accumulated at the bottom of the engine room space causing stagnancy due to the inability of the engine room ventilation to withdraw the CO_2 contaminants. The increment of CO_2 concentrations will surplus the total velocity inside the engine room causing the indoor air velocity to increase. This condition created a velocities mixture of indoor air and CO_2 contaminants causing the increment of room velocity. An increase in CO_2 contaminants might lead to drowsiness and concentration problems. From Fig. 6, the incoming air from ventilators seemed unable to exhaust the CO_2 contaminants from the room.

The relationship between CO_2, air velocity, and engine speed in Table 2 during maneuvering for berthing and at engine speed decreases as shown in Fig. 7 and

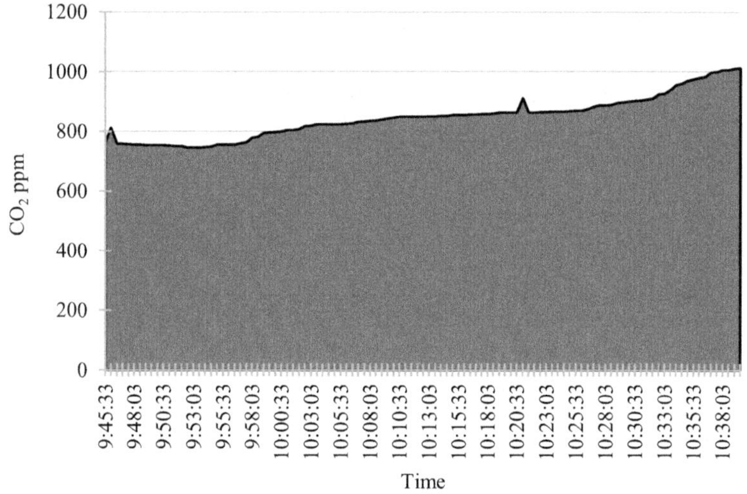

Fig. 5 CO_2 versus time compared to standard value 1000 ppm during full away

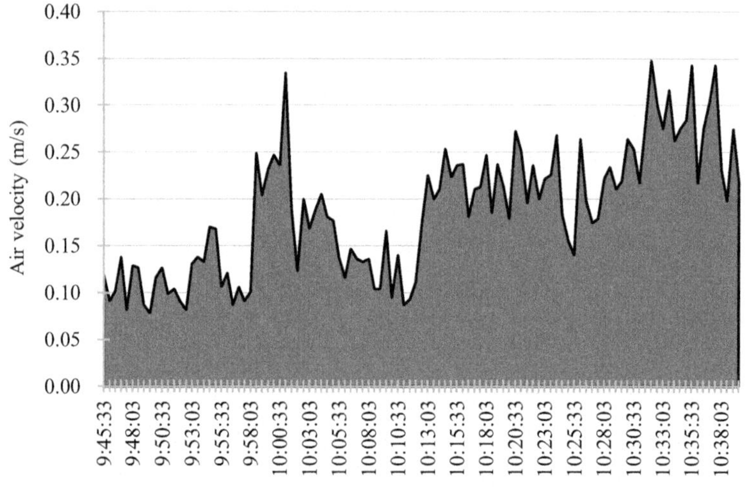

Fig. 6 Air velocity versus time compared to standard value 0.5–0.8 m/s during full away

Fig. 8. The engine started to lower the speed gradually. The output power required will reduce as well as the reduction of the intake air required for the combustion's. The concentrations of CO_2 in the room started to decrease but the CO_2 contaminants level was still higher than the recommended value with an average of 1077.58 ppm. Even though the room's air velocity was at the required value but the incoming air from ventilators seemed unable to exhaust the high-density CO_2 contaminants from the room.

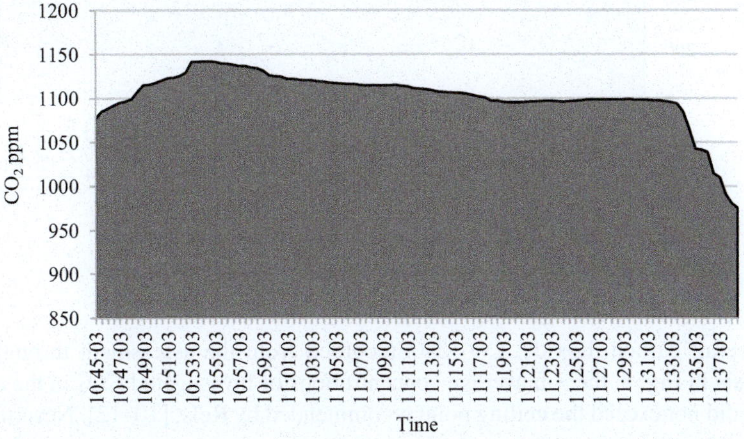

Fig. 7 CO_2 versus time compared to standard value 1000 ppm during maneuvering for berthing and engine speed decreasing

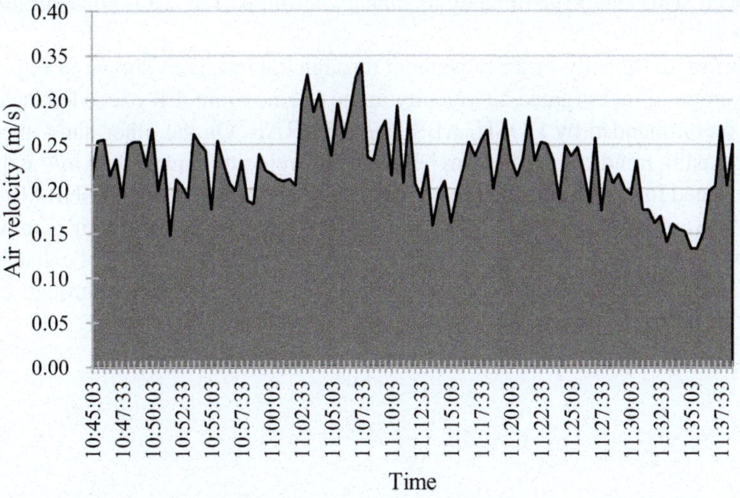

Fig. 8 Air velocity versus time compared to standard value 0.5 m/s–0.8 m/s during maneuvering for berthing and engine speed decreasing

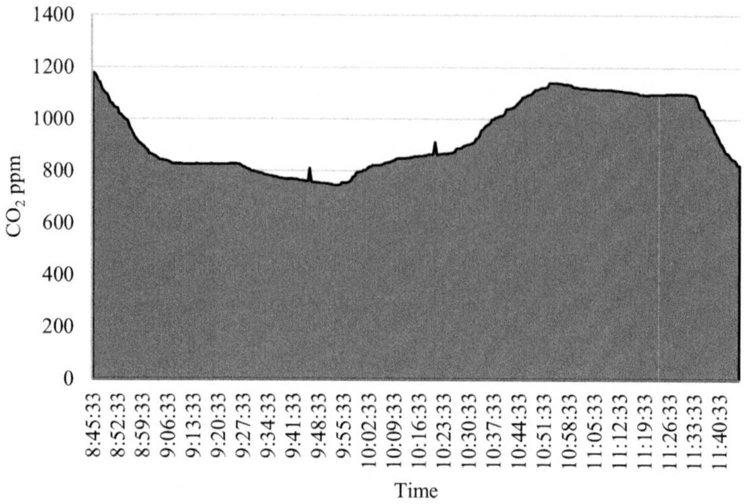

Fig. 9 Overall three-hour CO_2 concentrations level

Overall carbon dioxide, CO_2 data recorded from the assessment through data analysis in Fig. 8, the 3-h average concentrations of investigated CO_2 in the engine room did not exceed the ceiling point recommended by Refs. [10–12]. Nevertheless, during the study, there was a growing trend pointing to greater amounts of carbon dioxide pollutants above the 1000 ppm maximum limit. The increments are due to the leakages of exhaust gases through exhaust pipe fittings in the engine room. The longer the running hours of the engines, the more CO_2 contaminants will be produced. This can be justified by the linear condition of the CO_2 concentration line in Fig. 9.

Overall air velocity or air movement through data analysis shown in Fig. 10, the 3-h average of investigated air velocity in the engine room, ER exceeded the ceiling point recommended by DOSH, ABS, and ASHRAE. On the other hand, there was an increasing trend condition from below the air velocity limit of 0.15 m/s at the start of the speed increment during the voyage to a velocity advised by ASHRAE, 0.5 m/s by ABS, and 0.8 m/s by DOSH. Even though it seems sufficient air is moving in the space, the increasing velocity is due to the additional turbulence of exhaust gases including CO_2 contaminants, from the leakages of exhaust gases through exhaust pipe fittings in the engine room.

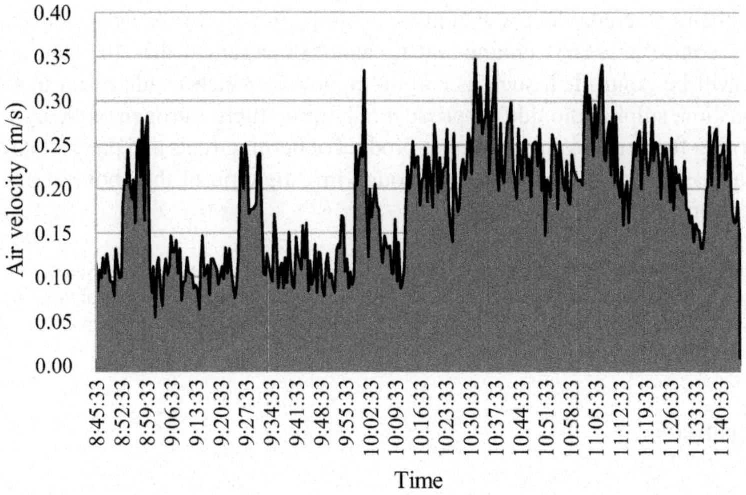

Fig. 10 Overall three-hour air velocity level

5 Conclusion

Investigating the engine room indoor air quality onboard ships has two basic purposes. First, to provide an environment that permits the machinery and equipment to function properly with dependable service life. Second, to provide an environment in which personnel can work comfortably and effectively. The level of CO_2 in a room shows the level of ventilation efficiency, especially in the workplace. The rate of CO_2 removal from an area is related to the rate of air being removed from the space as well as the rate of fresh air entering the ventilation system from outside. An engine room has been chosen, and the indoor air has been characterized to assess its quality.

A conclusion may be drawn here, there were increments of CO_2 concentrations in the engine room due to the effects of leakage in the exhaust emission system related to low preventive maintenance and insufficient ventilation that could remove the accumulated CO_2. The exchange of air, the process of replacing indoor air with outside air, and the dilution of airborne pollutants, such as carbon dioxide, in an enclosed space, are all important aspects of ventilation design that are often overlooked. This evaluation is valuable as a gauge of the engine room's air quality, which is crucial for determining whether workplace health and safety regulations are being followed and has a significant impact on the effectiveness of the combustion process for the ship's main propulsion systems.

While investigating indoor air quality situations, many parameters need to be considered and checked as it may be contributing to some sort of problem in the future. It is typical to encounter multi-layered issues; addressing and resolving one problem might not address the underlying cause. Maintaining CO_2 levels below the ceiling limit does not mean that the space is in compliance with the standard. There are many chemical contaminants that always will be present in the indoor environment and by controlling one contaminant it cannot be assumed that all other contaminants will be controlled, such as carbon monoxide which results from incomplete combustion, sulphur dioxide by-product of burning fuels, nitrogen oxide by-product of burning fuels, nitrogen dioxide by-product of burning fuels and particulate matter. This assessment can be continued through investigation of the above elements by using appropriate research tools.

Acknowledgements This assessment was conducted onboard a patrol vessel. Sincere thanks are expressed to the officers and engine room staff for their contribution to the results of this assessment.

References

1. Yaakob, O., Nee, L.S.: Application of ergonomics in ship design. In: Proceedings of the International Conference on Mechanical Engineering 2003, Dhaka, Bangladesh, 26–28 Dec 2003
2. Veenstra, A.W., Ludema, M.W.: The relationship between design and economic performance of ships. Maritime Policy Manag. **33**(2), 159–171 (2006)
3. Lundh, M., Lützhöft, M., Rydstedt, L., Dahlman, J.: Working conditions in the engine department—A qualitative study among engine room personnel on board Swedish merchant ships. Appl. Ergon. **42**(2), 384–390 (2010)
4. Kim, S.S., Lee Y.G.: Field measurements of indoor air pollutant concentrations on two new ships. Build. Environ. **45**(10), 2141–2147 (2010)
5. Elsafty, A.F.: Ventilation efficiency inside a ship's cabin based on air diffusion performance index (ADPI). Eur. J. Sci. Res. **40**(1), 102–114 (2010)
6. Yahaya, R., Arif, H.M., Kamel, S.N.M.: Preliminary study on indoor air quality (IAQ) on a Royal Malaysian Navy (RMN) ship. Defence S T Tech. Bull. **5**(1), 11–20 (2012)
7. Che Isa, M., Yunus, Z.: An evaluation of indoor air quality (IAQ) in a machinery room of a floating vessel. J. Defence Security **1**(2), 258–274 (2010)
8. Spengler, J.D., Samet, J.M., McCarthy, J.F.: Indoor Air Quality Handbook. Mcgraw-hill (2001)
9. Persily, A.K.: The relationship between indoor air quality and carbon dioxide. In: Indoor Air '96 Proceedings of the 7th International Conference on Indoor Air Quality and Climate, Nagoya, Japan, 21–26 July, 961–966 (1996)
10. Mokhtar, A.M., Rahman, M.A., Amin, F.M.: Department of Occupational Safety & Health (DOSH), Industry code of practice of indoor air quality 2010. Ministry of Human Resources, Malaysia (2010)
11. American Bureau of Shipping (ABS), Guide for crew habitability on ships. American Bureau of Shipping (ABS), Houston, Texas (2016)
12. American Society of Heating, Refrigerating, and Air-Conditioning Engineers, ASHRAE. Ventilation for acceptable indoor air quality. Peachtree Corners, Georgia 30092 US (2016).

Preliminary Development of a Software for Ship Maneuvering Ability Prediction

Yuda Apri Hermawan, Erlangga Zuhdi Aulia Rijal, I. Made Ari Cahyadi, Muhamad Riduwan, Siska Arifiani, Rizky Chandra Ariesta, and Wardiah Mohd Dahalan

Abstract The maneuvering ability is one of the main capabilities which is compulsory possessed by a ship. Inadequate ship maneuverability can lead into marine accidents, especially ship collisions. The International Maritime Organization (IMO) has regulated ship maneuverability criteria, however, in practical and especially in the design stage, analysis related to ship maneuvers is often ignored due to the complexity and the cost of maneuvering calculations as well as maneuvering experiments, respectively. Therefore, the development of a software for ship maneuvering ability prediction is necessary. In this paper, the preliminary development of a software for ship maneuvering ability prediction is conducted. The software is constructed by using a set of computer program codes based on Mathematical Maneuvering Modeling Group (MMG) model integrated with computer user interface and three-dimensional

Y. A. Hermawan (✉) · E. Z. A. Rijal · I. M. A. Cahyadi · R. C. Ariesta
Department of Naval Architecture, Faculty of Marine Technology, Institut Teknologi Sepuluh Nopember, Surabaya, Indonesia
e-mail: yuda.apri@its.ac.id

E. Z. A. Rijal
e-mail: erlangga.19041@mhs.its.ac.id

I. M. A. Cahyadi
e-mail: cahyadi.19041@mhs.its.ac.id

R. C. Ariesta
e-mail: chandra@its.ac.id

M. Riduwan
Department of Marine Transportation Engineering, Faculty of Marine Technology, Institut Teknologi Sepuluh Nopember, Surabaya, Indonesia
e-mail: riduwan@its.ac.id

S. Arifiani
Department of Informatics, Faculty of Intelligent Electrical and Informatics Technology, Institut Teknologi Sepuluh Nopember, Surabaya, Indonesia
e-mail: siska@its.ac.id

W. M. Dahalan
Department of Marine Electrical Engineering, Universiti Kuala Lumpur Malaysian Institute of Marine Engineering Technology, Lumut, Perak, Malaysia
e-mail: wardiah@unikl.edu.my

A. Ismail et al. (eds.), *Engineering Frontiers*, Advanced Structured Materials 208,
https://doi.org/10.1007/978-3-031-56844-2_31

simulation. The developed software is tested to predict a turning and zig-zag motion of a ship, and it works well. The maneuvering ability including ship trajectory, heading, and drift angle can be shown in the software. A three-dimensional view of the ship movement and trajectory is also displayed. The maneuverability IMO criteria can also be examined. The developed software can be used for investigating the maneuvering ability prediction of a ship.

Keywords Ship maneuvering ability · Maneuverability · Ship maneuvering software · Maneuvering ability prediction

1 Introduction

In designing a ship, the technical characteristics and performance capabilities of the ship are crucial aspects and those must comply some acceptable requirements, which are regulated on statutory regulations or stated in technical references. One of the foremost required ship's performance capability is the ship's maneuverability that is being compulsory to fulfill the International Maritime Organization (IMO) standard [1, 2]. A ship must be able to respond accurately to various ship maneuver patterns at sea either for arbitrary movement or avoiding objects/other ships. IMO [1, 2] recommends a set of acceptable criteria for investigating ship and other marine vehicle maneuverability.

Maneuverability refers to the ability of a ship to move or change over from one point to another, encompassing the ability to maintain a specific heading/course, change ships heading/course direction, and ships stopping motion ability. The maneuvering capability of a ship is essential and vital towards the safety of the vessel in the waters, especially in narrow navigational areas such as harbors. It also affects to the safety of marine transportation and has sufficient enough contribution for human error on marine accident [3].

According to the Korean Institute of Marine Science and Technology (KIMST), there were recorded 55 ship accidents resulting from errors or inaccuracies in ship maneuvering during 2013 to 2017 [4]. Among the numerous accidents caused by maneuvering failures, the majority were due to collisions, including both vessel-to-vessel collisions and collisions with other floating objects or docks. As instance, a significant accident that shocked the world occurred in March 2021 when the container ship Ever Given got stuck in the Suez Canal for six days due to maneuvering failure, disrupting maritime traffic and resulting in an estimated economic loss of approximately USD 10 billion [5]. Therefore, efforts to predict the ship performance, especially in terms of maneuverability become crucial in reducing accidents caused by maneuvering failures, considering that 89–96% of accidents at sea occur due to ship collisions [3] where human error contributes more than 80% in the occurrence of the ship collision [6].

A careful examination on ship's maneuverability is necessary at preliminary design stage to ensure the maneuvering performance of a ship [7]. Moreover, an attentive and continuous control to the ship maneuvering ability should be conducted on each design stage as well as at the end of the ship production process to ensure the maneuvering ability of the ship [8]. Thereby, an appropriate approach, technique, and tool are necessary to predict and examine the ship maneuverability during design and production stages.

Ship maneuvering capabilities can be predicted through model tests, mathematical models, or empirical formula. The model test can be performed by a free running model test or a captive model test. Mathematical models can be divided into three types i.e. hydrodynamic models by expressing hydrodynamic forces and moment based on Taylor expansion called as whole ship/Abkowitz model [9], hydrodynamic models by modular mathematical models such as Mathematical Maneuvering Modeling Group (MMG) model [10] and response models that investigate the course control problem by the relationship of rudder action and ship motion response called as Nomoto model [11]. Meanwhile, the empirical formula is developed based on accumulated or empirical data of existing ships [12].

The maneuvering characteristics (expressed by hydrodynamic coefficient) and maneuvering ability resulted from model tests are more accurate comparing with others, however, the model tests have limitations especially for equipment/facility availability, cost, and time [13]. The results coming from empirical formulas are also considered to be not sufficient precise due to limitations in the applicability [7] and the different ship hull types against ship data used for developing the empirical formula.

In the practical design process, the maneuvering capabilities of a ship are rarely calculated or predicted in detail during the preliminary design phase. The maneuvering capabilities of a ship are usually not to be done in the design stage due to the complex calculation of ship maneuvering ability prediction process, or at least it is only focused to meet the IMO criteria without any effort for reaching the optimum level of maneuverability. Moreover, software for maneuvering ability prediction is also limited aggravating the attempt to conduct the ship maneuvering ability prediction in the design stage. A software for predicting ship maneuvering ability is necessary to ensure that the maneuvering performance of a ship meets the IMO criteria. This software is required to enable ship designers to evaluate the ship maneuvering performance in the design stage, thereby, it can minimize the risk of accidents resulting from maneuvering failures.

Based on the mentioned study, this paper attempts to propose a preliminary design of a software to be used for ship maneuvering ability prediction. The software is constructed by using a set of computer program systems integrated with a computer user interface system as well as three-dimensional simulation. The ship maneuvering simulation is constructed by adopting the MMG model since this model has a good accuracy, it is easy to figure out the physical understanding of each its parameters, and easy to add/remove external forces acting to the ship according to the ship condition. The results for maneuvering motions including ship trajectory, heading angle, and drift angle are figured on the software results while the ship trajectory results can also be displayed by three-dimensional simulation. Moreover, the IMO criteria for maneuvering ability can be investigated based on the maneuvering prediction results.

2 Mathematical Model

2.1 Maneuvering Motion Equations

Ship's dynamic motion in horizontal plane is commonly described by three degrees of freedom: surge, sway, and yaw. A coordinate system is employed to represent the ship's motion, consisting of an earth-fixed coordinate system ($o - x_0y_0$) and a body-fixed coordinate system ($G - xy$) with the ship's center of gravity (G) as origin. In the body-fixed coordinate system, the x-axis points towards the ship's fore perpendicular, y-axis extends towards starboard side, and the z-axis is oriented downwards. Forces acting in the x- and y- axes are denoted as X and Y, respectively, while the moment around the vertical axis passing through G is represented by N.

The ship's speed is expressed as U, with its components u and v corresponding to the x and y directions, respectively. The incident angles of the current and wind are denoted as α and ν, respectively, while V_C and V_W represent the speeds of the current and wind. The heading angle ψ, measured as the angle between the x_0 and x axes, is assumed to be positive in the clockwise direction. Angle β is indicating the ship's drift angle from its intended course, with its positive direction as shown in Fig. 1.

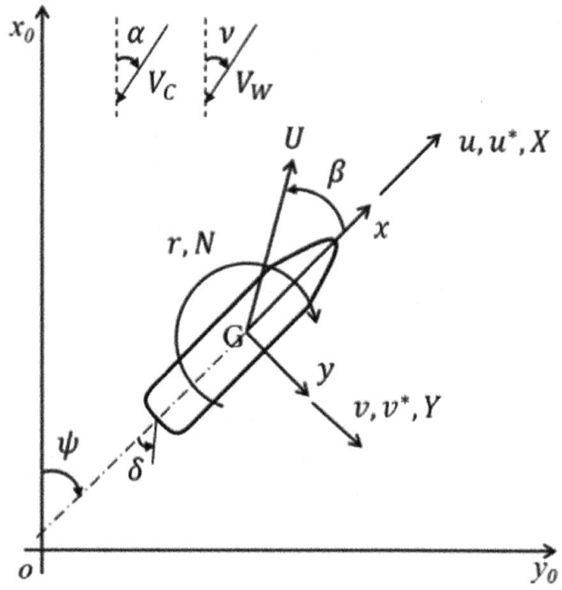

Fig. 1 Coordinate systems

2.2 Hull Forces and Moment

The forces acting on the ship hull affect the ship in three main axes: x-axis, y-axis, and z-axis (yaw). Therefore, there are three different equations that need to be solved to calculate the hull force on each respective axis. Thus, the designed programming system will be able to provide comprehensive information about the forces acting on the ship hull, enabling researchers to analyze and design ships more effectively within the research context. Hull forces and moment action on the ship are expressed as follows:

$$\left.\begin{array}{c} X_{H'} = X_{\beta r'} \sin\beta + X_{uu'} cos^2\beta, \\ Y_{H'} = Y_{\beta'}\beta + Y_{r'}r' + Y_{\beta\beta'}\beta|\beta| + Y_{rr'}r'\left|r'\right| + \left(Y_{\beta\beta r'}\beta + Y_{\beta rr'}r'\right)\beta r', \\ N_{H'} = N_{\beta'}\beta + N_{r'}r' + N_{\beta\beta'}\beta|\beta| + N_{rr'}r'\left|r'\right| + \left(N_{\beta\beta r'}\beta + N_{\beta rr'}r'\right)\beta r' \end{array}\right\} \quad (1)$$

$X_{H'}$ and $Y_{H'}$, are non-dimensional hull forces in x-axis and y-axis respectively, while $N_{H'}$ is the non-dimensional hull moment with reference to the z-axis. $X_{\beta r'}$, $X_{uu'}$, $Y_{\beta'}$, $Y_{r'}$, $Y_{\beta\beta'}$, $Y_{rr'}$, $Y_{\beta\beta r'}$, $Y_{\beta rr'}$, $N_{\beta'}$, $N_{r'}$, $N_{\beta\beta'}$, $N_{rr'}$, $N_{\beta\beta r'}$, and $N_{\beta rr'}$ are known as hydrodynamic derivatives or maneuvering coefficients, determined by conducting a captive model test or estimated from adequate empirical formulas such as Kijima et al. [14].

2.3 Propeller Forces and Moment

The propeller forces acting on the ship hull affect to ship in the in x-axis. The surge force due to propeller is shown as follow:

$$X_{P'} = \left(1 - t_p\right)n^2 D_P^4 K_T J_P / \tfrac{1}{2}\rho L d U^2 , \quad (2)$$

with the thrust coefficient (K_T) and advanced ratio (J_P):

$$K_T = C_1 + C_2 J_P + C_3 J_P^2 , \quad (3)$$

$$J_P = U \cos\beta(1 - w_P)/(nD_P) . \quad (4)$$

Wake fraction of the propeller (w_P) and geometrical inflow angle ($\beta_{P'}$) are:

$$w_P = w_{po}\exp\left(-4\beta_P^{'2}\right) , \quad (5)$$

$$\beta_P^{'2} = \beta - x_{P'} . \quad (6)$$

2.4 Rudder Forces and Moment

The forces and moments associated with the rudder (X_R, N_R, Y_R) are determined by considering the rudder's normal force (F_N), rudder angle (δ), and the hull-rudder interaction coefficients (t_R, a_H, x_H). The terms on rudder force are assumed as follows:

$$\left.\begin{aligned} X_{R'} &= -(1 - t_R)F_{N'} \sin\delta, \\ Y_{R'} &= -(1 - a_H)F_{N'} \cos\delta, \\ N_{R'} &= -\left(x'_R - a_H x'_H\right)F_{N'} cos\delta. \end{aligned}\right\} \quad (7)$$

Parameters required for prediction of rudder normal force(s) (F_N):

$$F_{N'} = \left(\tfrac{A_R}{Ld}\right) C_N U_R'^2 \sin\alpha_R \ , \quad (8)$$

in which, the coefficient of rudder normal force (C_N) and rudder aspect ratio (K_R) are:

$$C_N = 6.13K_R/(K_R + 2.25) \ , \quad (9)$$

$$K_R = H_R^2/A_R \ . \quad (10)$$

The effective inflow speed (U_R) is as follows,

$$\left.\begin{aligned} U_R &= (1 + C_{lr}g)(1 - w_r)^2, \\ g &= \eta K(2 - (2 - K)s)s/(1 - s)^2, \\ K &= 0.6\left(1 - w_p\right)/(1 - w_R), \end{aligned}\right\}. \quad (11)$$

while slip ratio (s) can be expressed as follow,

$$s = 1 - \left(1 - w_p\right)U \cos\beta/\eta_P \ . \quad (12)$$

2.5 Maneuvering Equations

In this stage, the software utilizes the method of solving differential equations to calculate the desired values, such as the ship's position (x_0, y_0), heading angle, and other parameters. These differential equations are based on the physical relationships between the forces acting on the ship, and ship movement. The differential maneuvering equations used for modelling maneuvering calculation is expressed as follows,

$$\begin{aligned}
\frac{d\beta}{dt} &= \frac{U}{L}\left\{\frac{m' + m_{x'}}{m' + m_{y'}}\right\} r' |\cos^2| \beta + \left\{\frac{m' + m_{y'}}{m' + m_{x'}}\right\} r' |\sin^2| \beta \\
&\quad - \frac{1}{m' + m_{x'}} \sin \beta X' - \frac{1}{m' + m_{y'}} \cos \beta Y', \\
\frac{dr'}{dt} &= \frac{U}{L}\left\{\frac{N'}{I_{zz'} + i_{zz'}}\right\} - r' \frac{U}{L}\left\{\frac{X'}{m' + m_{x'}} \cos \beta - \frac{Y'}{m' + m_{y'}} \sin \beta\right\} \\
&\quad - \left(\frac{m' + m_{y'}}{m' + m_{x'}} - \frac{m' + m_{x'}}{m' + m_{y'}}\right) r' \sin \beta \cos \beta, \\
\frac{dU}{dt} &= \frac{U^2}{L}\left\{\frac{X'}{m' + m'_x} \cos \beta - \frac{Y'}{m' + m_{y'}} \sin \beta\right\} \\
&\quad - \left(\frac{m' + m_{y'}}{m' + m'_x} - \frac{m' + m'_x}{m' + m_{y'}}\right) r' \sin \beta \cos \beta, \\
\frac{d\psi}{dt} &= \frac{Ur'}{L}, \\
\frac{dx_o}{dt} &= U \cos(\psi - \beta), \\
\frac{dy_o}{dt} &= U \sin(\psi - \beta).
\end{aligned} \tag{13}$$

Details of the calculation for the MMG model can be seen in [10].

3 Results and Discussion

3.1 *Subject Ship Data*

To ensure that the developed software can perform a maneuvering simulation of a ship, a ship "X" data which has 1:75.5 scale is used for the simulation. The principle particular of the ship models is shown in Table 1.

The simulation is conducted for both turning circle and zig-zag motion. The turning circle motions are performed for 10° and 20° of rudder angle. The effect of environmental loads is not considered in this simulation while the turning circle

Table 1 Ship "X" data

Particular	Value
L [m]	3.046
B [m]	0.427
d [m]	0.143
Cb	0.651

motion with 10° rudder angle are presented in the three dimensional (3D) simulation to show the performance of the 3D simulation software.

3.2 Simulation for Turning Circle Motion with 10° of Rudder Angle

A simulation of turning circle motion with 10° of rudder angle for the model ship, taking into account the hydrodynamic derivatives, rudder, and propeller parameters is conducted in the developed software. The results of maneuvering simulation presented in the software interface is shown in Fig. 2. According to the figure, the ship trajectory, heading angle, and drift angle can be evaluated.

The maneuvering performance of the ship was evaluated by comparing the results obtained from simulating a 10° turning circle with the criteria established by the International Maritime Organization (IMO). The criteria used in this software is the maximum tactical diameter that should be less than 5.0 times of ship's length (5.0 *L*). According to the examination results, the calculated tactical diameter for the model ship exceeds the accepted threshold of 5.0 *L*, which means that the maneuverability of the ship does not meet the criteria defined by the IMO. The developed software can

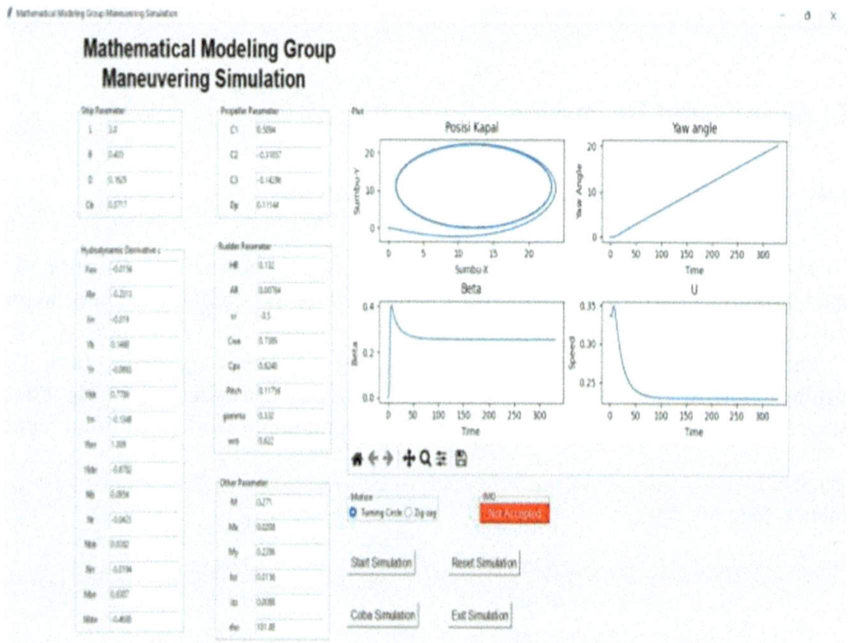

Fig. 2 Turning circle results for 10° rudder angle

evaluate the maneuverability prediction of the ship in this turning circle simulation by calculating the discrepancy of calculated results and required IMO criteria.

3.3 Simulation for Turning Circle Motion with 20° of Rudder Angle

Similar with the previous turning circle motion simulation, the software is also performed for turning circle with 20° of rudder angle. The result is shown in Fig. 3. According to the figure, the trajectory of turning circle can be drawn by the software. The diameter of turning circle is smaller than the turning circle which is in the previous simulation. Through the same manner, the maneuverability criteria are investigated. Upon examining the graph generated by this software, the calculated tactical diameter for the model ship is less than the threshold of 5.0 L, which means the maneuvering performance of the model ship, as predicted by this software, meets the criteria defined by the IMO. Again, the developed software can perform maneuvering simulation and evaluate the maneuverability prediction.

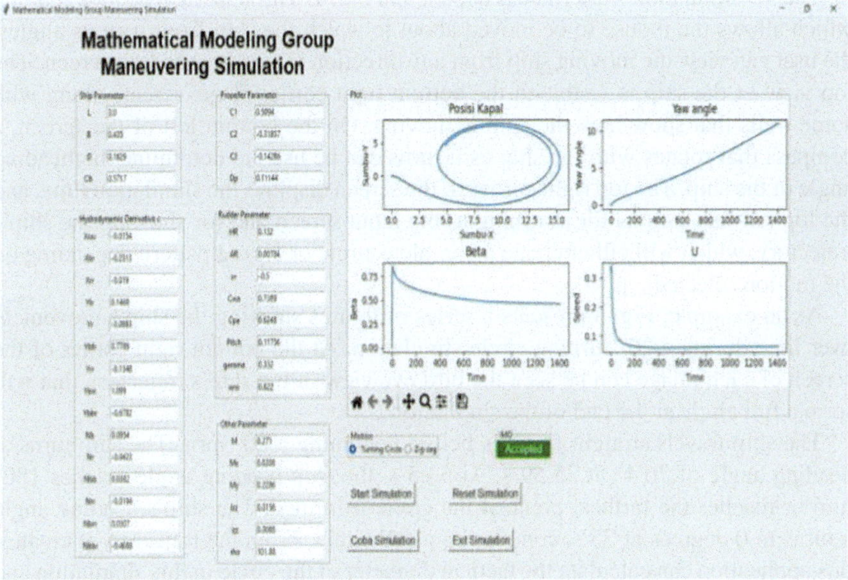

Fig. 3 Turning circle results for 20° rudder angle

Fig. 4 Display of 3D maneuver simulation

3.4 Simulation of Ship Maneuvering Motion in 3D View

From the ship maneuver calculations that have been carried out previously, it will produce coordinates in the x- and y-axes (horizontal plane) which will be used as reference points in the 3D simulation of ship maneuvering simulation. In this 3D view, the user will be shown how the ship moves during her maneuver as well as one can also observe the ship's heading angle and simulation time.

The 3D simulation's interface is depicted in Fig. 4. Through the free-look feature, which allows the mouse to be moved about to watch the ship from various angles, the user can view the moving ship from any direction in the center of the screen. The top view of the ship is visible in the bottom right corner of the screen, along with some trails that show how the ship is moving. On the bottom left of the screen, a compass that rotates with the ship as it turns can be used to determine the heading angle of the ship. The top right corner of the screen displays the simulation time, and the top left has buttons for returning to the input screen and for showing the ship's trajectory, which will either create a complete circle or zigzag pattern depending on the motion selected.

As an example, Fig. 5 presents a series of figures showing the ship's movement over time during a 20° turning circle simulation. At the bottom right corner of the screen of each image can be seen the bird eye view of the ship's trajectory that will form a full circle at the end of the simulation.

The ship travels straight for 20 s before beginning a 20° turn. The ship turns at heading angle of 76.4° at 35.59 s. At 53.5 s, the ship heading angle reaches 180° turn or reaches the farthest point of the circle it formed. The ship's heading angle returns to 0 degrees at 75 s, completing a full circle. From the trajectory it creates, this application can calculate the tactical diameter of the circle in this simulation and compare it to the requirement of IMO.

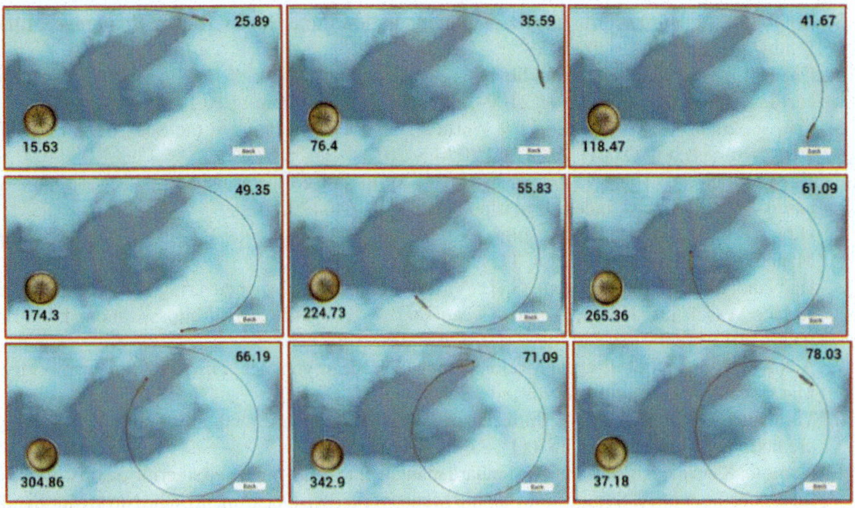

Fig. 5 Ship trajectory over time

4 Conclusion

A preliminary development of a software for predicting the ship maneuvering ability is conducted. The software is developed based on MMG maneuvering model integrated with a computer user interface and 3D simulations. Two turning circle simulations with 10° and 20° rudder angle degree was performed to examine the performance of the developed software. The software can perform the maneuvering simulation displaying the ship trajectory and maneuvering ability data. The investigation of ship maneuvering against IMO criteria can also be calculated by the software by examining the result maneuvering simulation towards required IMO criteria. Moreover, the three-dimensional simulation of ship trajectory can be simulated and displayed properly according to ship maneuvering simulation results.

Acknowledgements The works in this paper is supported by Institut Teknologi Sepuluh Nopember (ITS) through Research, Innovation, and Entrepreneurship Project Higher Education for Technology and Innovation (HETI) ADB Fund with contract numbers: 2178/PKS/ITS/2022.

References

1. The International Maritime Organization: Resolution Standards for Ship Maneuverability MSC.137 (76). IMO, London (2002)
2. The International Maritime Organization: Explanatory Notes to the Standards for Ship Maneuverability MSC/Circ 1053. IMO, London (2002)

3. Rothblum, A.: Human error and marine safety. In: Maritime Human Factors Conference, MD, Linthicum, 13–14 Mar 2000
4. Chan, S.R., Hamid, N.A., Mokhtar, K.: A theoretical review of human error in maritime accidents. Adv. Sci. Lett. **22**(9), 2109–2112 (2016). https://doi.org/10.1166/asl.2016.7058
5. Yee, V., Glanz, J.: How One of the World's Biggest Ships Jammed the Suez Canal. https://www.nytimes.com/ (2021). Accessed 20 May 2022
6. Hu, Y., Zhang, A., Tian, W., Zhang, J., Wu, H.: Multi-ship collision avoidance decision-making based on collision risk index. J. Marine Sci. Eng. **8**, 640 (2020)
7. Hirano, M., Takashina, J., Fukushima, M., Moriya, S.: A Computer Program System for Ship Maneuvering Motion Prediction. TU Delft (1975)
8. Hajivand, A., Mousavizadegan, S.H.: Virtial simulation of maneuvering captive test for a surface vessel. Int. J. Nav. Archit. Ocean Eng. **7**, 848–872 (2015). https://doi.org/10.1515/ijnaoe-2015-0060
9. Abkowitz, M.A.: Stability and Motion Control of Ocean Vehicles. The M.I.T and Cambridge, Massachusetts and London (1969)
10. Yasukawa, H., Yoshimura, Y.: Introduction of MMG standard method for ship maneuvering predictions. J. Mar. Sci. Tech. **20**(1), 37–52 (2015)
11. Nomoto, K.: Analysis of Kempf's standard maneuver test and proposed steering quality indices. In: Proceedings of 1st Symposium on Ship Maneuverability. Department of the Navy, Maryland, United State of America, pp. 275–304, 24–25 May 1960
12. Lewis, E.V.: Principle of Naval Architecture Second Revision Vol. III Motion in Waves and Controllability. The Society of Naval Architects and Marine Engineers, Jersey City (1989)
13. Khan, O.U., Mansoor, A., Zai, B.A., Saqib, N.U., Hussain, M., Jamil, T.: Virtual captive model test for maneuvering prediction. J. Maritime Res. **XIX**(III), 47–51 (2022)
14. Kijima, K., Nakiri, Y., Katsuno, T., Furukawa, Y.: On the Maneuvering Performance of a Ship with the Parameter of Loading Condition. The Society of Naval Architect of Japan (1990)

Review on Coconut Shell Charcoal Thermal Decomposition Under Low Heat Temperature

Mahani Mohd Zamberi, Rawaida Muhammad, Norsheila Buyamin, and Shaharin Anwar Sulaiman

Abstract Charcoal has been among the alternative energy sources that are used worldwide. The famous feedstock in producing charcoal is normally wood. Yet, due to certain circumstances varieties of raw materials have been used to make charcoal. Even so, suitable processing methods need to be considered so that the feedstock is fully utilized. Major drawbacks that need to be considered for charcoal are in terms of low-density, low fixed carbon, high ash, contamination, moisture content and many more. These problems are addressed and being partially solved by using coconut shell as the feedstock and pyrolysis method as the process of producing coconut shell charcoal. This paper focuses on the works that have been carried out in the last decade for producing charcoal. Different types of material sources are presented with a focus on approaches in developing charcoal. Pyrolysis or thermal decomposition of feedstock under low temperature is believed to prove better characteristics of coconut shell charcoal. This is because by applying indirectly the low temperature heating process could enhance the charcoal benefit. Nevertheless, it is important to be aware of any beneficial nutrient from coconut shell after pyrolysis process or any technologies which can offer better solutions to charcoal problems.

Keywords Charcoal · Coconut shell · Thermal decomposition · Pyrolysis

M. M. Zamberi (✉) · R. Muhammad · N. Buyamin
Politeknik Ungku Omar, Jalan Raja Musa Mahadi, 31400 Ipoh, Perak, Malaysia
e-mail: mahani@puo.edu.my

R. Muhammad
e-mail: rawaida@puo.edu.my

N. Buyamin
e-mail: nsheila@puo.edu.my

S. A. Sulaiman
Universiti Teknologi Petronas, 32610 Seri Iskandar, Perak, Malaysia
e-mail: shaharin@utp.edu.my

A. Ismail et al. (eds.), *Engineering Frontiers*, Advanced Structured Materials 208,
https://doi.org/10.1007/978-3-031-56844-2_32

1 Introduction

In fact, 24 million tonnes of charcoal were consumed worldwide in 1992 and half of the world's production is consumed by Africa alone [1]. According to the alternative energy development and efficiency department, the trend of charcoal consumption has increased in the last ten years [2]. Charcoal is a solid fuel or other meaning is lightweight black carbon used for heating, cooking, adsorber or catalyst through a slow heating process into carbon or other chemical compounds. Charcoal is also mainly used in the household sector, with a total of around 4.2 million households [3]. While 97% of the total charcoal consumption is represented in the rural area [2]. There are lots of raw materials that can be processed to make charcoal. Generally, the raw materials containing carbon are more suitable such as wood, coconut shell, biomass, bamboo, nutshell, and others.

For traditional production, wood material is commonly used to produce charcoal. Nevertheless, the output of the traditional method is lower production while nowadays people are more likely to use new mechanisms according to its wide range of raw material and high carbon rate. Based on The Food and Agriculture Organization (FAO), about 100 million tons of wood are annually cut for charcoal production and official harvest level in forest are often exceeded [1]. Apart from that, many agricultures waste by product are used to replace wood as the carbon source such as cherry stones, macadamia nut shells, coconut shells and palm husks [4, 5].

Several renewable energy sources, abundantly available biomass can contribute around 10% of annual energy needs and reduce waste disposal [6, 7]. Biomass is classified as clean energy, the highest carbon rate [8], potentially sustainable and environmentally friendly. In fact, a good quality charcoal contains about 82% fixed carbon and about 54% is the maximum charcoal yield from cellulose [9]. According to the previous research, Table 1 shows the comparison of the properties of high-yield charcoals with those of commercial products of biomass. Table 1 shows that the volatile matter content is below 30% and the related fixed carbon contents are all high with a good quality.

Interest and research of new raw material are needed to ensure the production of chemicals and the reduction in chemicals and the energy crisis shortly [10]. Agriculture waste by products has been successfully used in production of activated carbon is coconut shell [11]. Thus, the main part of this review study is focused on coconut shell charcoal and its production. The world coconut production accounts about ten million hectares in 92 countries and 75% of the mass production are from Asian countries [12]. Mainly, coconut is used to produce oil, otherwise it is also used for health, beauty, energy, foods, water, fresh milk, and others. Consequently, abundant coconut waste is being produced from coconut industries such as coconut shell. Therefore, the coconut shell is ready to produce charcoal or activated carbon. It shows in the Thai Ministry of Commerce (1998) statistics, about 1.4 million tons of coconut (*Cocos nucifera L*) are produced annually where unused resources from coconut such as coconut shells were exchanged to the activated carbon which can be used as an adsorbent in water purification [13]. Besides, it helps reduce the cost

Table 1 Comparison of the properties of high-yield charcoals with those of commercial products [9]

Feedstok	Ash (%)	Volatile matter (%)	Fixed carbon (%)	Calorific value^d (MJ/kg)
Eucalyptus	1.29	23.3	75.4	32.0
Kiawe	2.65	21.4	75.9	30.6
Leucaena	2.85	14.4	82.8	32.8
Kukui nutshell	1.79	20.1	78.1	31.2
Macadamia nutshell	0.81	28.7	70.5	31.4
Palm nutshell	3.48	18.8	77.7	31.2
Commercial Briquette charcoal	17.2	36.4	46.5	22.8
Commercial Kiawe charcoal	1.96	28.3	69.8	29.7
Commercial Mesquite charcoal	4.17	9.04	86.8	31.8

^{d}Refers to higher heating value

of waste disposal and provide inexpensive alternative commercial activated carbon (CAC) from the existing CAC. Coconut shells are also suitable for producing microporous activated carbons due to its good natural structure and low ash content [14]. For the next section, the advantages and disadvantages of coconut shell charcoal are addressed for discussion.

2 Pros and Cons Coconut Shell Charcoal

As reviewed earlier, there are various types of charcoal in the market. This study focuses on coconut shell charcoal that is made of shells of coconuts. Various pros and cons of coconut shell have been listed in Table 2. One of the benefits is in terms of the burning process coconut shell charcoal that is much longer than other types of natural charcoal. For certain industries, it is crucial to have longer burning sessions as an example for those having hookah smoking to smoke the tobacco. This is due to its unique bone structure, heterogeneous shapes with macro-pores scales that have higher volatile content [12]. By having these kinds of properties, granulated carbon with constant and big pore volume can be produced with slow burning process and lesser smoke [15].

Instead of considering its combustion process, the ability to emit more energy also gives more advantage to coconut shell as a charcoal. The activation step for this coconut shell produced fixed high composition carbons and volatiles that

Table 2 Advantages and disadvantages of coconut shell charcoal with particular application

Advantages	Disadvantages	Uses	References
Detoxification support—good adsorption characteristics	Less carbonized than coconut husks	Food industry	[4]
Cleansing agent—low heavy metal	Still having low percentage of ash	Teeth whitener	[16]
Reduces gas & bloating	Still having fewer gas emission	Animal feed	
Treats poisoning	Must go through chemical activation step	Pharmaceutical	[4, 12, 16]
Renewable resource of energy—high lignin content	Needed binder for better combustion medium	Fuel, Charcoal	[4, 12, 17]
Decrease waste		Cooking purpose	[4]

stored the energy. This shows an advantage for the feedstock when undergoing the devolatilization process becoming energy storage too [18, 19].

This type of charcoal would have less side effects due to its influence in water gas reaction. Thus, possess higher value of hydrogen that lowers the composition of nitrogen and sulfur. By that, it means that the emission from sulfur and nitrogen due to combustion process is lower making coconut shell as one of environmentally friendly renewable energy source [18].

There are many ways that can be used as methods in optimizing the uses of coconut shells. As for this study, pyrolysis is chosen as the best method to better utilize the coconut shells [17]. For the next section, various procedures of charcoal production are studied for better understanding of the differences for all methods. Then, better detailing pyrolysis technique of future study would be discussed and concluded.

3 Methodology

3.1 Charcoal Technologies Principles

Charcoal production in various countries depends on their natural sources such as from plants, natural materials, and biomass from wastes to produce energy processes. Besides mass charcoal production, regardless sources are also based on geographical demography and climates exchanges to meet production demand. Since that there are two major methods used in producing charcoal: modern technology and traditional technology. Different kinds of batch kilns create the largest portion of the world's total production of charcoal. Using modern industrial methods, charcoal is produced at which most of them use wood as their primary fuel source and operate under atmospheric pressure. Meanwhile the pit or mound turf kilns are the oldest of all, then the pyrolysis gas was then somewhat modified to Casamance kilns by adding exhaust pipes. Other than that brick or concrete kiln and steel kiln were considered as traditional technologies.

In general, three methods of heating are utilized to start the carbonization process and sustain high temperatures throughout the procedure [20]: internal heating where internal heating the raw material is burnt under controlled air flow, external heating in which oxygen is not allowed to enter the reactor and the retort is heated from the outside, and the third heating with recirculated gas where pyroligneous vapours are burned in an external combustion chamber and then sent into the reactor which is in close proximity to the raw material.

3.2 *Internal Heating Principle*

Based on partial combustion of the fuel, more than 90% of charcoal production systems use internal heating. The kilns made of brick or concrete can be found in this category. The kiln's design is straightforward, and the upfront expenses are frequently modest and simple in design.

Conventional Wood Carbonization Technologies

Due to their simplicity and inexpensive cost, especially for small companies, these kilns are the most common around the world [1]. Their construction commonly needs open space and flat sites due to its design structure with several kilns made of baked bricks, sand mortar and clay placed, or built in the allocated area [21]. Besides there are also kilns made of stake of wood covered with sawdust or rice husk for mound turf depending on the country's plant-based or agricultural waste products. A brief explanation of certain charcoal processing techniques using these heating principles is provided in the section that follows. The constraints of mechanising the loading of firewood and the unloading of charcoal are one of the limitations of the old methods for producing charcoal due to the fact that charcoal fragmentations occur during unloading.

In Malaysia, due to the higher demand from local and international markets the area planted to coconut has progressively increased from 84,609 ha in 2016 to 85,630 ha in 2020 with the production of 542,000 MT when compared with 504,780 MT. In the northern and southern regions, charcoal producers mainly used clay brick kilns. Malaysia has mangrove trees with the surrounding salt water under provision and authorized by Matang Mangrove Forest Reserve situated in Perak providing optimum raw materials for high quality charcoal and the largest sustainable mangrove forest in Malaysia. Most companies employ brick clay kilns that operate on the internal heating principle. The kiln's structure is up to 4.5-to-5-m height and a 5 to 5.5 m diameter equipped with one main door opening for loading and unloading charcoal purpose. The next step of loading is to close the door with a simple brick over and sealed with mud during carbonization process. Air infiltration is regulated by the vent surrounding the kiln's base. Only one route of exhausting smoke from the kiln is through the "eye hold" on the top of the kiln [22].

Modern Industrial Technologies

According to research on the supply-side economics of scale, businesses that produce in large quantities profit because they can spread their fixed expenses across more units, which lowers their cost per unit, and because they can use more efficient technologies [23]. The efficiency of kilns could be increased by 30% or more by adopting improved kiln preparation techniques. In Brazil, the Missouri kiln, which is mechanized to increase its capacity of producing charcoal, is the technique that is most frequently utilized. Its technologies are made of poured concrete and designed into rectangular or square shapes. Doors at either end allow the use of front-end loaders for charging and discharging, and the kilns are appropriately filled with round wood, primarily from eucalyptus plantations. Other designs of kilns to improve the cooling phase were developed by the Veradas system. It is made up of 16 kilns that are arranged in 4 connected buildings. In order to shield the kilns from damage brought on by the sun, rain, and other weather conditions, each integrated building has a roof over it. In order to convey hot smoke, the kilns in each integrated building are connected to one another by outside pipes.

Most charcoal kilns in Europe is steel retorts. The CML carbonization system [20] is one of these technologies that has gained the most traction in France. A manufacturing plant is a standardized building made up of 4 to 12 production kilns connected to a depollution system that incinerates waste gases. A top lid and a base discharge hatch are features of the cylindrical kilns used to make charcoal. Manual valves placed at the bottom and all around the lower edges of the kilns are used to regulate the airflow.

3.3 External Heating Principle

A part of the pyroligneous vapours used in these procedures are burned in an external combustion chamber before being sent into the reactor, where they encounter the raw material. Compared to earlier technologies, these are scarce and costly. Even though there are only a relatively small number of units now, the Charbon engineering technology is the most often used. This dual retort system, which is a semi-continuous production module that has one module made up of two retorts that are each housed in their own insulated oven and installed on a concrete floor that is armed. This module is housed in a hallway.

3.4 Heating with Recirculated Gas

Most people are familiar with the continuous Lambiotte carbonisation process. With a feeding entrance at the top and a discharge cone at the bottom, the retort is shaped like a vertical cylinder. Wood enters the retort through a lock-hopper after being hoisted to the top of the device by a hoist. The drying, carbonization, and chilling zones are the three areas through which the wood traverses as it descends.

3.5 Pyrolysis Process Under Low Temperature

The thermochemical conversion process of pyrolysis is the most prevalent and straightforward. It has a great potential to transform using thermochemical conversion techniques that in the presence of little or no oxygen, may be able to convert large amounts of lignocellulosic biomass into valuable solid, liquid, and gas products. An environmentally friendly approach for producing charcoal is one that uses a downdraft reactor with controlled pyrolysis variables. Hence, some parameters need to be considered such as temperature and heating rate. Temperature in this matter is the key point to investigate and measure important physical characteristic of biomass including effects on the percentage weight loss, density, moisture content, ash content, carbon content, yield, carbon yield, and energy content. There are two conditions of the temperature settings applied in the previous studies but most of them were on the higher temperature process. Some studies resulted as when temperature was increased the percentage of mass of solid char decreased, while gas and oil products increased, and the other studies result contrary in the decreased temperature [24]. The coconut shell was used in the purification industry as well as other sectors like pharmaceuticals and cosmetics as well as a feedstock to produce chemicals. The pyrolysis process in the study for producing coconut shell charcoal uses a downdraft reactor electric heater due to its primary difference between more sophisticated pyrolysis technologies and conventional kilns where process heat is not supplied by burning a portion of the input biomass material; rather, the process is sustained utilizing the pyrolysis gas created during the pyrolysis process itself. The temperature correlated to residence time is being studied for this experiment to assure the carbonization process works well to produce better performance with favorable qualities of charcoal.

4 Results and Discussion

The pyrolysis of coconut shell process to produce charcoal has been done by many researchers [25]. In this study the investigation needs to be done on how the pyrolysis method affected the quality parameters with elevated temperatures range from 250 to

500 °C (50 °C intervals) and longer residence time of 15–75 min (15 min intervals) per process being used to obtain the possible value of the biomass's moisture content, ash content, and calorific value. Some data obtained from previous researchers showed that temperature during pyrolysis has an impact on the product's composition by causing the volatiles to volatilize and produce more solids. The impact of carbonization residence time on charcoal calorific value, moisture content, and ash content at a constant temperature by observation show that the calorific value and the ash content increased. Some studies found that the longer residence time of the pyrolysis process the moisture content tends to be low, while the calorific value and ash content increased due to elevated temperature [25]. Some other study also found in the case of biochar used in agricultural production affects the chemical and physical characteristics of biochar at the relative high pyrolysis temperature and long-term oxidation increased nutrient concentrations as well as producing negative charges on the biochar surface [26].

5 Conclusion

The longer duration of residence time with the lower temperature applies to the thermal decomposition process may prove the better performance and quality of coconut shell charcoal. This is because the benefits from the process by using low temperature could enhance the charcoal reusability and quality. This paper also suggests studying the beneficial nutrient from coconut shell charcoal after the pyrolysis process in contributing to better quality of charcoal. Further work and study need to be carried out to meet the requirements.

Acknowledgements The authors would like to thank the *Centre of Air Conditioning & Refrigeration* (CARe), Politeknik Ungku Omar for supporting this review paper to the conference of International Conference on Marine and Advanced Technologies (ICMAT 2023). The authors are also grateful to Prof. Dr. Shaharin Anwar for guiding the information of this research.

References

1. Kammen, D.M., Lew, D.J.: Review of Technologies for the Production and Use of Charcoal. Berkeley (2005)
2. Kajina, W., Junpen, A., Garivait, S., Kamnoet, O., Rousset, P.: Charcoal Produced from Mangrove in Center Thailand (2018)
3. Kajina, W., Junpen, A., Garivait, S., Kamnoet, O., Keeratiisariyakul, P., Rousset, P.: Charcoal Production Processes: An Overview
4. Cobb, A., Maurer, E.P., Chiesa, S.: Low-tech coconut shell activated charcoal production. IJSLE **7**(1), 93–104 (2012)
5. Adpour, A.A., Do, D.D.: The Preparation of Activated Carbon from Macadamia Nutshell by Chemical Activation (1997)

6. Höök, M., Tang, X.: Depletion of fossil fuels and anthropogenic climate change-a review. Energy Policy **52**, 797–809 (2013). https://doi.org/10.1016/j.enpol.2012.10.046
7. Sun, J.X., Sun, X.F., Sun, R.C., Fowler, P., Baird, M.S.: Inhomogeneities in the chemical structure of sugarcane bagasse lignin. J. Agric. Food Chem. **51**(23) (2003). https://doi.org/10.1021/jf034633j
8. Rashidi, N.A., Yusup, S.: Biochar as potential precursors for activated carbon production: parametric analysis and multi-response optimization. Environ. Sci. Pollut. Res. **27**(22), 27480–27490 (2020). https://doi.org/10.1007/s11356-019-07448-1
9. Antal, M.J., et al.: High-yield biomass charcoal. Energy Fuels **10**(3) (1996). https://doi.org/10.1021/ef9501859
10. Mahfouz, A., Gad, M. S., El Fatih, A., Emara, A.: Comparative study of combustion characteristics and exhaust emissions of waste cooking-diesel oil blends. Ain Shams Eng. J. **9**(4) (2018). https://doi.org/10.1016/j.asej.2018.03.004
11. Daud, W.M.N.S.M., Mustafa, S., Rahmadulla, S.R., bin Che Ghani, S.A., Hisyam, A.: Technical and economic feasibility study of coconut shell charcoal production as precursor to activated carbon in East Coast Malaysia. Int. J. Supply Chain Manage. **6**(2) (2017)
12. Kabir Ahmad, R., Anwar Sulaiman, S., Yusup, S., Sham Dol, S., Inayat, M., Aminu Umar, H.: Exploring the potential of coconut shell biomass for charcoal production. Ain Shams Eng. J. **13**(1) (2022). https://doi.org/10.1016/j.asej.2021.05.013
13. Babel, S., Kurniawan, T.A.: Cr (VI) removal from synthetic wastewater using coconut shell charcoal and commercial activated carbon modified with oxidizing agents and/or chitosan. Chemosphere **54**(7), 951–967 (2004). https://doi.org/10.1016/j.chemosphere.2003.10.001
14. Yang, K., Peng, J., Xia, H., Zhang, L., Srinivasakannan, C., Guo, S.: Textural characteristics of activated carbon by single step CO2 activation from coconut shells. J. Taiwan Inst. Chem. Eng. **41**(3) (2010). https://doi.org/10.1016/j.jtice.2009.09.004
15. Tam, M.S., Antal, M.J.: Preparation of activated carbons from macadamia nut shell and coconut shell by air activation. Ind. Eng. Chem. Res. **38**(11) (1999). https://doi.org/10.1021/ie990346m
16. Amidon, G.E., Meyer, P.J., Mudie, D.M.: Particle, powder, and compact characterization. J. Pharm. Policy Pract. (2017). https://doi.org/10.1016/B978-0-12-802447-8.00010-8
17. Rout, T.K.: Pyrolysis of Coconut Shell. M. Tech Thesis, no. 211 (2013)
18. Akhtar, A., Ivanova, T., Jiříček, I., Krepl, V.: Detailed characterization of waste from date palm (Phoenix dactylifera) branches for energy production: Comparative evaluation of heavy metals concentration. J. Renew. Sustain. Energy **11**(1) (2019). https://doi.org/10.1063/1.5027578
19. Inayat, M., Sulaiman, S.A., Naz, M.Y.: Thermochemical characterization of oil palm fronds, coconut shells, and wood as a fuel for heat and power generation. MATEC Web Conf. (2018). https://doi.org/10.1051/matecconf/201822501008
20. Grønli, M.: Industrial Production of Charcoal
21. de Oliveira Vilela, A., Lora, E.S., Quintero, Q.R., Vicintin, R.A., Pacceli da Silva e Souza, T.: A new technology for the combined production of charcoal and electricity through cogeneration. Biomass Bioenergy **69** (2014). https://doi.org/10.1016/j.biombioe.2014.06.019
22. Lehmann, J., Stephen, J.: Biochar for Environmental Management: Science, Technology and Implementation, vol. 1 (2015)
23. Porter, M.E.: The five competitive forces that shape strategy. Harvard. Bus. Rev. **86**(1) (2008)
24. Williams, P.T., Besler, S.: The influence of temperature and heating rate on the slow pyrolysis of biomass. Renew. Energy **7**(3) (1996). https://doi.org/10.1016/0960-1481(96)00006-7
25. Ahmad, R.K., Sulaiman, S.A., Yusuf, S., Dol, S.S., Umar, H.A., Inayat, M.: Influence of Pyrolysis Process Conditions on the Quality of Coconut Shells Charcoal (2020)
26. Gaskin, J.W., Steiner, C., Harris, K., Das, K.C., Bibens, B.: Effect of low-temperature pyrolysis conditions on biochar for agricultural use. Trans. ASABE **51**(6), 2061–2069 (2008). https://doi.org/10.13031/2013.25409

Numerical Investigation on the Bulkhead Structure Due to Sloshing Loads

Septia Hardy Sujiatanti, Wasis Dwi Aryawan, Dandy Fathullah Zain, Zulzamri Salleh, Satriyo Rahmadianto, Dian Purnamasari, and Dony Setyawan

Abstract The present study focuses on the bulkhead strength due to sloshing loads in the cargo oil tank of tanker ships. Sloshing simulation is examined according to the ship's motion during its operation period at sea. Ship motion of surge and pitch were considered in the longitudinal sloshing issue in a certain range of rotation. The maximum pressure on a cargo oil tank's bulkhead is obtained using computational fluid dynamics simulation. Various cases of the filling level or liquid heights are employed to investigate the bulkhead strength due to the pressure of the sloshing effects. Several cases of the filling level including 10%, 30%, 50%, 70%, and 90% are considered. The simulation results show that the maximum pressure on the bulkhead due to the sloshing load is obtained at the filling level of 50% of the tank height. Furthermore, the maximum stress and deformation are calculated by using numerical analysis for all various filling levels. It is found that the maximum stress and deformation occur on the bulkhead subjected to maximum pressure. The maximum stress increased by 33.8% and 18.7% for the filling level 10% and 30% against 50% of tank height, respectively. On the other hand, the maximum stress decreased by 25.6% and 38.8% for the filling level 70% and 90% against 50% of tank height, respectively. In addition, the maximum deformation shows a similar trend to the maximum stress for all various filling levels.

S. H. Sujiatanti (✉) · W. D. Aryawan · D. F. Zain · D. Setyawan
Department of Naval Architecture, Sepuluh Nopember Institute of Technology, Surabaya, Indonesia
e-mail: septia.rahmadianto@gmail.com

D. Setyawan
e-mail: dony@na.its.ac.id

Z. Salleh
Universiti Kuala Lumpur Malaysian Institute of Marine Engineering Technology, Lumut, Perak, Malaysia
e-mail: zulzamri@unikl.edu.my

S. Rahmadianto
PT Surabaya Maritim Konsultan, Surabaya, Indonesia

D. Purnamasari
National Research and Innovation Agency, Cibinong, Indonesia
e-mail: dian.purnama.sari@brin.go.id

A. Ismail et al. (eds.), *Engineering Frontiers*, Advanced Structured Materials 208,
https://doi.org/10.1007/978-3-031-56844-2_33

Keywords Sloshing · Finite element analysis · Ship structure · Computational fluid dynamic · Transverse bulkhead

1 Introduction

Various waveforms due to sloshing motion in the tank produce different values. It is highly dependent on the depth of the liquid and the oscillation frequency. The waves that may occur include standing waves, traveling waves, hydraulic jumps, and combinations of the three waves [1]. Standing waves will form when the liquid level is shallow and oscillate at a frequency much lower than the resonant frequency. When the frequency increases, standing waves will turn into traveling waves with very short wavelengths. The hydraulic jump wave type will occur when there is a small disturbance and appears above the frequency range near the resonant frequency. If the frequency increases, a solitaire wave will occur [2].

Several methods of determining the movement of the liquid which is called sloshing have been studied [3–5]. One of the methods used is a quasi-static method. The method is used to simplify the free surface of the liquid by describing it as a straight line related to the rolling or pitching motion and resulting lateral acceleration [5].

One of the components in the cargo hold of a ship that is subjected to sloshing loads is the bulkhead structure. A bulkhead is part of a ship that divides one compartment from another [6]. Longitudinal bulkheads are often subjected to sloshing loads due to the rolling movement and transverse bulkheads are mostly subjected to sloshing loads due to pitching motion. An analytical approach is quite difficult to estimate the force that arises due to sloshing loads around the resonant frequency, especially when the resonance occurs with a large excitation amplitude [7]. The linear sloshing theory is very limited to low amplitude and excitation frequencies. Besides that, the linear theory is merely used to estimate the response in the resonance area [8]. On the other hand, the non-linear sloshing theory can be used to estimate the force of dynamic pressure due to sloshing loads on the tank or cargo hold. The non-linear sloshing theory is very common to apply on small excitation amplitudes and general tank shapes [9, 10]. In this study, CFD and numerical analysis are performed to estimate the structural behavior of the bulkhead structure of the tanker due to sloshing motions. In the first part, CFD simulation is carried out to obtain the maximum pressure on the bulkhead structure due to the sloshing motion with various filling levels in the cargo tank [11]. The maximum pressure is taken from the 20-s first of the sloshing motion. The second part, a numerical examination, is carried out using the CFD software ANSYS. The finite element analysis is very widely used to estimate structural strength by evaluating the maximum structural response behavior to retain the loads due to sloshing motion [12–16]. In order to obtain the appropriate structural behavior, the overall bulkhead structures are modeled by using a suitable finite element model. The maximum pressure resulting from CFD simulation is then applied to the bulkhead structure to obtain the maximum stress and deformation on the structure.

2 Methodology

Several filling levels including 10%, 30%, 50%, 70%, and 90% are considered in the CFD simulation in order to obtain the maximum pressure due to sloshing motion on the bulkhead structure. The bulkhead structure to be studied is bulkhead no. 2 at frame number 78. The position of the bulkhead is between cargo tank (A) and cargo tank (B) as shown in Fig. 1 bordered by the red line. The bulkhead structural design on the tanker ship is shown in Fig. 2. The ship's principal dimension and the cargo tank's dimension are given in Tables 1 and 2, respectively.

For the sloshing case of the tank, the most dominant motions of the ship to be considered were only surge, heave, and pitch. In the CFD analysis, the operational conditions such as acceleration of gravity, fluid density, and the temperature of fluid were determined according to fluid characteristics. The fluid boundary condition of each bulkhead was specified with zero velocity conditions. Meshing was modeled into the dynamic mesh using a layering method and the dynamic mesh zone was set up on the rigid body of the bulkhead. The sloshing analysis for all various filling levels is illustrated in Fig. 3. Sloshing simulation results were presented in terms of the pressure value per unit of time step on the bulkhead. A sloshing characteristic can be seen from the pressure transition on the bulkhead for each time step. The value of the maximum pressure and its location on the bulkhead might be appropriately predicted. The pressure value on bulkhead no. 2 taken from the 20-s first of sloshing motion on the filling level of 30% is shown in Fig. 4.

In the structural analysis, a numerical model for the plate and stiffener on the bulkhead is generated by using shell and beam elements, respectively. Figure 5 shows the finite element model of the bulkhead structure with an element size of 100 mm. The pressure value resulting from the sloshing analysis is then employed in the

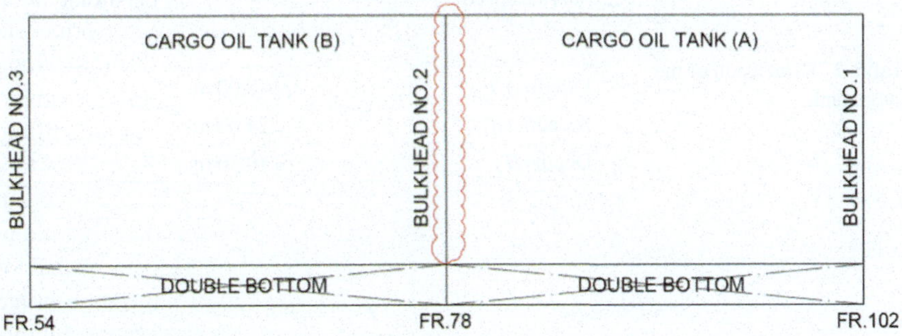

Fig. 1 Bulkhead position

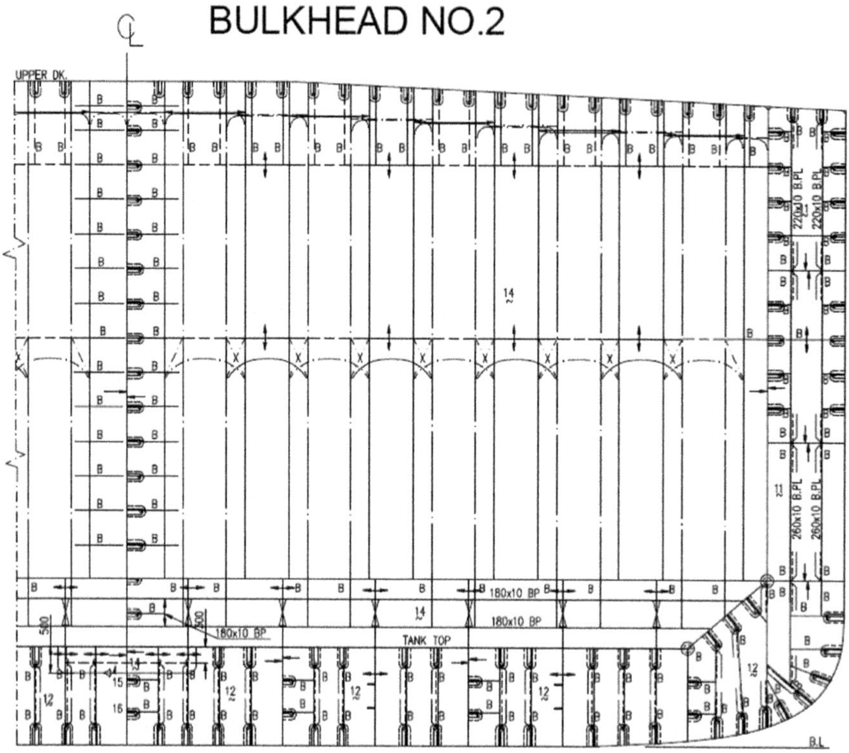

Fig. 2 Bulkhead structural design

Table 1 Principal dimension of the ship

Length overall (*Loa*)	157.50 m
Length between perpendicular (*LPP*)	149.50 m
Breadth (*B*)	27.70 m
Depth (*H*)	12.00 m
Draught (*T*)	7.00 m
Service speed	13.00 knot

Table 2 Dimension of the cargo tank

Length (*L*)	18.00 m
Breadth (*B*)	24.00 m
Depth (*H*)	10.50 m

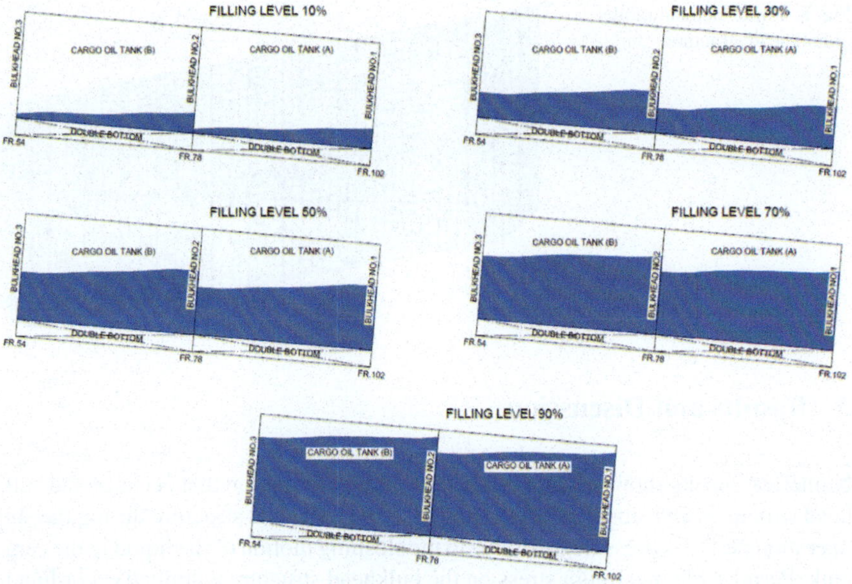

Fig. 3 Sloshing analysis for all various filling level

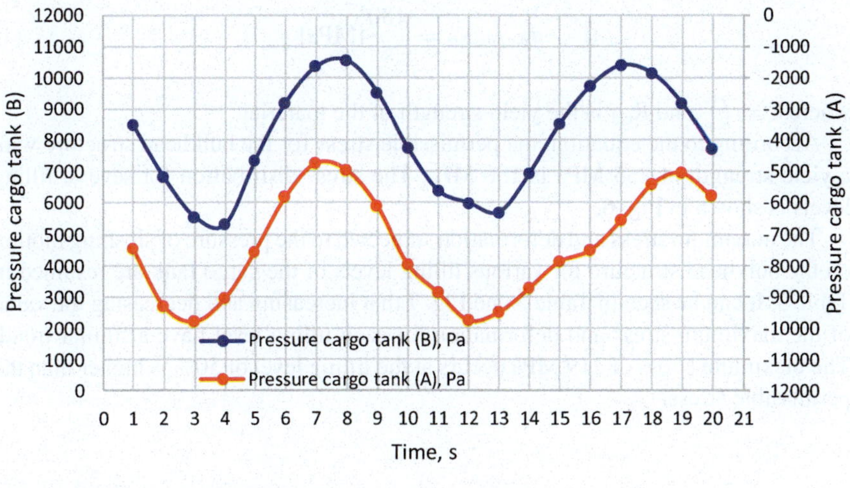

Fig. 4 Pressure value on bulkhead no. 2 for filling level of 30%

bulkhead structural analysis. In this analysis, the pressure value from the sloshing load in the cargo tank (A) and cargo tank (B) is used together since bulkhead no. 2 is subjected to the sloshing load from those cargo tanks.

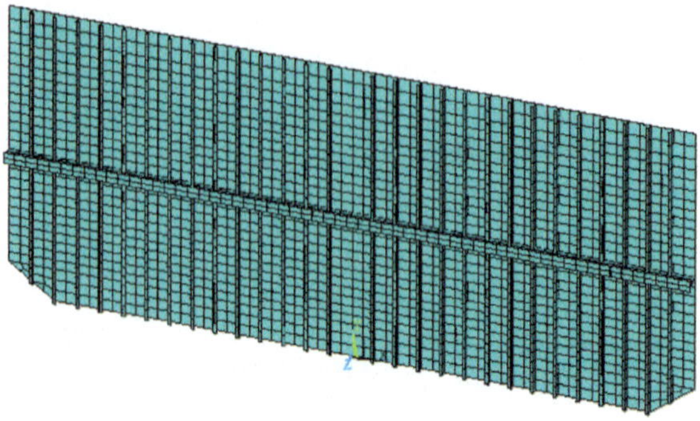

Fig. 5 Finite element model of bulkhead structure

3 Results and Discussion

Numerical results show the maximum stress and deformation that occur on the bulkhead structure. The sloshing effect applied to the bulkhead structure during the ship operation period is pressure loads due to the sloshing motion of the liquid in the cargo tank. Permissible maximum stress on the bulkhead structure is limited according to the material properties and regulation of the strength criteria [17]. The permissible stress to be used is expressed in Eq. (1).

$$\sigma_{permissible} = \frac{180}{k}[\text{MPa}] \tag{1}$$

where $k = \frac{235}{R_{eH}}$ and R_{eH} is the yield strength of the material.

According to the equation, the permissible stress for the bulkhead structure with a yield strength of 235 MPa is 180 MPa. The stress distribution for several filling levels is shown in Fig. 6.

The maximum stress and deformation subjected to the pressure of sloshing motion on the bulkhead structure for various filling levels of the cargo tank are resumed in Table 3. It can be seen in Table 3 and Fig. 7 that increasing and decreasing the value of the maximum stress and deformation for each filling level have a similar trend. The maximum stress of 219 MPa occurs at the filling level of 50% is higher than the permissible stress.

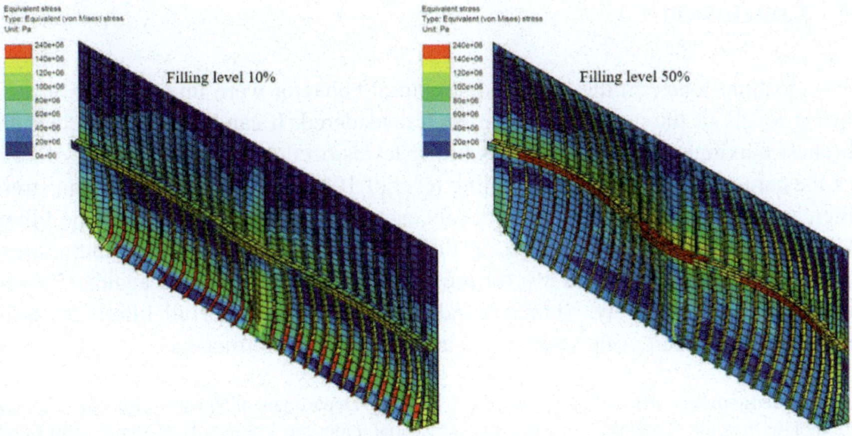

Fig. 6 Stress distribution for filling levels 10 and 50%

Table 3 Maximum stress and deformation for all filling levels

Filling level (%)	Stress (MPa)	Deformation (mm)
10	145	1.84
30	178	4.52
50	219	6.17
70	163	2.90
90	134	1.25

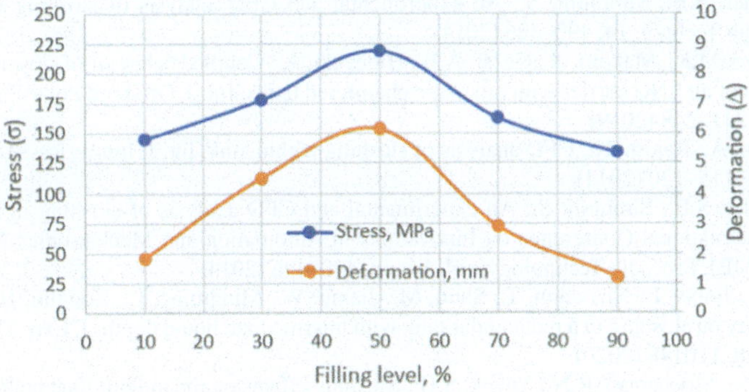

Fig. 7 Maximum stress and deformation for all filling levels

4 Conclusion

The sloshing loads on the bulkhead structural behavior were investigated. Several filling levels of the cargo tank have been considered. It can be concluded that the highest maximum stress for various filling levels occurs at the filling level of 50% of the cargo tank height. For the filling level of 10% and 30%, the maximum stress increased by 33.8% and 8.7, respectively, against the stress maximum at the filling level of 50%. On the other hand, at the higher filling level, the maximum stress decreased by 25.6% and 38.8% for the filling level 70% and 90% against 50% of tank height, respectively. Similar trends have been found for all filling levels in increasing and decreasing stress value and maximum deformation.

Acknowledgements The author would like to thank the Department of Naval Architecture, Institut Teknologi Sepuluh Nopember, for supporting financial under the scheme of the Department Fund.

References

1. Won, M.P., Dae, K.C., Kyungsoo, K., Sung, M.S., Se, H.O., Kang, H.L., Heung, S.K., Choengryul, C.: Simple analytical method for predicting the sloshing motion in a rectangular pool. Nucl. Eng. Technol. **52**, 947–955 (2020)
2. Vijayan, A., Nisthul, G.: Experimental and numerical modal analysis of sloshing in a rectangular tank. Int. J. Eng. Res. Technol. **5**(08), 227–230 (2016)
3. Yue, M.Y., Ning, M., She, M.F., Xie, C.G.: Experimental and numerical studies on sloshing in a membrane-type LNG tank with two floating plates. Ocean Eng. **129**, 217–227 (2017)
4. Gurinder, S.B., Simranjit, S.: An experimental and CFD analysis of sloshing in a tanker. Procedia Technol. **14**, 490–496 (2014)
5. Gustavo, M.K., Makoto, A., Cezar, A.B., Hideyuki, A.: Coupled behavior of sloshing and ship motion of an LNG carrier with prismatic chamfered tanks. Int. J. Offshore Polar Eng. (ISOPE) **29**(3), 241–248 (2019)
6. Roshan, A., Rajesh, K.: CFD analysis of sloshing within tank. Int. J. Innov. Res. Technol. Sci. **02**(03), 115–120 (2014)
7. Gurinder, S.B., Simranjit S.: An experimental and CFD analysis of sloshing in a tanker.In: 2nd International Conference on Innovations in Automation and Mechatronics Engineering (ICIAME), Procedia Technology, vol. 14, pp. 490–496 (2014)
8. Xin, J., Jinbo, T., Xiaochun, T., Shuo, M., Jiaxin, W., Mingming, L., Zongliu, H.: Effect of viscosity on sloshing in a rectangular tank with intermediate liquid depth. J. Exp. Therm. Fluid Sci. **118**, 110148 (2020)
9. Aries, S., Mochamad, R.N., Yoyok, S.H.: Sloshing effects on the longitudinal tank type C due to motions of the LNG Ship. J. Appl. Eng. Sci. **18**(670), 140–146 (2020)
10. Wenyuan, W., Qi, Z., Qianli, M., Li, R.: Sloshing effects under longitudinal excitation in horizontal elliptical cylindrical containers with complex baffles. J. Waterway, Port, Coastal, Ocean Eng. **144**(2), 04017044 (2018)
11. Chrismianto, D., Zakki, A.F., Arswendo, B., Abar, I.A.C.: Comparison of propeller type B-series and Au-outline Gawn series for improving on submarine propulsion performance using CFD. Int. J. Adv. Res. Eng. Technol. **10**(2), 647–651 (2019)
12. Sujiatanti, S.H., Tanaka, S., Shinkawa, S., Setoyama, Y., Yanagihara, D.: Experimental and numerical studies for buckling and collapse behaviors of a cracked thin steel panel subjected to sequential tensile and compressive loading. Thin-Walled Struct. **157**, 107059 (2020)

13. Syahab, H., Ariesta, R.C., Misbah, M.N., Zubaydi, A., Sujiatanti, S.H., Putra, W.H.A., Setyawan, D.: Structural design evaluation for underwater remotely operated vehicle (ROV), case study: Madura straits. IOP Conf. Ser. Earth Environ. Sci. **1198**(1), 012007 (2023)
14. Misbah, M.N., Sujiatanti, S.H., Setyawan, D., Ariesta, R.C., Rahmadianto, S.: Structural analysis on the block lifting in shipbuilding construction process. MATEC Web Conf. **177**, 01027 (2018)
15. Zakki, A.F., Hadi, E.S., Windyandari, A., Ilham, R.: Numerical analysis of unidirectional GFRP composite mechanical response subjected to tension load using finite element method. Int. J. Autom. Mech. Eng. **19**(2), 9790–9799 (2022)
16. Zakki, A.F., Sisworo, S.J.: Hull girder ultimate strength analysis of 17500 LTDW oil tanker under the combined global and local load using explicit finite element analysis. IOP Conf. Ser. Earth Environ. Sci. **1198**(1), 012003 (2023)
17. Biro Klasifikasi Indonesia: Rules For Classification and Construction, Rules for Hull, vol. II, Part 1: 11-8 (2022)

An Investigation of Lifting Cranes for Loading and Offloading at Offshore Terminals

Muhammad Adli Mustapa, Nur Nadhirah Rosli, and Rohaizad Hafidz Rozali

Abstract This study focuses on the investigation of lifting cranes used for loading and unloading operations at offshore terminals. The primary objectives encompass scrutinizing the specific types of lifting cranes employed in offshore terminals, conducting an in-depth analysis of stress distributions during lifting and loadout procedures, reviewing essential lifting criteria, assessing crane design standards, and evaluating provisions for dynamic loads within open water environments. To achieve these goals, a structural analysis computer system (SACS) software was employed as a design tool to develop the deck and jacket drawings. Furthermore, this software facilitated the execution of lifting and loadout analyses. The lifting analysis was configured to include scenarios involving one spreader bar, two spreader bars, and a spreader frame, all designed for the utilization of the ringing system. The paper delves into examining four rigging configurations featuring three distinct angles 40°, 60°, and 80°for lifting a deck structure. In the context of loadout analysis, the GAP element report was utilized to ascertain the number of iterations, and load cases for each structural member, and verify the deflected shape of the jacket. The project's comprehensive methodology involves an exploration of lifting cranes for offshore terminal operations, the application of SACS software for designing and analyzing lifting and loadout procedures, and a comprehensive study of crane design standards and their adaptations to account for dynamic loads within the challenging open water environment.

Keywords Loading and offloading · SACS · Lifting and loadout · Four rigging configurations · Crane design standards

M. A. Mustapa (✉) · N. N. Rosli · R. H. Rozali
Maritime Engineering Technology, University Kuala Lumpur Malaysian Institute of Marine Engineering Technology, Lumut, Perak, Malaysia
e-mail: adli@unikl.edu.my

N. N. Rosli
e-mail: nnadhirah.rosli@s.unikl.edu.my

R. H. Rozali
e-mail: rohaizad@unikl.edu.my

A. Ismail et al. (eds.), *Engineering Frontiers*, Advanced Structured Materials 208,
https://doi.org/10.1007/978-3-031-56844-2_34

1 Introduction

Carrying out lifting operations to load and unload equipment from supply boats is a routine task at offshore terminals. Hence, meticulous planning and thorough investigations of lifting cranes become paramount to ensure safe and cost-effective project execution. Offshore cranes, as fixed structures, cannot move alongside their loads. The surge in oil and gas demands has propelled offshore activities, often accompanied by intricate challenges. Making an inappropriate choice of offshore crane type can lead to crane failures. Neglecting proper inspection, maintenance, and usage protocols can result in crane accidents, potentially causing fatalities and severe injuries.

The primary consequence of crane accidents is improper lifting practices. This occurs when loads are raised at an inclined angle, excessive loading at the end of the crane arm, and the utilization of components that fall short of required technical specifications. Despite governmental regulations in place since 2000, the frequency of crane accidents remains concerning. Globally, there have been 1125 recorded crane accidents resulting in 780 fatalities. This underscores the significance of this research, which compiles comprehensive investigations into the lifting cranes utilized for loading and unloading operations at offshore terminals. Various types of offshore cranes are discussed, including board offshore cranes, compact board offshore cranes, offshore heavy lift cranes, crane around the leg, mast-type cranes, luffing cranes, knuckle boom cranes offshore, and offshore pedestal cranes.

1.1 Lifting Criteria

Numerous lifting criteria and precise specifications have been formulated exclusively for demanding offshore heavy lifts. These include API-RP2A (2000) [1], DNV Marine Operation Part 2 Recommended Practice RP5 (1996), Phillips Petroleum (1989), Heerema (1991), Noble Denton & Associates (NDA) (1996), Health and Safety Executive UK (HSE) (1992), and Shell (1990) [2].

In the case of a single crane, the overall lift point design factor (K) prescribed by API RP 2A (2000) is 2.00, as presented in Table 1. Similarly, for a double hook lift, the overall lift point design factor (K) outlined in API RP 2A (2000) is also 2.00, as detailed in Table 2 [1].

1.2 Lifting Components

Lifting operations involve a range of crucial components. Cranes are divided into static and mobile variants, each serving distinct purposes. Lifts and cranes rely on slings, which can be crafted from cable, chain, rope, or webbing, to ensure the proper

Table 1 Lifting criteria comparison—single crane lift [2]

	Noble Denton	DNV	Heerema	Shell	BP	Oxy	Amoco	Chevron
Range of module weight	>2500	>500	>2500	>1000	>2500	>2500	>2500	>2500
A. Weight factor (Pre-AFC)	1.125	1.10	1.15	1.15	1.15	1.15	1.15	1.15
B. DAF (Slings)	1.10	1.10	1.10	1.10	1.10	1.10	1.10	1.10
C. Skew load factor	1.25	1.25	1.50	1.50	1.50	1.50	1.25	1.50
D. CG Shift factor	1.05	1.00	1.00	1.05	1.00	1.00	1.05	1.00
E. Tilt factor	1.00	1.00	1.00	1.00	1.00	1.00	1.00	1.05
$F = A \times B \times C \times D \times E$	1.62	1.51	1.90	1.99	1.90	1.90	1.66	1.99
G. Rigging weight factor	1.03	1.03	1.03	1.03	1.03	1.03	1.03	1.03
H. Lift point design factor	1.35	1.35	1.00	1.00	1.25	1.30	1.30	1.30
I. Load member design factor	1.15	1.15	1.00	1.00	1.10	1.15	1.15	1.15
J. Sling design $= (F \times G)$	1.67	1.56	1.95	2.05	1.95	1.95	1.71	2.05
K. Lift point design $= (F \times H)$	2.19	2.04	1.90	1.99	2.37	2.47	2.16	2.59
L. Load member design $= (F \times I)$	1.87	1.74	1.90	1.99	2.09	2.18	1.91	2.29

elevation and equilibrium of weights. Lifting shackles act as essential connection links, enabling a secure linkage between the load-lifting equipment and the designated load. Additionally, padeyes function as attachment points, with applications extending to both general lifting activities and specialized contexts like oil and gas projects.

Table 2 Lifting criteria comparison—double hook lift [2]

	Noble Denton	LOC	Heerema	Chevron	BP	Amoco
Range of module weight	>2500	>1000	>2500	>2500	>8000	>2500
A. Weight factor (Pre-AFC)	1.125	1.15	1.15	1.25	1.15	1.15
B. DAF (Slings)	1.10	1.10	1.10	1.10	1.10	1.10
C. CG shift factor	1.03	1.05	1.05	1.05	1.08	1.05
D. Tilt factor	1.03	1.03	1.03	1.03	1.03	1.03
E. Yaw factor	1.05	1.05	1.05	1.00	1.00	1.05
F. Torsion factor	1.00	1.00	1.00	1.00	1.00	1.10
G. Skew factor	1.00	1.10	1.00	1 10	1.10	1.00
$H = A \times B \times C \times D \times E \times F \times G$	1.38	1.58	1.44	1.64	1.55	1.58
I. Rigging weight factor	1.03	1.03	1.03	1.03	1.03	1.00
J. Lift point design factor	1.35	1.00	1.10	1.30	1.25	1.35
K. Load member design factor	1.15	1.00	1.10	1.15	1.10	1.15
L. Sling design = $(H \times I)$	1.42	1.63	1.48	1.68	1.59	1.58
M. Lift point design = $(H \times J)$	1.86	1.58	1.58	2.13	1.93	2.13
N. Load member design = $(H \times K)$	1.59	1.58	1.58	1.88	1.70	1.82

1.3 Crane Design Standards and Allowances Made for Dynamic Loadings in Open Water Environments

Design Guidelines

In accordance with design guidelines established by various organizations, a clear distinction is maintained between shipboard cranes and offshore cranes. The design factor offers insight into the divergence across different codes and functions, ranging from a peak of 1.17 for API to a minimum of 0.77 for DNV (OCIMF) [3]. A concise overview of the design guidelines for API, ABS, DNV, and Lloyd's is presented in Table 3.

Design Factors

The design parameters for offshore hose cranes consider the specific duty and stress factors outlined in the relevant guidelines. When evaluating shipboard harbour criteria, the hose crane design factor used for offshore operations corresponds to the

Table 3 Summary of design guidelines

Summary of design guidelines	API	ABS	DNV	Lloyd's
	API is offshore only			
Shipboard cranes				
Minimum vertical dynamic factor	–	1.00	1.15	1.10
Basic allowable stress/yield stress				
Without wind forces	–	NA	0.67	0.67
With wind forces	–	0.50	0.75	0.75
Duty factor	–	None (1.00)	None (1.00)	1.05
Heel/trim assumed, deg	–	By designer	5/2	5/2
Overall design factor				
Without wind forces	–	NA	0.86	0.86
With wind forces	–	1.00	0.77	0.77
Shock loads	–	Specified by user	Required	Required
Fatigue analysis	–	Not mentioned	Required	Trade-off w/duty factor
Offshore cranes		ABS adopts API		
Minimum vertical dynamic factor	1.40	1.40	1.30	1.10
Basic allowable stress/yield stress				
Without wind forces	NA	NA	0.67	0.67
With wind forces	0.60	0.60	0.75	0.75
Duty futor	None (1.00)	None (1.00)	None (1.00)	1.20
Heel/trim assumed, deg	By designer	By designer	5/2	5/2
Overall design factor				
Without wind forces	NA	NA	0.97	0.99
With wind forces	1.17	1.17	0.87	0.88
Shock loads	Required	Required	Required	Required
Fatigue analysis	Required	Required	Required	Trade-off w/duty factor

DNV (harbour) design factor from OCIMF [3]. The DNV factors align with the DNV (harbour) design factor. A comparative analysis of design factors among API, ABS, DNV, and Lloyd's is presented in Table 4.

2 Methodology

The overall approach employed to carry out this study is outlined in Fig. 1.

2.1 *Design Aspect*

The design process encompasses two distinct phases: the development of the deck structure and the jacket structure. Similarly, the analysis process is divided into two stages: conducting the lifting analysis and performing the loadout analysis. The design of the deck, jacket, and lifting configurations is executed to accurately ascertain intricate stress distributions through meticulous lifting and loadout analysis. This design and analysis procedure was facilitated by the utilization of the structural analysis computer system (SACS) software.

In the literature review, data extracted from previous research papers, theses, and relevant journals is employed as secondary information to establish connections, contextualize the study, and obtain conclusive results.

- **Deck Design**

  1. Main Deck elevate (+) 22 m
  2. Wellhead Mezzanine Deck elevate (+) 17 m
  3. Cellar Deck elevate (+) 14 m
  4. Subcellar Deck Elev. (+) 11 m.

The image of the deck platform structure constructed in SACS Software is shown in Fig. 2.

- **Jacket Design**

  1. Water depth: 79.5 m
  2. Working point elevation: 4.0 m
  3. Pile connecting elevation: 3.0 m
  4. Number of legs: 4 Leg spacing at working point: X1 = 15 m, Y1 = 10 m.

The image of the jacket structure constructed in SACS Software is shown in Fig. 3.

Table 4 Comparison of design factors

Comparison of design factors from various design guidelines	API API is offshore only	ABS	DNV	Lloyd's
Shipboard cranes				
Minimum vertical dynamic factor	–	1.00	1.15	1.10
Basic allowable stress/yield stress				
Without wind forces	–	NA	0.67	0.67
With wind forces	–	0.50	0.75	0.75
Duty factor	–	None (1.00)	None (1.00)	1.05
Heel/trim assumed, deg	–	By designer	5/2	5/2
Overall design factor				
Design factor without wind forces	–	NA	0.86	0.86
Design factor with wind forces	–	1.00	0.77	0.77
Hose crane design factor assuming boom tip acceleration = 1.15 g				
Vertical dynamic factor	1.15	1.15	1.15	1.15
Design factor without wind forces	–	NA	0.86	0.9
Design factor with wind forces	–	1.15	0.77	0.81
Offshore cranes		ABS adopts API		
Minimum vertical dynamic factor	1.40	1.40	1.30	1.10
Basic allowable stress/yield stress				
Without wind forces	NA	NA	0.67	0.67
With wind forces	0.60	0.60	0.75	0.75
Duty Factor	None (1.00)	None (1.00)	None (1.00)	1.20
Overall design factor				
Without wind forces	NA	NA	0.97	0.99
With wind forces	1.17	1.17	0.87	0.88
Hose crane design factor assuming boom tip acceleration = 1.15 g				
Vertical dynamic factor	1.15	1.15	1.15	1.15
Design factor without wind forces	NA	NA	0.86	1.03
Design factor with wind forces	0.96	0.96	0.77	0.92

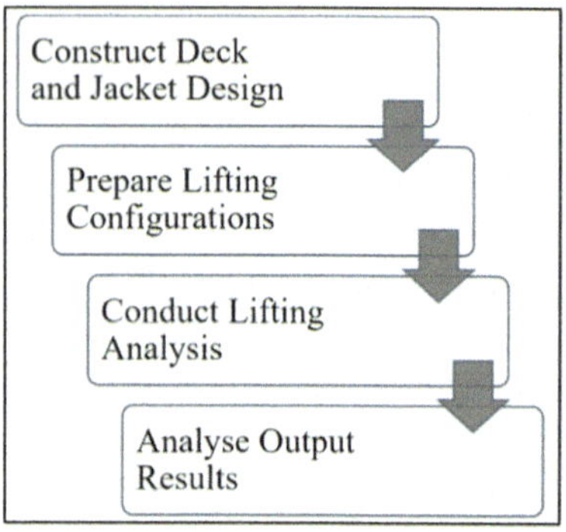

Fig. 1 Research flowchart

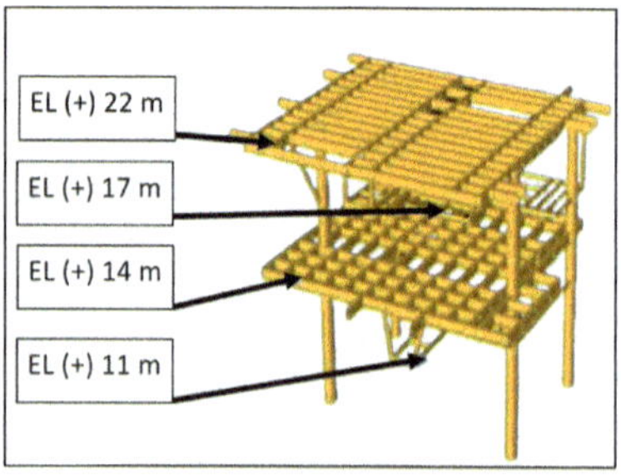

Fig. 2 Deck platform structure in SACS software

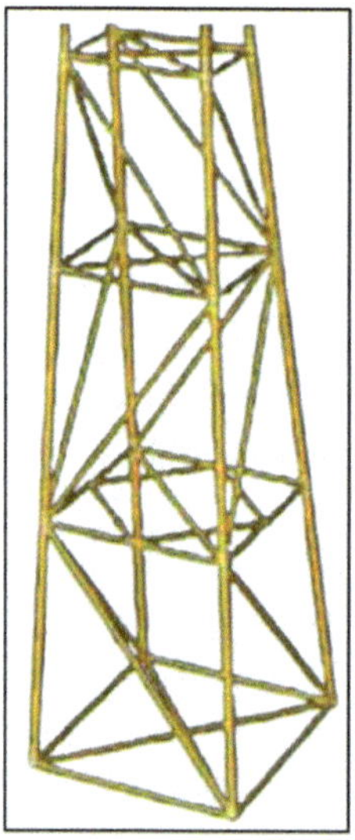

Fig. 3 Jacket structure in SACS software

2.2 *Lifting Analysis*

The lifting analysis serves the purpose of establishing the maximum permissible design sea state within which the structure can be securely installed. Essential lifting calculations, including sling angle and load distribution across each leg, are depicted in Figs. 4 and 5, respectively.

In adherence to the DNV marine lifting standard and Noble Denton code, a minimum lifting angle of 60° is employed [4]. This study introduces three distinct lift angles—namely 40°, 60°, and 80°—to comparatively assess the most suitable angle for the lifting procedure. Furthermore, four lifting configurations are investigated, encompassing lifting without a spreader bar, lifting with one spreader bar, lifting with two spreader bars, and lifting with a spreader frame. The lifting configurations for the 60° angle, simulated using the SACS Software, are illustrated in Figs. 6, 7, 8 and 9.

- **The Maximum Unity Check Ratio for The Spreader Bar for 60°**.

Figure 10 shows the spreader bar arrangement.

H / L = Sling Angle Factor

Sling Angle (A)	Sling Angle Factor	Sling Angle (A)	Sling Angle Factor
90°	1.000	55°	0.819
85°	0.996	50°	0.766
80°	0.985	45°	0.707
75°	0.966	40°	0.643
70°	0.940	35°	0.574
65°	0.906	30°	0.500
60°	0.866	-	-

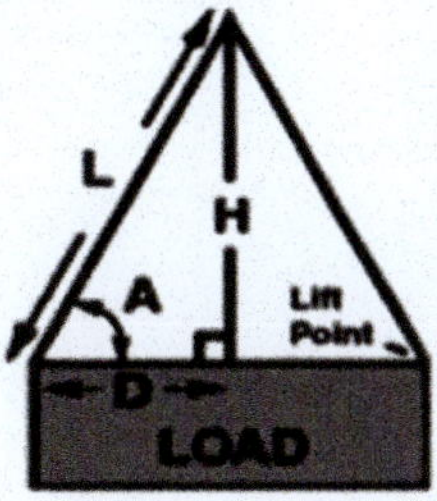

Fig. 4 Determine sling angle

Fig. 5 Determine load on each leg with unequal legs

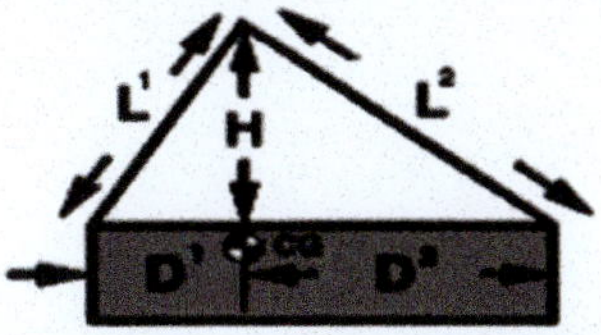

$$\text{Load on Leg } L^1 = \frac{\text{Load (lbs)} \times D^2 \times L^1}{H \times (D^1 + D^2)}$$

$$\text{Load on Leg } L^2 = \frac{\text{Load (lbs)} \times D^1 \times L^2}{H \times (D^1 + D^2)}$$

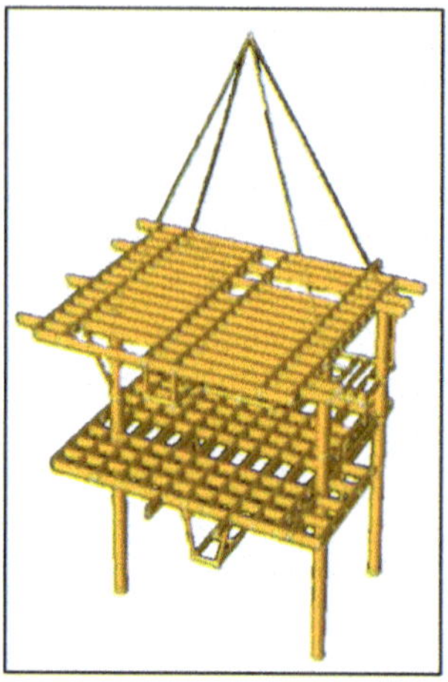

Fig. 6 Without spreader bar

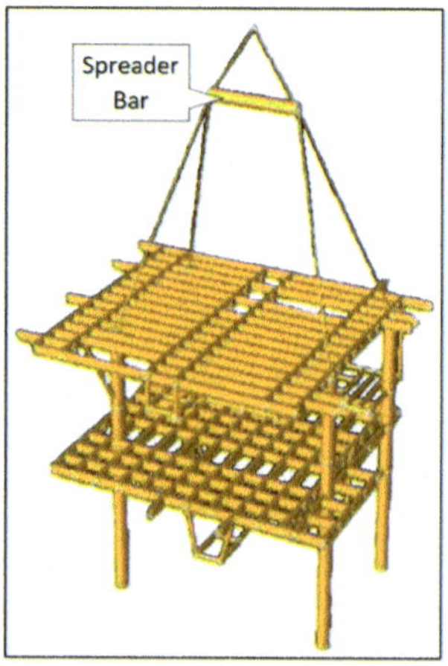

Fig. 7 With one spreader bar

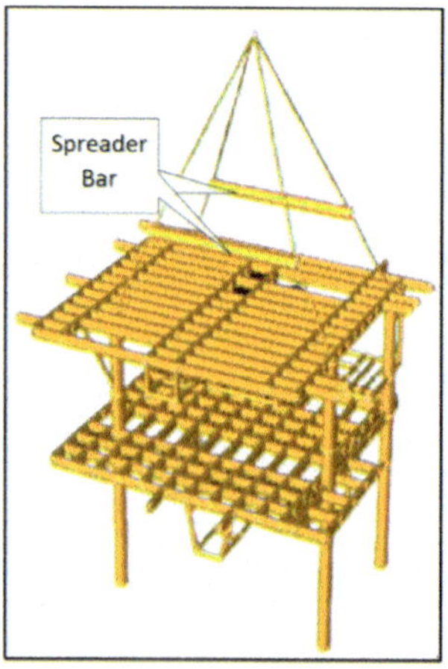

Fig. 8 With two spreader bars

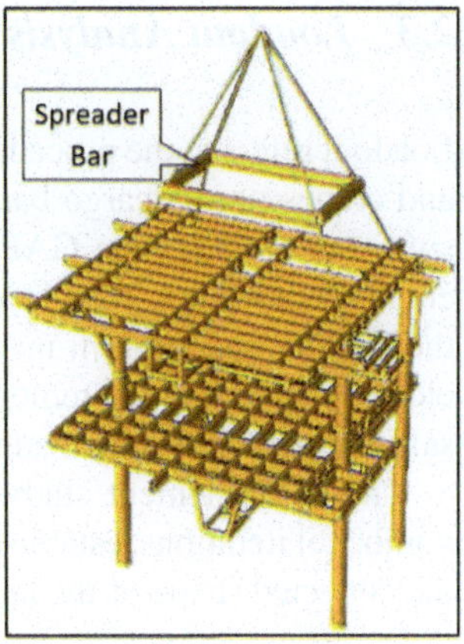

Fig. 9 With spreader frame

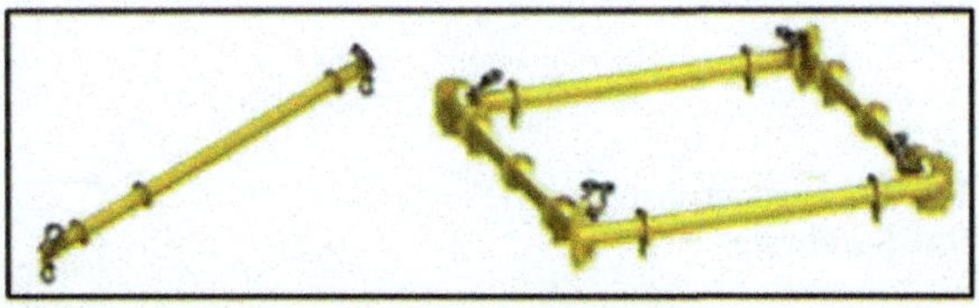

Fig. 10 Spreader bar

- **The Maximum Sling Force on Padeye (kN) for 40°, 60°, and 80°.**

Figure 11 shows the padeye arrangement.

- **The Maximum Sling Force on Hook Point (kN) for 40°, 60°, and 80°.**

Figure 12 shows the hook point arrangement.

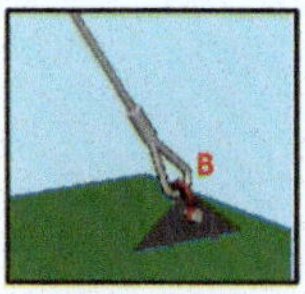

Fig. 11 Padeye

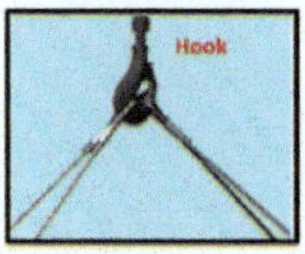

Fig. 12 Hook point

2.3 Loadout Analysis

Loadout refers to the procedure of transferring a structure from land-based skidways and cranes onto a cargo barge. The loadout analysis, conducted within the SACS software, employs the GAP element method, illustrated in Fig. 13. This analysis entails defining non-load-bearing elements to replicate the absence of supports during the loadout operation. It involves evaluating the stress condition within a structural element caused by the removal of one or more supports, along with determining the safest approach for the loadout, as depicted in Fig. 14.

The GAP element study encompasses key aspects, including determining the number of iterations, establishing load cases for individual members, and scrutinizing the deflected shape of the jacket as part of the loadout process.

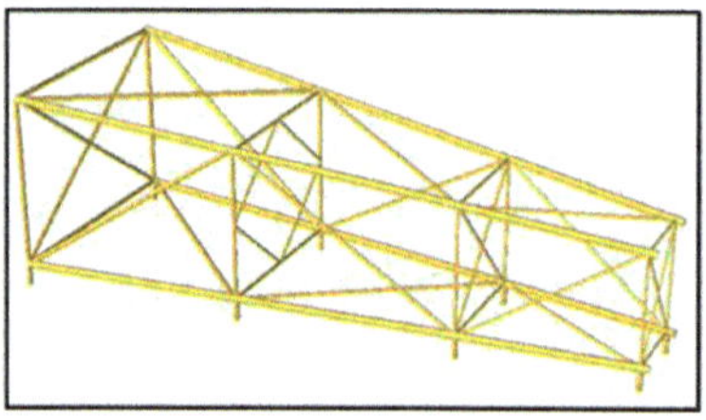

Fig. 13 Loadout analysis in SACS software

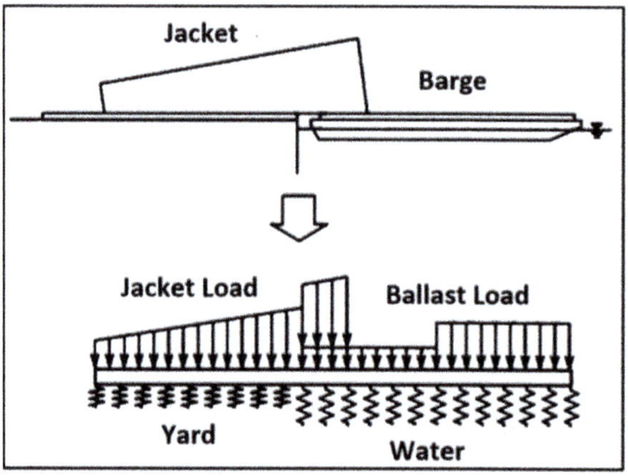

Fig. 14 Loadout process

3 Results and Discussion

3.1 Lifting Analysis

- **The Maximum Unity Check Ratio for The Spreader Bar for 60°**

The unity check assesses the capacity of deck structure members to endure stresses arising during installation via the lifting method. The maximum unity check at a 60° angle is detailed in Table 5 and Fig. 15. Analysis of the graph leads to the observation that the unity check value is lower for the one spreader bar configuration in comparison to the spreader frame and two spreader bars configurations. Furthermore, all unity check values for various rigging configurations remain below one, indicating that tension generated in each lifting scenario does not surpass the permissible stress limit of the structure.

- **The Maximum Sling Force on Padeye (kN) for 40°, 60° and 80°**

Comprehensive findings regarding the maximum sling force applied to the padeye (in kN) for the 40°, 60°, and 80° angles are elaborated in Table 6 and visually depicted in Fig. 16.

From the depicted graph in Fig. 16, it can be inferred that the lifting process lacking a spreader bar yields a comparatively lower sling force measurement in contrast to the

Table 5 Maximum unity check ratio for 60°

	Maximum unity check ratio for 60°
With one spreader bar	0.0600
With two spreader bars	0.1008
With spreader frame	0.0955

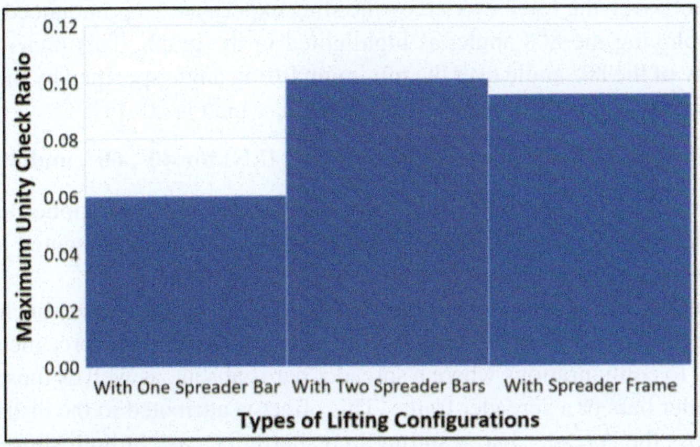

Fig. 15 Maximum unity check ratio for 60°

Table 6 Maximum sling force on padeye (kN) for 40°, 60°, and 80°

	40°	60°	80°
Without spreader bar	4547.0	3909.5	3591.2
With one spreader bar	4040.3	3379.2	3153.9
With two spreader bars	3345.3	3277.8	3157.4
With spreader frame	3772.2	3628.9	3403.4

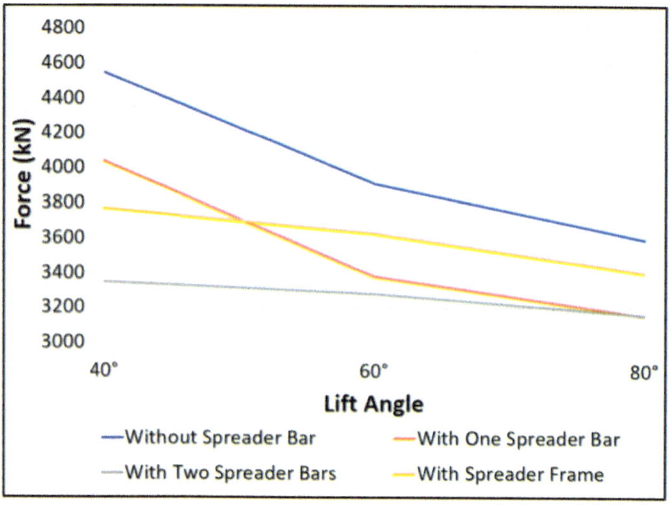

Fig. 16 Maximum sling force on padeye (kN) for 40°, 60°, and 80°

configurations involving one spreader bar, two spreader bars, and a spreader frame. This discrepancy can be attributed to the manner in which force is distributed across the spreader bar, resulting in a significant reduction in sling force as well as tension values. Moreover, the force exerted on the sling experiences a pronounced reduction when employing the 80° angle, as highlighted in the graph. This underscores the superiority of the 80° angle over the minimum lifting angle specified by the standard codes of DNV marine lifting and Noble Denton, which is 60° [4].

- **The Maximum Sling Force on Hook Point (kN) for 40°, 60°, and 80°**

Comprehensive outcomes pertaining to the maximum sling force applied at the hook point (in kN) for the 40°, 60°, and 80° angles are meticulously presented in Table 7, accompanied by a graphical representation in Fig. 17.

The interpretation of the graph presented in Fig. 17 leads to the conclusion that the lifting process involving a spreader bar yields a reduced sling force measurement compared to configurations where a spreader bar is absent, as well as those utilizing two spreader bars or a spreader frame. This effect is attributed to the distribution of force across the spreader bar, resulting in a sharp decrease in both sling force and tension values. Notably, when employing the 80° angle, the force exerted on the sling

Table 7 Maximum sling force on hook point (kN) for 40°, 60°, and 80°

	40°	60°	80°
Without spreader bar	4547.0	3909.5	3591.2
With one spreader bar	9190.0	6404.5	5382.6
With two spreader bars	4208.0	3443.5	3166.9
With spreader frame	4764.0	3535.5	2980.5

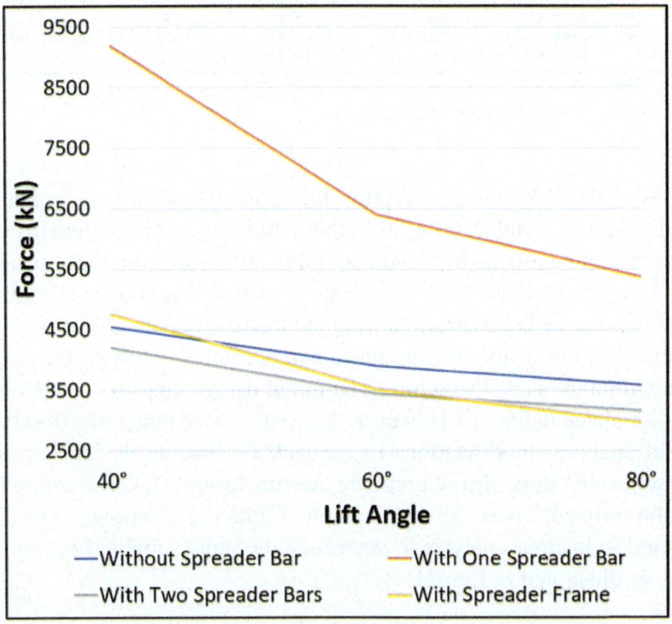

Fig. 17 Maximum sling force on hook point (kN) for 40°, 60°, and 80°

experiences a significant reduction, as visually evident in the graph. This observation underscores the superior performance of the 80° angles over the minimum lifting angle stipulated by the standard codes of DNV marine lifting and Noble Denton, which is set at 60° [4].

3.2 *Loadout Analysis*

Loadout analysis involves defining non-load-bearing elements to replicate the loss of supports during the loadout procedure. It encompasses the evaluation of stress conditions within a structural element during the loadout process, considering the absence of one or more supports, and identifying the safest loadout approach. The results derived from the GAP element report are depicted in Fig. 18.

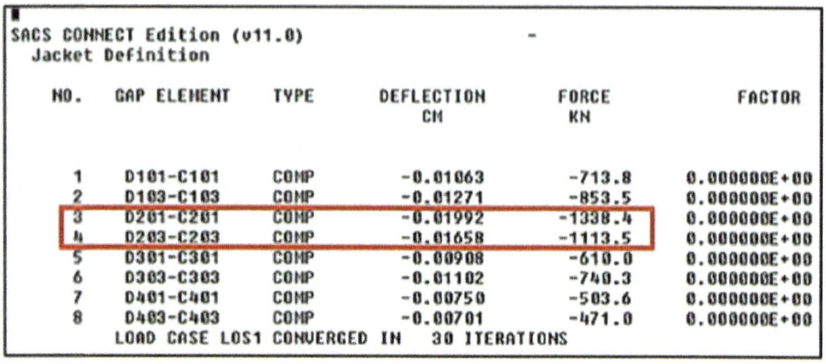

SACS CONNECT Edition (v11.0)
Jacket Definition

NO.	GAP ELEMENT	TYPE	DEFLECTION CM	FORCE KN	FACTOR
1	D101-C101	COMP	-0.01063	-713.8	0.000000E+00
2	D103-C103	COMP	-0.01271	-853.5	0.000000E+00
3	D201-C201	COMP	-0.01992	-1338.4	0.000000E+00
4	D203-C203	COMP	-0.01658	-1113.5	0.000000E+00
5	D301-C301	COMP	-0.00908	-610.0	0.000000E+00
6	D303-C303	COMP	-0.01102	-740.3	0.000000E+00
7	D401-C401	COMP	-0.00750	-503.6	0.000000E+00
8	D403-C403	COMP	-0.00701	-471.0	0.000000E+00

LOAD CASE LOS1 CONVERGED IN 30 ITERATIONS

Fig. 18 Result from GAP element report

Analysis of the GAP element report highlights that within the jacket structure, members D201–C201 and D203–C203 exhibit the highest force values in comparison to other structural members, as illustrated in Fig. 19. Moreover, these structures also exhibit the most significant deflection values, which can potentially influence the deformation of the jacket structure during the loadout process.

The configuration of the bracing has a notable impact on both force values and deflection within the jacket structure. The initial design employs a rhombus-shaped bracing, as depicted in Fig. 20. However, due to increased deflection observed during the loadout analysis, modifications to the jacket's bracing design become necessary. Consequently, three distinct bracing designs for the jacket structure have been selected and utilized. These designs, namely X-bracing, Z-bracing, and K-bracing, are subjected to loadout analysis to determine the most suitable bracing design for the jacket, as illustrated in Fig. 21.

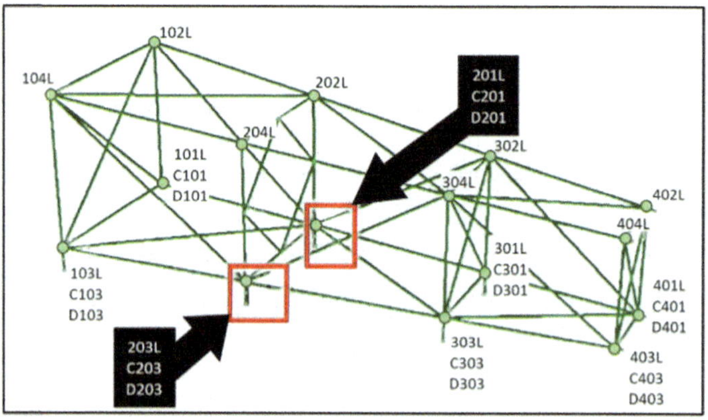

Fig. 19 D201–C201 and D203–C203 in jacket structure

Fig. 20 Rhombus bracing (initial design) in SACS software

Fig. 21 X, Z and K-bracing in SACS software

- **Result from GAP element study for every bracing.**

Comprehensive findings regarding the correlation between the type of bracing and the corresponding force values (in kN) for the D201–C201 section are meticulously presented in Table 8, accompanied by a graphical representation in Fig. 22.

Based on the graph in Fig. 22, it can be inferred that the deflection experienced by the jacket structure when employing a K-bracing design yields lower force and deflection values compared to the Rhombus, X, and Z-bracing alternatives. Notably, the force value exerted on member D201-C201 witnesses a significant reduction when the K-bracing design is utilized, as vividly illustrated in Fig. 22. This phenomenon

Table 8 Types of bracing versus force (kN) for D201–C201

Type of bracing	Force (kN)
(1) Rhombus bracing (initial design)	1338.4
(2) X-bracing	1305.2
(3) Z-bracing	1199.1
(4) K-bracing	1181.5

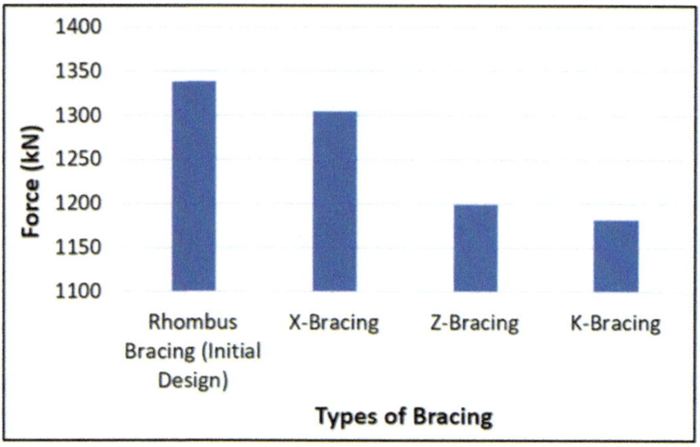

Fig. 22 Types of bracing versus force (kN) for D201-C201

can be attributed to the dispersion of force across the structure, leading to a marked decrease in force and tension values.

4 Conclusion

The study explores the investigation of lifting cranes' utilization for loading and unloading operations at offshore terminals. The primary aim of this research is to examine the application of lifting cranes in the context of cargo handling at offshore terminals. To address the study's objectives, a three-phase approach was undertaken. The initial phase involves an in-depth exploration of lifting cranes to pinpoint challenges encountered in the offshore industry concerning loading and unloading activities at offshore terminals. The subsequent phase entails the development of both the deck and jacket structures, achieved through an extensive literature review and research, culminating in lifting and loadout analyses facilitated by the SACS Software. The final phase is dedicated to evaluating crane design standards and their responsiveness to dynamic loadings within open water environments. Based on the obtained results, the study provides conclusions and recommendations. The conclusion is bifurcated into two key components. Firstly, it encompasses the findings derived from the lifting analysis and loadout analysis. The second component presents an overarching summary of the study's overall conclusions.

Acknowledgements The author would like to thank the Universiti Kuala Lumpur-Malaysian Institute of Marine Engineering Technology (UniKL-MIMET) for supporting the equipment and support.

References

1. API: Recommended Practice for Planning, Designing and Constructing Fixed Offshore Platforms—Working Stress Design. API recommended practice 2A-WSD (RP 21-WSD) (2000)
2. Liang, L.: Heavy lift installation study of offshore structures. Dissertation, National University of Singapore (2004)
3. OCIMF: A study into crane loads associated with hose handling at offshore terminals. https://www.ocimf.org/publications/information-papers/a-study-into-crane-loads-associated-with-hose-handling-at-offshore-terminals (2015). Accessed 8 Feb 2023
4. DNV: Standard for certification of lifting appliances, 1-114 (2011)

New Opening Inlet Design of the Pico-Hydropower System

Muhammad Adli Mustapa, Khairul Azhar Norazman, and Iwan Mustaffa Kamal

Abstract This research focuses on designing and assessing an innovative inlet opening for the pico hydro power (PHP) system, aimed at enhancing its overall performance. The proposed approach involves adapting a similar inlet slope design with a 45° angle while introducing modifications to the inlet configuration. Computational fluid dynamics (CFD) software was utilized to conduct the investigation, with the primary goal of optimizing energy extraction from a Malaysian river. The study examines four distinct incoming velocities: 0.1, 0.3, 0.6, and 2.0 m/s. The study's outcomes are presented in terms of velocity magnitude at a critical point, velocity reduction percentage, power estimation, and design efficiency. The findings reveal that among the evaluated designs, design 3 exhibits superior performance compared to designs 1 and 2 in terms of velocity magnitude at the critical location. Notably, design 3 outperforms the others in power production estimation and device efficiency, achieving an efficiency range of 16–26%.

Keywords Pico hydro power (PHP) · CFD software · Computational fluid dynamics (CFD) simulation · Opening inlet of PHP · Velocity magnitude

1 Introduction

1.1 General

Malaysia's energy production still relies on non-renewable sources such as crude oil, coal, and natural gas. With global energy consumption projected to outpace population growth, every nation needs to transition towards renewable energy sources.

M. A. Mustapa (✉) · K. A. Norazman · I. M. Kamal
Maritime Engineering Technology, Universiti Kuala Lumpur Malaysian Institute of Marine Engineering Technology, Lumut, Perak, Malaysia
e-mail: adli@unikl.edu.my

I. M. Kamal
e-mail: iwanzamil@unikl.edu.my

A. Ismail et al. (eds.), *Engineering Frontiers*, Advanced Structured Materials 208,
https://doi.org/10.1007/978-3-031-56844-2_35

Renewable energy, often referred to as clean energy, originates from continually replenishing natural sources or processes [1]. On April 2, 2019, the International Renewable Energy Agency published a report titled 'Renewable Capacity Statistics 2019,' highlighting a substantial growth of 171 GW in renewable power. This surge was primarily propelled by the addition of solar and wind energy, contributing 84% of the increase [2]. Malaysia has effectively harnessed hydro power (HP) installations of varying scales for electricity generation. Pico hydro power (PHP) systems have emerged as an optimal choice for electricity production in rural areas and have gained notable attention [3]. Additionally, PHP stands as the most compact option for renewable energy devices, boasting small infrastructure and operating costs, setting it apart from other available off-grid power generation methods [4]. PHP holds the potential to elevate the living standards of individuals residing in remote regions across the globe [5].

1.2 Hydropower

Hydropower technology showcases impressive versatility and adaptability, capable of providing electricity to individual households on a small scale, while also addressing the energy needs of diverse sectors, the general populace, and even entire nations at the national or regional level. Among the top ten power stations worldwide in terms of generating capacity, eight are installations utilizing hydropower. Currently, there exist four primary categories of hydropower: river-based, storage-based, pumped storage, and offshore installations [6]. As outlined in [7], critical components of a hydroelectric plant encompass the forebay, intake structure, penstock, surge chamber, hydraulic turbines, powerhouse, draft tube, and tailrace.

1.3 Gravitational Water Vortex Power Plant

A groundbreaking method of generating hydropower involves utilizing an ultra-low head turbine, which demonstrates efficient performance even with as little as a 0.7 m head. This innovative technology gives rise to a gravitational water vortex power plant, as outlined in [8]. Presently categorized as micro hydropower due to its capacity falling below 100 kW, the gravitational water vortex mechanism operates by guiding water through an inlet channel into a circular basin, where it acquires tangential momentum and forms a potent vortex, as explained in [9]. Activation of the device is initiated as water enters the inlet channel and is directed tangentially into the circular basin, resulting in the creation of a forceful vortex. The vortex is subsequently channelled through an outlet hole at the base of the basin for release. Importantly, the turbine's functionality is dependent on the dynamic force generated by the vortex, rather than differential pressure, a distinguishing feature discussed in [8]. This unique characteristic contributes to relatively lower expenses for development and power generation compared to other hydropower technologies. While hydropower plants are generally recognized as environmentally friendly in terms of air pollution, their impact primarily pertains to water quality and wildlife habitats, often obstructing fish migration. The gravitational water vortex power system, serving as a horizontal variant of a hydroelectric dam, holds the potential for addressing these environmental concerns.

1.4 Water Head

Basar underscores the need for the PHP generation scheme to consider two pivotal elements: water head and flow [4]. Water head encompasses either water pressure or the vertical descent of water sources, both of which are measurable. Traditionally, this head is quantified as a vertical distance (H, in meters) or as pressure (N/m^2). Conversely, water flow rate pertains to the volume of water passing through.

1.5 Power from Water

The generated output power is contingent upon the vertical water flow drop (H) and the rate of water flow (Q) [4]. The below formulas embody fundamental equations that underpin the advancement of hydropower.

$$P_{net} = \eta P_{gross} \tag{1}$$

$$P_{net} = \eta Q g H_{gross} \tag{2}$$

The weight of descending water is attributed to the gravitational force exerted by the water. This downward force, which is influenced by gravity (g), can be articulated in terms of mass (m) and acceleration (a). The ensuing equation unveils that the energy released (measured in Joules) due to water descent can be determined by the product of the vertical distance H_{gross} and the combination of mass (m) and gravitational acceleration (g), referred to as weight (mg). The formula for computing the released energy is as follows:

$$Energy\ Released = mgH_{gross} \tag{3}$$

Subsequently, the density (ρ) and the volume of water (V) substitute for the water's mass (m), resulting in the formulation presented below:

$$Energy\ Released = \rho V g H_{gross} \tag{4}$$

From formula (4), the volume (V) could be replaced with volume flow rate, Q (in m^3/s). Consequently, the generated energy can be denoted as gross power (in J/s or W). The calculation for estimating the hydro system's energy output or total power is as follows:

$$P_{gross} = \rho Q g H_{gross} \tag{5}$$

1.6 Inlet and Outlet Configuration

The incoming flow rates encompass the water that is released into a stream linked to the basin of the gravitational water vortex. The channel has the role of guiding the water into the basin with a tangential direction and can be positioned either horizontally or at an angle, depending on the necessity [9]. The width of the channel can vary or remain uniform at its two extremities. The height of the inlet carries a twofold significance: it denotes both the water's elevation and the gap between the bottom of the channel and the base of the basin.

1.7 Basin Configuration

The arrangement of the developed vortices demonstrates a range of variations contingent upon the design of the basin within the gravitational water vortex system. Earlier explorations involved the simulation of three distinctive configurations: (1) A cylindrical central outlet basin, (2) a rectangular pre-rotation basin, and (3) a cylindrical inlet basin. The results underscored that the most effective basin design was the cylindrical basin featuring an inlet guide, which yielded consistently higher and superior flow velocities compared to alternative basin layouts [10].

In the realm of historical research, the focus has predominantly centered on conical basin designs in contrast to the more conventional cylindrical basin setups. Across various basin diameters, notch angles, notch inlet widths, cone angles, and channel heights, the velocity of the vortex were quantitatively assessed [10]. Importantly, the width of the inlet, the cone angle, and the channel height wield notable influences over vortex velocity. Suggestions were made to decrease the notch inlet width while maximizing the cone angle and channel height to optimize performance. Additionally, it is advisable to sustain a longer notch length (the length of the inlet channel) to alleviate undesired losses [10].

1.8 Efficiency of Device

Achieving the optimal PHP device necessitates attaining the highest efficiency. The device's output efficiency can be computed using the subsequent formula [11]:

$$Efficiency(\eta) = P_{out} P_{in} x 100\% \tag{6}$$

The equation for Pin and Pout are;

$$P_{in} = \rho g H Q \tag{7}$$

$$P_{out} = \eta \rho g H Q \tag{8}$$

where P_{in} is the input power (PHP), ρ is the water density (kg/m^3), g is the gravitational acceleration (9.81 m/s^2), H is the effective head or pressure (m), Q is the water flow rate (m^3/s), P_{out} is the output power (Watt) and η is the hydraulic efficiency of water turbine (%).

1.9 Motion Equation

The computational fluid dynamics (CFD) software is specifically designed to computationally solve equations governing fluid motion, providing three-dimensional solutions for flow challenges that span different scales, are transient in nature, and involve multiple physical phenomena. The central focus of fluid dynamics lies in identifying and calculating the fluid's velocity within a stationary reference frame. Achieving this entails working with two fundamental equations:

- The continuity equation.
- The momentum equation.

Mass Continuity Equation

The inherent attributes of a fluid frequently display notable non-uniformity, comprising distinct individual molecules. When scrutinized on a larger scale, where each "small" fluid component continues to encompass numerous molecules, the fluid is regarded as a continuum. Conversely, an option exists to attribute a local bulk flow velocity v(x, t) to the element at point x by means of averaging, akin to the local mean density ρ(x, t). Consequently, this leads to the comprehensive formulation for mass continuity, which can be articulated as follows:

$$V_F\frac{\partial_\rho}{\partial t}+\frac{\partial}{\partial x}(\rho u A_x)+R\frac{\partial}{\partial y}(\rho v A_y)+\frac{\partial}{\partial z}(\rho w A_z)+\xi\frac{\rho u A_x}{x}=R_{DIF}+R_{SOR} \tag{9}$$

V_F indicates the volume fraction designated for flow, with ρ representing fluid density. R_{DIF} represents the turbulent diffusion factor, while R_{SOR} signifies a mass source. A_x, A_y, and A_z denote the fractional flow area in the x, y, and z directions, respectively. The velocity components in the x, y, and z directions are denoted by (u, v, w). R_{DIF} stands for turbulent diffusion factor, and ξ is disregarded in the Cartesian coordinate system.

$$R_{DIF}=\frac{\partial}{\partial x}\left(v_p A_x\frac{\partial\rho}{\partial x}\right)+R\frac{\partial}{\partial y}\left(v_p A_y R\frac{\partial\rho}{\partial y}\right)+\frac{\partial}{\partial z}\left(v_p A_z R\frac{\partial\rho}{\partial z}\right)+\xi\frac{\rho v_p A_x}{x} \tag{10}$$

The coefficient of v_ρ corresponds to S_c μ/ρ. Here, μ represents the coefficient of momentum diffusion, and S_c stands as the constant reciprocal, commonly referred to as the turbulent Schmidt number. Dealing with a compressible flow entails solving the density transport equation derived from Eq. (9) comprehensively. The equation describing incompressibility becomes simplified due to the constant nature of fluid density in incompressible conditions. Consequently, the subsequent equation demonstrates the reduction of Eq. (9).

$$\frac{\partial}{\partial x}(u A_x)+R\frac{\partial}{\partial y}(v A_y)+\frac{\partial}{\partial z}(w A_z)+\xi\frac{u A_x}{x}=\frac{R_{SOR}}{\rho} \tag{11}$$

However, a circumstance will arise where the propagation of acoustic pressure waves is necessary, and the rate of change of density with respect to time is determined by:

$$\frac{\partial\rho}{\partial t}\approx\frac{1}{c^2}\frac{\partial p}{\partial t} \tag{12}$$

Here, c^2 represents the square of the speed of sound, and p stands for pressure. This approximation holds valid within the range of:

$$\left|\frac{\delta\rho}{\rho}\right|<0.1 \tag{13}$$

Hence, the continuity equation has been modified into;

$$\frac{V_F}{\rho c^2}\frac{\partial\rho}{\partial t}+\frac{\partial u A_x}{\partial x}+R\frac{\partial v A_y}{\partial y}+\frac{\partial w A_z}{\partial z}+\xi\frac{u A_x}{x}=\frac{R_{SOR}}{\rho} \tag{14}$$

Momentum Equation

Navier–Stokes equation is the fluid velocity component of the equation of motion for (u, v, w) with additional terms;

$$\begin{aligned}
&\frac{\partial u}{\partial t}+\frac{1}{V_F}\left\{uA_x\frac{\partial u}{\partial x}+vA_y\frac{\partial u}{\partial y}+wA_z\frac{\partial u}{\partial z}\right\}-\xi\frac{A_y v^2}{xV_F}\\
&=-\frac{1}{\rho}\frac{\partial p}{\partial x}+G_x+f_x-b_x-\frac{R_{SOR}}{\rho V_F}(u-u_w-\delta u_s)\\
&\frac{\partial v}{\partial t}+\frac{1}{V_F}\left\{uA_x\frac{\partial v}{\partial x}+vA_y\frac{\partial v}{\partial y}+wA_z\frac{\partial v}{\partial z}\right\}+\xi\frac{A_y uv}{xV_F}\\
&=-\frac{1}{\rho}\left(R\frac{\partial p}{\partial y}\right)+G_y+f_y-b_y-\frac{R_{SOR}}{\rho V_F}(v-v_w-\delta v_s)\\
&\frac{\partial w}{\partial t}+\frac{1}{V_F}\left\{uA_x\frac{\partial w}{\partial x}+vA_yR\frac{\partial w}{\partial y}+wA_z\frac{\partial w}{\partial z}\right\}\\
&=-\frac{1}{\rho}\frac{\partial p}{\partial z}+G_z+f_z-b_z-\frac{R_{SOR}}{\rho V_F}(w-w_w-\delta w_s)
\end{aligned} \tag{15}$$

In this context, (fx, fy, fz) denote viscous accelerations, (Gx, Gy, Gz) represent body accelerations, and (bx, by, bz) indicate flow losses in porous baffle plates or porous media. The source components' velocity is denoted by Uw (uw, vw, ww), and the fluid's velocity at the source's surface, represented as Us (us, vs, ws), is computed within each control volume as follows:

$$U_s=\frac{dQ}{\rho QdA}n \tag{16}$$

1.10 Equation of Motion

The computational fluid dynamics (CFD) software is crafted to numerically solve fluid motion equations, yielding transient 3D solutions for flow problems that span multiple scales and involve various physical phenomena [12]. These equations describe the nonlinear, transient, and differential secondary equations governing fluid motion. In fluid dynamics, a pivotal objective is discerning the position and velocity of the liquid within the body-fixed frame, necessitating the utilization of a continuity equation.

2 Methodology

There are two phases required to achieve the research objective which are to develop a new opening inlet for the pico hydro power (PHP) device corresponding to the Malaysian rivers and to optimize reservoir system parameters to harness maximum energy from the river.

2.1 *Assessment of New Opening Inlet for PHP*

The first phase required to design a new opening inlet for a PHP device. The software used is Autodesk Inventor. The new design of the opening inlet will have a certain angle and shape. The new design will have a floor with an angle of 45° combined with a different inlet shape. The main idea is to find a design with optimum efficiency to harvest energy in the Malaysian River. The three proposed designs are shown in Figs. 1, 2 and 3.

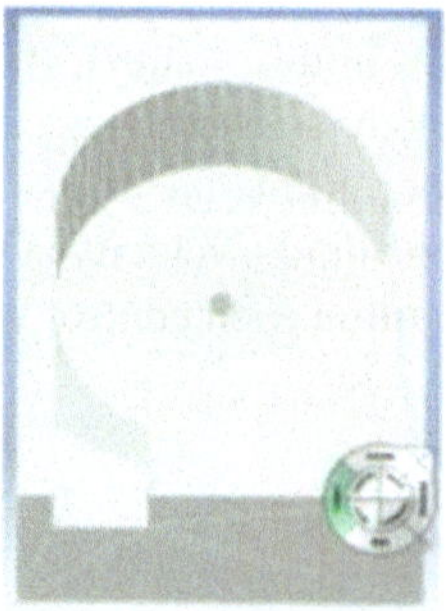

Fig. 1 Design 1

Fig. 2 Design 2

Fig. 3 Design 3

2.2 *CFD Simulation*

Simulation Works

The precision-demanding design process necessitates continuous drafting, for which Auto-desk Inventor software is adopted. This software aids in creating 3D designs of PHP devices. Subsequently, the 3D model is imported into the CFD software, and the 3D model's grid is generated using the software's inherent meshing tools. The meshing employs the nested mesh technique, concentrated in the area of interest to refine critical regions, thereby enhancing computational efficiency and reducing simulation time. Notably, this approach underscores the significance of meshing quality to augment simulation outcomes, as disregarding meshing quality could compromise the reliability of simulation results.

Computational Fluid Dynamics

Computational fluid dynamics serves as a tool for simulating fluid flow systems by employing the standard Navier–Stokes and continuity equations to address individual computational cells. The configuration and settings applied within the CFD software mirror experimental approaches; accurate outcomes arise when both factors are appropriately configured. However, this study does not involve experimental work. The software employs the volume of fluid (VOF) method to track fluid surface locations. Offering a multi-physics solver, the CFD software is adept at resolving a diverse array of flow problems. It incorporates mass and momentum conservation formulas, essential for solving dynamic flow issues. The cut cell formulation, encompassing mass and momentum, is integrated into the conservation equations. This technique enhances overall simulation efficiency by reducing integration demands compared to body-fitted meshing, consequently aiding in moderating mesh resolution precision.

Mesh Generation

This process aims to divide the framework model proposed into a small rectangular grid. In this study, only an optimum mesh will be used. Thus, it requires an additional grid-independent study to test several mesh element sizes. The critical areas for this study are the inlet and also basin. Hence, the area needs to be covered with a fine mesh.

Simulation Goal

The objective of this simulation endeavor is to determine the velocity magnitude (V_{mag}) generated by the proposed inlet configuration. The outcomes are assessed at specific probes positioned at the PHP device's inlet. The findings can be extracted in the format of velocity magnitude, documented at the designated probe locations.

Probe Location

The location of the probe for designs 1, 2 and 3 are shown in Figs. 4, 5 and 6, respectively.

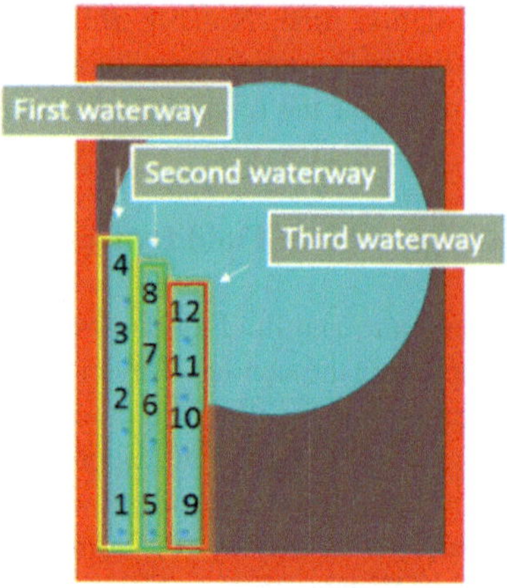

Fig. 4 Probe location for design 1

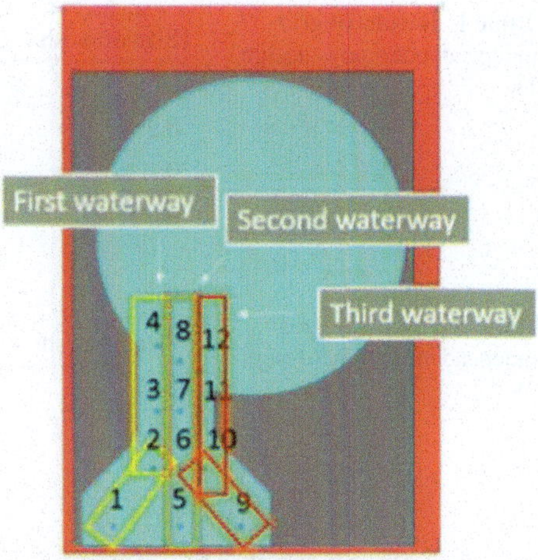

Fig. 5 Probe location for design 2

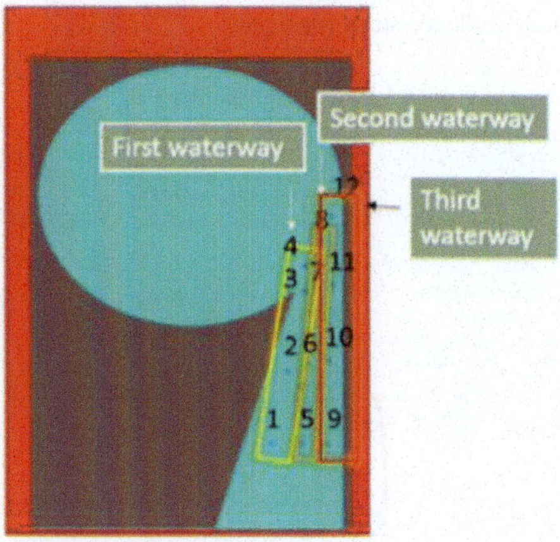

Fig. 6 Probe location for design 3

3 Results and Discussion

3.1 Grid Independence Study

Conducting a grid independence study represents a crucial stage in ascertaining the ideal count of mesh elements for each envisioned design. As mesh quality becomes finer, simulation duration extends. Consequently, determining the optimal mesh element count becomes imperative to curtail simulation time while maintaining result integrity. Elaborate outcomes for designs 1, 2, and 3 are outlined in Tables 1, 2, and 3, correspondingly. These findings culminate in a grid independence study graph illustrated in Fig. 7.

Table 1 Results of grid independence study (design 1)

Size of mesh	Number of mesh element	Vmag (m/s)
0	0	0
0.2	55,440	1.2159
0.1	454,050	1.3481
0,05	3,590,400	1.3458

Table 2 Results of grid independence study (design 2)

Size of mesh	Number of mesh element	Vmag (m/s)
0	0	0
0.2	58,590	2.1697
0.1	476,010	2.1505
0.05	3,763,800	2.1478

Table 3 Results of grid independence study (design 3)

Size of mesh	Number of mesh element	Vmag (m/s)
0	0	0
0.2	66,822	0.9403
0.1	547,304	1.0390
0.05	4,302,360	1.043S

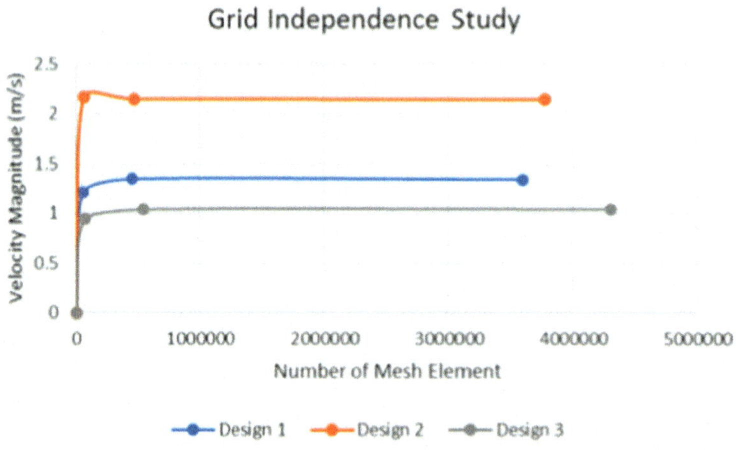

Fig. 7 Graph of grid independence study

In conclusion, it is safe to choose mesh elements between 0.5 (million) to 4.3 (million). It is equivalent to a mesh size of 0.05 to 0.1.

3.2 *Percentage of Velocity Reduction*

It is important to analyse velocity reduction percentage data as it helps to determine which models are the best in producing high-velocity magnitude of water based on incoming water velocity. The data used in this section are obtained from velocity magnitude produced at probe 8 for each model which is classified as a critical area right before the water enters the vortex pool basin.

i. Design 1:

Detailed velocity magnitude, velocity differences and velocity reduction for design 1 are illustrated in Table 4 and plotted as shown in Fig. 8.

ii. Design 2:

Detailed velocity magnitude, velocity differences and velocity reduction for design 2 are illustrated in Table 5 and plotted in Fig. 9.

iii. Design 3:

Detailed velocity magnitude, velocity differences and velocity reduction for design 3 are illustrated in Table 6 and plotted in Fig. 10.

Table 4 Results for design **1**

Vin (m/s)	Vmag (m/s)	Velocity differences (m/s)	Velocity reduction (%)
0.1	0,0552	0.0448	45
0.3	0.1470	0.1530	51
0.6	0.2503	0.3497	58
2	0.4578	1.5422	77

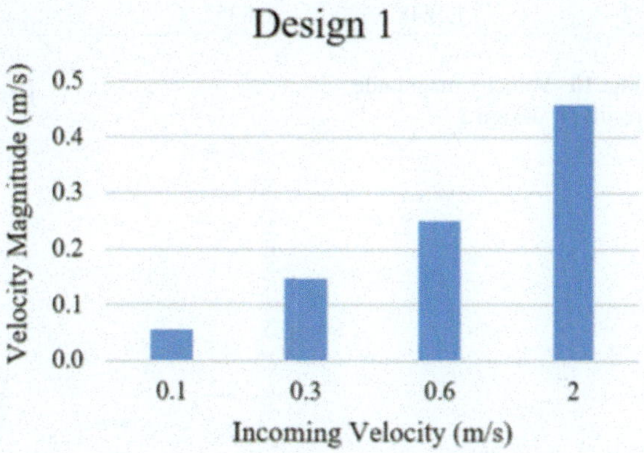

Fig. 8 Velocity magnitude result for design 1

Table 5 Results of design 2

Vin (m/s)	Vmag (m/s)	Velocity Differences (m/s)	Velocity Reduction (%)
0.1	0.0789	0.0211	21
0.3	0.2092	0.0908	30
0.6	0.3320	0.2630	45
2	0.5293	1.4707	74

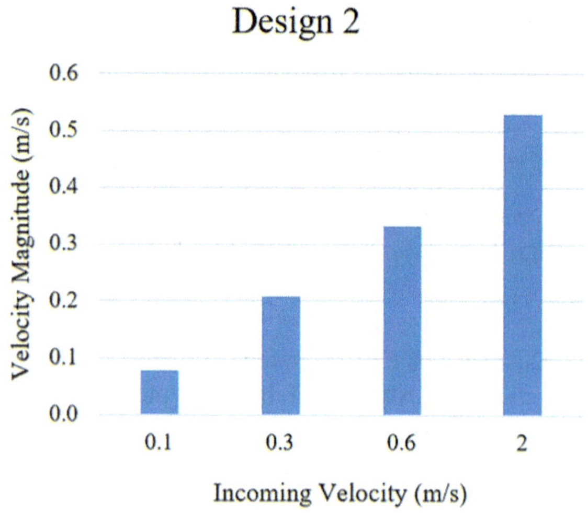

Fig. 9 Velocity magnitude result for design 2

Table 6 Results for design 3

Vin (m/s)	Vmag (m/s)	Velocity differences (m/s]	Velocity reduction (%)
0.1	0.0516	0.0484	48
0.3	0.1862	0.1138	38
0.6	0.5108	0.0892	15
2	1.1648	0.8352	42

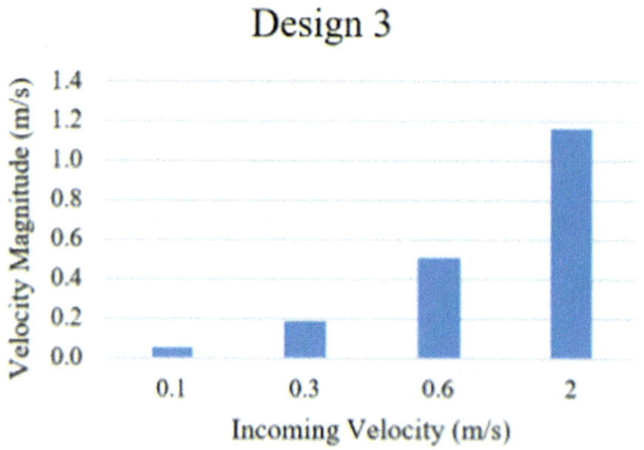

Fig. 10 Velocity magnitude result for design 3

Table 7 Results for each design

Vin (m/s)	Velocity Magnitude, Vmag (m/s)		
	Design 1	Design 2	Design 3
0.1	0,05,521	0.07894	0,0516
0.3	0.14696	020,921	0.1362
0.6	0.25030	033,193	0.5108
2	0.457S4	0.52933	1.1648

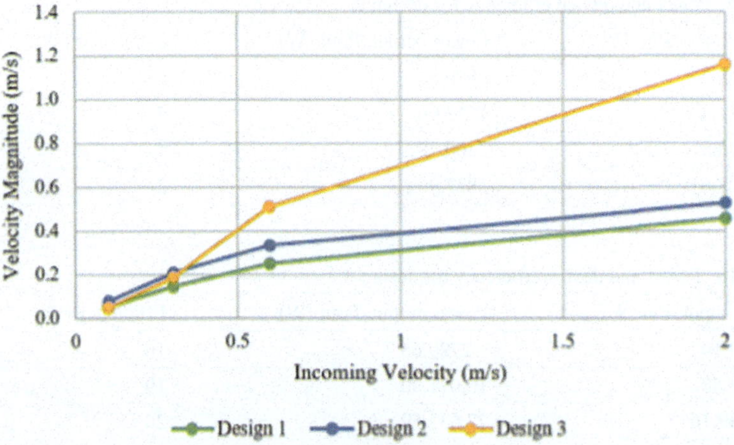

Fig. 11 Comparison of velocity magnitude for each design

3.3 *Comparison Between Designs*

A comparison between each design is illustrated in Table 7 and Fig. 11. Based on the results, at low incoming water velocities which are 0.1 and 0.3 m/s, design 2 has dominated other designs as it produces the highest velocity magnitude. However, at high incoming water velocities which are 0.6 and 2 m/s, design 3 showed a significant improvement by showing drastic changes in velocity magnitude leaving design 1 and design 2 behind with almost twice the result.

3.4 *Power and Efficiency Estimation*

Detailed power output and efficiency of designs 1–3 are illustrated in Tables 8, 9 and 10, respectively. Based on the results, there are three significant results. First, at the low incoming water velocity, the result is dominated by design 1 with 21% of device efficiency. However, at medium incoming water velocity, the result was dominated both by design 1 and design 3 with 19% of device efficiency while at high

Table 8 Power and device efficiency for design 1

Power input, Pin (W)	Power output, Pout (W)	Efficiency (%)
5636.19	1206.83	21
16,908.57	3212.33	19
33,817.14	5471.27	16
112,723.79	10,007.87	9

Table 9 Power and device efficiency for design 2

Power input, Pin (W)	Power output, Pout (W)	Efficiency {%)
9506.28	570.86	6
28,518.84	1512.96	5
57,037.68	2400.78	4
190,125.61	3827.96	2

Table 10 Power and device efficiency for design 3

Power input, Pin (W)	Power output, Pout (W)	Efficiency (%)
17,523.19	2789.65	16
52,569.58	10,071.90	19
105,139.16	27,633.59	26
350,463.86	63,012.65	18

incoming water velocity, the results are dominated by design 3 with 26% and 18% for incoming water velocity of 0.6 and 2 m/s respectively. From these results, the highest efficiency was recorded by design 3 at an incoming water velocity of 0.6 m/s with an efficiency of 26%.

4 Conclusion

To achieve all of the three objectives, the research has been separated into three stages. The valuation needs to be done to optimize opening inlet parameters to harness maximum energy from the river. Most of the rivers in Malaysia have a low water velocity and hence the stages mentioned earlier are important before implementing PHP device construction in the Malaysian river. The final stage of the research is to evaluate the performance of the new opening inlet of the PHP device.

An answer given based on simulation work has answered the first objective of this research which is to develop a new opening inlet for PHP devices corresponding to the Malaysian river. After analysing the results, design 3 was the best design among the others as it produced high-velocity magnitude results which were recorded at probe

8 for every different incoming velocity used. The inlet opening area and the height of the head for design 3 are the key factors that contribute to the improvement of the performance. By comparing the head height and also the opening inlet area for design 3 with other designs, it could be seen that design 3 produces better performance. Next, the researcher proceeded to the next step which is calculating the power and efficiency of the model. These answer the objectives of this study. Again, in this matter, an observation made by the researcher noticed that the area of the inlet and also the head height are important to produce a higher power output. The velocity output in the inlet area is also important since it will determine the value of the water flow rate. Finally, the best design recorded in this study is design 3 which produced a range of efficiency between 16 and 26%. Thus, design 3 was chosen as the most optimum design to harness river energy in Malaysia.

Acknowledgements Special thanks to Universiti Kuala Lumpur (UniKL), for funding this project under the Short Term Research Grant (STRG) No: 19055.

References

1. Denchak, M.: Natural Resources Defense Council, Flint Water Crisis: Everything You Need to Know, pp. 1–29 (2018)
2. Global Forum on Sustainability Energy.: Renewable Capacity Statistics 2019. https://www.gfse.at/news/renewable-capacity-statistics-2019 (2019). Accessed 5 Mar 2023
3. Kadier, A., Kalil, M.S., Pudukudy, M., Hasan, H.A., Mohamed, A., Hamid, A.A.: Pico hydropower (PHP) development in Malaysia: potential, present status, barriers and future perspectives. Renew. Sustain. Energy Rev. **81**(2), 2796–2805. Elsevier (2018)
4. Basar, M.F., Ahmad, A., Hasim, N., Sopian, K.: Introduction to the Pico Hydro Power and the Status of Implementation in Malaysia, pp. 283–288. IEEE (2011)
5. Güney, M.S., Kaygusuz, K.: Hydrokinetic energy conversion systems: a technology status review. Renew. Sustain. Energy Rev. **14**(9), 2996–3004 (2010)
6. International Hydropower Association: Types of hydropower: https://www.hydropower.org/iha/discover-types-of-hydropower (2022). Accessed 5 Mar 2023
7. Anupoju, S.: Components of a Hydropower Plant and Their Functions. https://theconstructor.org/structures/hydropower-plant-components-functions/19705/ (2017). Accessed 5 March 2023
8. Dhakal, S., Timilsina, A.B., Dhakal, R., Fuyal, D., Bajracharya, T.R., Pandit, H.P.: Effect of dominant parameters for conical basin: gravitational water vortex power plant. In: Proceedings of IOE Graduate Conference (2014)
9. Rahman, M.M., Tan, J.H., Fadzlita, M.T., Muzammil, A.W.K.: A review on the development of gravitational water vortex power plant as alternative renewable energy resources. IOP Conf. Ser. Mater. Sci. Eng. **217**(1), 012007 (2017)
10. Wanchat, S., Suntivarakorn, R.: Preliminary design of a vortex pool for electrical generation. Adv. Sci. Lett. **13**(1), 173–177 (2012)
11. Yadav, G., Chauhan, A.K.: Design and development of pico micro hydro system by using house hold water supply. Int. J. Res. Eng. Technol. **3**(10), 114–119 (2014)

SPRINGER NATURE

GPSR Compliance

The European Union's (EU) General Product Safety Regulation (GPSR) is a set of rules that requires consumer products to be safe and our obligations to ensure this.

If you have any concerns about our products, you can contact us on ProductSafety@springernature.com

In case Publisher is established outside the EU, the EU authorized representative is:

Springer Nature Customer Service Center GmbH
Europaplatz 3
69115 Heidelberg, Germany

The manufacturer's authorised representative in the EU is Springer Nature Customer Service Centre GmbH, Europaplatz 3, 69115 Heidelberg, Germany. If you have any concerns regarding our products, please contact ProductSafety@springernature.com

Printed and bound by CPI Group (UK) Ltd, Croydon, CR0 4YY
30/09/2026
02225586-0015